Mathematics Education in the Early Years

Tamsin Meaney • Ola Helenius
Maria L. Johansson • Troels Lange
Anna Wernberg
Editors

Mathematics Education in the Early Years

Results from the POEM2 Conference, 2014

Springer

Editors
Tamsin Meaney
Bergen University College
Bergen, Norway

Maria L. Johansson
Luleå Technical University
Luleå, Sweden

Anna Wernberg
Malmö University
Malmö, Sweden

Ola Helenius
National Centre for Mathematics
Gothenburg University
Gothenburg, Sweden

Troels Lange
Bergen University College
Bergen, Norway

ISBN 978-3-319-23933-0 ISBN 978-3-319-23935-4 (eBook)
DOI 10.1007/978-3-319-23935-4

Library of Congress Control Number: 2015958765

Springer Cham Heidelberg New York Dordrecht London

Printed on acid-free paper

Springer International Publishing AG Switzerland is part of Springer Science+Business Media (www.springer.com)

Judy Patterson (1948–2015) was an inspiring teacher educator and it is to her and her family that this book is dedicated. One of the last academic tasks that she undertook was to complete a review of a chapter that is included in this book. Her knowledge, skills, and willingness to give to the community of mathematics education researchers will be sorely missed, as will her infectious laughter.

Vale, Judy

Acknowledgement

Funding to bring the keynote speakers, Professor Alan Bishop, Monash University, and Professor Helen May, University of Otago, to Malmö for the conference was obtained from the Swedish Research Council. Without this funding and the keynotes, the conference would not have prompted the in-depth discussions which led to the development of the chapters included in this book.

Contents

Contributors

Paul Andrews Stockholm University, Stockholm, Sweden

Kerstin Bäckman University of Gävle, Gävle, Sweden

Ruthi Barkai Tel Aviv University, Tel Aviv, Israel

Ergi Acar Bayraktar Goethe-Universität Frankfurt am Main, Frankfurt, Germany

Melanie Beck University of Frankfurt/IDeA Center, Frankfurt, Germany

Christiane Benz University of Education Karlsruhe, Karlsruhe, Germany

Alan J. Bishop Faculty of Education, Monash University, Melbourne, VIC, Australia

Camilla Björklund University of Gothenburg, Gothenburg, Sweden

Manon Boily Université du Québec à Montréal, Gatineau, QC, Canada

Lisa Björklund Boistrup Stockholm University, Stockholm, Sweden

Nathalie Silvia Anwandter Cuellar Université du Québec à Outaouais, Gatineau, QC, Canada

Laurence Delacour Malmö University, Malmö, Sweden

Ann Gervasoni Monash University, Clayton, VIC, Australia

Ola Helenius National Centre for Mathematics, Gothenburg University, Gothenburg, Sweden

Maria L. Johansson Luleå Technical University, Luleå, Sweden

Ulrich Kortenkamp Universität Potsdam, Potsdam, Germany

Götz Krummheuer Goethe-Universität Frankfurt am Main, Frankfurt, Germany

Silke Ladel Universität des Saarlandes, Saarbrücken, Germany

Troels Lange Bergen University College, Bergen, Norway

Dorota Lembrér Malmö University, Malmö, Sweden

Geneviève Lessard Université du Québec à Outaouais, Gatineau, QC, Canada

Esther Levenson Tel Aviv University, Tel Aviv, Israel

Danielle Mailhot Université du Québec à Outaouais, Gatineau, QC, Canada

Helen May University of Otago, Dunedin, New Zealand

Tamsin Meaney Bergen University College, Bergen, Norway

Marcus Nührenbörger TU Dortmund University, Dortmund, Germany

Hanna Palmér Linnaeus University, Växjö, Sweden

Bob Perry Charles Sturt University, Albury, NSW, Australia

Andrea Peter-Koop Faculty of Mathematics, University of Bielefeld, Bielefeld, Germany

Eva Riesbeck Malmö University, Malmö, Sweden

Judy Sayers Stockholm University, Stockholm, Sweden

Marcus Schütte Technical University Dresden, Dresden, Germany

Anna Susanne Steinweg University of Bamberg, Bamberg, Germany

Julia Streit-Lehmann Faculty of Mathematics, University of Bielefeld, Bielefeld, Germany

Lovisa Sumpter Dalarna University, Falun, Sweden

Michal Tabach Tel Aviv University, Tel Aviv, Israel

Dina Tirosh Tel Aviv University, Tel Aviv, Israel

Pessia Tsamir Tel Aviv University, Tel Aviv, Israel

Dorothea Tubach TU Dortmund University, Dortmund, Germany

Anna Wernberg Malmö University, Malmö, Sweden

Johanna Zöllner University of Education Karlsruhe, Karlsruhe, Germany

List of Reviewers

Vigdis Flottorp, Oslo and Akershus University College of Applied Sciences, Oslo, Norway
Naomi Ingram, University of Otago, Dunedin, New Zealand
Barbara Clarke, Monash University, Melbourne, Australia
Lisser Rye Ejersbo, Danish School of Education, Aarhus University, Denmark
Sue Dockett, Charles Sturt University, Albury, Australia
Gert Monstad Hana, Bergen University College, Norway
Toril Eskeland Rangnes, Bergen University College, Norway
Ingvald Erfjord, Agder University, Norway
Marit Johnsen Høines, Bergen University College, Norway
Anita Wager, University of Wisconsin-Madison, USA
Ann Anderson, University of British Columbia, Canada
Per Sigurd Hundeland, Agder University, Norway
Martin Carlsen, Agder University, Norway
Rose Vogel, Goethe-University Frankfurt am Main, Germany
Jill Cheeseman, Monash University, Melbourne, Australia
Ewa Swoboda, Rzeszów University, Poland
Patti Barber, University College London
Jody Hunter, Massey University, Palmerston North, New Zealand
Dianne Siemon, RMIT University, Melbourne, Australia
Amy Parkes, Michigan State University, USA
Pirjo Aunio, University of Helsinki, Finland
Judy Paterson, University of Auckland, New Zealand

Part I
Introduction

Introduction

Ola Helenius, Maria L. Johansson, Troels Lange, Tamsin Meaney, and Anna Wernberg

Abstract In this chapter, we introduce the conference from which the chapters in this book originated. Then we discuss some of the background issues in order to describe the need for discussing issues to do with construction and instruction in early childhood mathematics. In particular, we talk about the evidence whether children can learn mathematics through play and the role of the teacher in this discussion. Finally, we discuss the chapters in relation to the "dance of construction with instruction".

The POEM Conferences

In 2012, a group of German mathematics educators, under the leadership of Götz Krummheuer, instigated the first conference on *A Mathematics Education Perspective on Early Mathematics Learning between the Poles of Instruction and Construction* (POEM) in Frankfurt am Main. This conference and the following one held in Malmö, Sweden, in 2014, were set up to discuss contentious issues around the provision of mathematics education in early childhood programmes and their possible contribution to children's mathematics achievement in school. In particular, POEM conferences are designed to allow researchers to link their work to the

O. Helenius
National Centre for Mathematics, Gothenburg University, Gothenburg, Sweden
e-mail: ola.helenius@ncm.gu.se

M.L. Johansson
Luleå Technical University, Luleå, Sweden
e-mail: maria.l.johansson@ltu.se

T. Lange • T. Meaney (✉)
Bergen University College, Bergen, Norway
e-mail: troels.lange@mah.se; tamsin.meaney@mah.se

A. Wernberg
Malmö University, Malmö, Sweden
e-mail: Anna.wernberg@mah.se

T. Meaney et al. (eds.), *Mathematics Education in the Early Years*,
DOI 10.1007/978-3-319-23935-4_1

question: In which way—and how much—should children be "educated" in mathematics before entering primary school?

The format for both conferences was the same. A group of researchers, many of whom had participated in earlier Congresses of European Research in Mathematics Education (CERME), were invited to attend and present a paper. A short version of each paper was made available a few weeks before the conference so that attendees could read them and be aware of the main ideas. At the paper presentation sessions, the focus was on discussing the main ideas. Several months after the conference, extended versions of the papers were submitted and peer-reviewed (see Kortenkamp et al. 2014 for the papers that arose from the first conference). The chapters that make up this book also come from the same extensive production process, and although in one sense they can be considered conference proceedings, they are in another sense much more than that. Not all conference presentations from the Malmö conference became book chapters, sometimes because of time constraints and at other times because they had been published elsewhere. The original conference presentations can be found at http://mah.se/poem.

As a result of funding from the Swedish Research Council, we were able to bring two keynote speakers to the second conference. These speakers, Alan Bishop and Helen May, provided the plenary talks at the beginning of each day, thus helping to set the scene and provide a historical background to the debate about construction and instruction or to what Norma Presmeg (2014) labelled in her keynote at the original conference, "the *dance* of instruction with construction" (p. 11).

In the following sections, we provide a background to the issues which the attendees at POEM brought up, before describing each of the five parts of the book and the individual chapters within them. Although some of the chapters in this book deal with issues to do with young children already attending schools, the research in these chapters is also informed by discussions about mathematics education in before-school settings. Therefore, it was important to provide some background to the current debates around young children engaging in mathematical ideas.

Mathematics and Early Childhood Education

Issues around mathematics education in early childhood education have become prominent in the last two decades, particularly in relation to young children not succeeding in school (Meaney 2014). For example, an analysis by Greg Duncan and colleagues of six longitudinal studies suggested that early mathematics knowledge is the most powerful predictor of later learning including the learning of reading (Duncan et al. 2007). At the same time, concerns have been raised about preschools inhibiting children from learning the deep mathematics of which they were capable (Clements and Sarama 2007). Many countries, such as New Zealand (Haynes 2000), face a tension of wanting to ensure that children begin school with stronger mathematical understandings whilst also wanting to adhere to the philosophy that preschool children should learn through play. This is a tension that some see as irreconcilable (Lee and Ginsburg 2009; Carr and May 1996).

In Sweden, play is considered the foundation for preschool children's learning experiences. However, in the last 10 years, this emphasis has begun to change in government documents highlighting the need to prepare children for school, through a focus on literacy, numeracy, and other school subjects (Broman 2010):

> An activity such as preschool, like most of the welfare institutions, is marked by its history. There is a clear relationship between a country's traditions in preschool and school system and its administration and integration of new challenges and demands. The former tradition and profession-driven activity is influenced by partially contradictory processes of late modernity. Among these are … globalization, democratization, segregation and marketisation as important influencing factors. (Broman 2010, p. 31; our translation)

The tension of expecting children to learn from playing whilst at the same time being prepared for school can be seen in the different emphases in the Swedish curriculum (Lembrér and Meaney 2014). A revised version of the curriculum for preschools was implemented in July 2011 (Skolverket 2010) with increased attention being given to mathematics, although the goals remain general. For example, "att orientera sig i … rum" (*to be able to orient themselves in … space*) (Skolverket 1998, p. 9; our translation) was expanded to "develop their understanding of space, shapes, location and direction" (Skolverket 2011, p. 11). At the same time, play retains a central role as the medium through which children are expected to learn:

> Play is important for the child's development and learning. Conscious use of play to promote the development and learning of each individual child should always be present in preschool activities. Play and enjoyment in learning in all its various forms stimulate the imagination, insight, communication and the ability to think symbolically, as well as the ability to co-operate and solve problems. Through creative and gestalt play, the child is given opportunities to express and work through his or her experiences and feelings. (Skolverket 2011, p. 6)

The presumption is that teachers will be able to support children to gain mathematical understandings whilst they play (Wernberg et al. 2010). Yet, this may not be simple to achieve (see, e.g., Lange et al. 2014). In the next section, we outline concerns about the mathematical learning that children can gain from play.

Mathematics, Play, and Direct Teaching

In many countries in the twentieth century, preschools or kindergartens were established in order to support children's learning through play (Meaney 2014). An adult watching or participating in child-initiated play is expected to develop children's mathematical ideas by stimulating their curiosity and language use (Doverborg 2006). Using results from several research projects, including one where a preschool teacher had designed activities around her children's interest in trolls, Doverborg and Samuelsson (2011) illustrated the importance of the teacher's role in encouraging children to think about mathematics. In the following extract, Nordahl (2011, p. 13; our translation)[1] provides an example from her preschool research where she monitored the free play of children aged between 1 and 3 years:

> Nancy, Minnie (2.5 years) and Jonna (3 years) build with wooden blocks. Minnie builds towers of as many blocks as she can, and when it collapses she laughs delightedly and then simply starts again. Jonna first builds a base and then continues on top of this.

Eva (förskollärare): Vad bygger du Jonna?	Eva (preschool teacher): What do you build Jonna?
Jonna: Jag bygger vårt hus, det har fyra våningar. Där bor jag (pekar) på trean.	Jonna: I build our house, it has four floors. I live there (points) on the third.
Eva: Oh, jag bor på ettan, mitt hus har bara en våning.	Eva: Oh, I live on the first, my house has only one floor.
Eva *vänder sig till Mimmi som balanserar upp ännu en kloss på sitt torn:* Du bygger riktigt högt.	Eva *turns to Minnie who balances yet a block on her tower:* You build really high.
Mimmi's *torn rasar och hon skrattar förtjust och utbrister:* Inte mer!	Mimmi's *tower collapses, and she laughs delightedly and exclaims:* No more!
Eva: Nä det har du rätt i nu är det inte högt längre (skrattande).	Eva: No you're right now, it is not high anymore (laughing).
Nancy *bygger bara ett lager och med " hålrum" emellan—nästan som en ritning.*	Nancy *builds only one layer and with "cavities" between—almost like a drawing.*
Eva: Det är ett stort hus du bygger, Nancy.	Eva: It's a big house you build, Nancy.
Nancy: Nej inte stort. Långt.	Nancy: No, not big. Tall.
Eva: Ja jättelångt. Lika långt som du nästan.	Eva: Yes very tall. As tall as you almost.
Nancy *blir förtjust och lägger sig ned bredvid och konstaterar samtidigt att hon behöver fylla på med klossar.*	Nancy is delighted and lies down next to it and acknowledges that she needs to fill it up with blocks.

In this example, the teacher encouraged Jonna's use of ordinal terms and supported Mimmi and Nancy to use a range of comparative terms about height. The teacher reinforced the use of these mathematical terms. In her research on similarly young children, Björklund (2008) also showed that adults were important in setting the parameters for children's opportunities to engage with mathematical ideas.

[1] As editors, we asked all contributors to provide their data in the original language as well as an English translation. We have done this for two reasons. The first is to encourage speakers of the original language to have opportunities to gain the nuances embedded in the original language which may be difficult to translate. The second is to contribute to discussions of the privilege of English in mathematics education research (Meaney 2013). Although not all authors were able to provide data in the original language, many did so or were willing to make them available to readers.

Acknowledgment of the importance of the teacher has led to investigations of the kind of learning that children can gain when left to their own devices:

> Children do indeed learn some mathematics on their own from free play. However, it does not afford the extensive and explicit examination of mathematical ideas that can be provided only with adult guidance. … Early mathematics is broad in scope and there is no guarantee that much of it will emerge in free play. In addition, free play does not usually help children to mathematise; to interpret their experiences in explicitly mathematical forms and understand the relations between the two. (Lee and Ginsburg 2009, p. 6)

However, if the adult's role is crucial in supporting preschool children's learning, there is a need for their teachers to be mathematically aware (Helenius et al. 2014). Other concerns are linked to preschool teachers' ability to support children's curiosity about mathematical ideas. In 2003, 100 preschool teachers in Sweden were surveyed about their teaching of mathematics. Only three teachers explicitly addressed the curriculum goals in their planning (Doverborg 2006). Many felt that learning occurred naturally as part of children's everyday lives and did not have to be planned for. Nordahl (2011) reported similar anecdotal experiences:

> My colleagues … often perceive mathematical development to only occur in the form of "learning to count". This has meant that they have not noticed when the children's mathematical development took place. Instead, they may even have impeded it by interrupting or trivialising the mathematical discoveries of preschool children, such as size perception. (p. 11, our translation)

Doverborg (2006) also felt that preschool teachers needed to see mathematics as more than "sifferskrivning och ramsrakning (*writing numerals and reciting counting rhymes*)" (p. 7; our translation) if children were to gain a deeper understanding of mathematics from play. In the book that resulted from the first POEM conference, Gasteiger (2014) highlighted both the demands on teachers for developing young children's mathematical understandings in play and some of the issues from providing teachers with professional learning opportunities about identifying potential mathematical teaching moments.

A consequence of these concerns about teachers' competency in identifying mathematical teaching moments, especially in English-speaking countries, has been the implementation of a number of mathematics teaching programmes in preschools. An American project, *Big Math for Little Kids*, was founded on the view that children needed to be presented with activities in a cohesive manner, but that these activities should be joyful and contribute to developing children's curiosity about mathematics (Greenes et al. 2004). Repetition of the activities provided opportunities for an extension of the mathematical ideas that were being introduced. For Greenes et al. (2004), the development of mathematical language was a key to helping children reflect on their learning.

Generally, preschool mathematics programmes of this type are sequenced so that children can move linearly through a development progression. For example, in another American project, *Building Blocks*, a set of activities were provided, based on learning trajectories for children (Sarama and Clements 2004). Teachers who understood the learning trajectories were better able to provide "informal, incidental mathematics at an appropriate and deep level" (p. 188). Papic et al. (2011) implemented an intervention programme on repeating and spatial patterning in one preschool over a 6-month period. Children were grouped according to how they performed on an initial diagnostic interview and then provided with tasks for their level. A combination of individual and group time was provided. Children progressed to the next level if they showed competency in their current level. Papic et al. (2011) found that, after 1 year in school, the children performed better on a general numeracy assessment than children from a control group.

Although these programmes support children to develop specific mathematical understandings, there are concerns about how formal instruction in early childhood settings could lead to "learned helplessness and a feeling of failure" (Farquhar 2003, p. 21). Many preschool and early school programmes, such as those described by Papic et al. (2011) and Clarke et al. (2006), include assessing children before, or as, they enter school on their mathematical knowledge. Such assessments risk children being labelled as "behind" or "at risk" when they are still very young. Although designed to support teachers to target their teaching to the children's levels, this also has the potential to lower teachers' expectations about children's capabilities, which may affect children's perception of themselves as learners of mathematics. When combined with stories about the value of mathematics in their adult lives, being marked out as needing extra support to learn mathematics could limit their willingness to persevere because there is too much risk of accepting themselves as failures if they persist and still do poorly (Lange 2008).

Another concern is whether a formal approach to mathematics teaching in preschool actually can produce a lasting impact on children's academic performances. In a study of children from three preschools with different pedagogical approaches, Marcon (2002) found that at different ages, children showed different academic proficiency. At the last stage of the study when children moved into their sixth year of school, children who had attended a preschool that was academically focused showed the least progress. "Grades of children from academically directed preschool classrooms declined in all but one subject area (handwriting) following the Year 6 transition" (Marcon 2002, p. 20).

Interventions about how to teach mathematics in preschools in English-speaking countries are more often targeted at preschools in low socio-economic areas, because of their enrolment of children perceived as being "at risk" of academic failure (Meaney 2014). For example, "the knowledge gap is most pronounced in the performance of U.S. children living in economically deprived urban communities" (Clements and Sarama 2007, p. 42). However, in a recent study, Clements et al. (2011) evaluated achievement gains of preschoolers from this targeted group who had been involved in *Building Blocks*. They found that after the first year in school, the gains in preschool were reduced, and after the second year of school, there was no substantial gain at all. They detailed other studies which showed similar results, suggesting that predictions that focus on preschool mathematics may not be the expected panacea for improving school results for children living in poverty.

Although a correlation may exist between mathematics knowledge on entering school and later learning, the circumstances of children's lives including the teaching that they receive in school will contribute to the knowledge that they show at all ages. In reporting on a longitudinal project, in New Zealand, which followed about 500 children till they were 10 years old, Wylie (2001) found that:

> Children who started school with low literacy and mathematics scores were much more likely to improve their scores if their parents were highly educated, or if their family had a high income. Good quality early childhood education and experiences at home, or later out-of-school activities using language, symbols, and mathematics, also made improvement more likely. (p. 11)

The circumstances that meant that young children did not have "good quality childcare" may be the same circumstances that did not provide them with rich in- or out-of-school activities. As Marcon (2002) warned, there are many variables that affect children's later school achievement, not just their preschool programmes.

The uncertainty around how mathematics in early childhood institutions, including preschools, kindergartens, and the first years of school should be provided, through making use of instruction or construction, requires that more research of the kind reported in this book should be undertaken and disseminated. It seems unlikely that children will ever again be considered too young to engage in mathematical activities and so what their capabilities and interests are needs to be better understood. However, it is also clear from the background, outlined above, that wider circumstances in which young children are engaged in mathematical activities will affect the outcomes. Therefore, this book, which provides information from projects in Australia, Canada, Israel, Germany, New Zealand, and Sweden, allows the possibility to better understand the contextual features which contribute to the projects being identified as important as well as how those features affect the outcomes of the projects. In the next sections, we provide an outline of the book and briefly describe each of the chapters and how they relate to construction or instruction.

Overview of the Book

The chapters in the book have been separated into five parts: "Introduction", "Transitions and Families", "Mathematical Processes", "Mathematical Content", and "Professional Learning". In each part, the chapters oscillate between being more or less focused on instruction or construction, with the majority of them dealing with different aspects of Presmeg's (2014) dance, by considering how instruction and construction are related.

In the introductory part, apart from this chapter, we have included the chapter that arose from Helen May's plenary for the conference. In this chapter, she charts the differences in approach over several centuries from a number of different geographical perspectives between looking at the present view of the child as a learner who is able to construct their own knowledge (construction) and the child as in need of specific knowledge and skills deemed as important by a particular society (instruction). May describes some of the political ideologies which affected the proponents of different early childhood programmes and how these proponents filled up the "socially empty spaces" that the education of young children represented.

The second part is about transitions and families and looks at the different contexts in which children operate and transverse. The initial chapter of this part is by Alan Bishop who gave the other plenary at the conference in Malmö. Through describing the six mathematical activities and the six sets of value clusters that he considers surrounding the cultural knowledge that is mathematics (Bishop 1988), Bishop discusses some of the transitions that young children may be engaged in. In doing so, he raises the need for more research to investigate how young children make sense of the changes in mathematical activities and values that occur as their transition between contexts.

In her chapter, Ergi Acar Bayraktar also focuses on the family but on the negotiation of mathematical meanings within family interactions. She adapts the "interactional niche in the development of mathematical thinking", developed at Goethe University, Frankfurt am Main, to understand the role of the interaction between family members when taking part in a geometry game in supporting the youngest child to think mathematically. Although the game is purposely designed for her study, suggesting that it is focused on instruction, it is the development of the young child's understanding as a consequence of the input received from family members which is analysed. This suggests that the point of the paper is actually about construction.

Similar to Bayraktar's chapter, Dorothea Tubach and Marcus Nührenbörger use purposely designed games to better understand the development of children's mathematical thinking. In their chapter, they consider how the same set of children play specific games in kindergarten and in their first year in school. The analysis of data indicates that the ways in which the children engage with the game change as they grew older. The authors conclude, amongst other things, that complementary learning environments can be used as a bridge to support children's developing mathematical thinking as they transition from kindergarten to school. Although their focus remains on the development of mathematical thinking (construction), how the different complementary learning environments change with the move to school suggests that how children can be taught socially valued knowledge about number (instruction) is also of interest.

The final two chapters in this part, by Julia Streit-Lehmann and Andrea Peter-Koop and by Ann Gervasoni and Bob Perry, document intervention studies designed to support children developing, in the home situation, appropriate mathematics knowledge for beginning school. Although recognising that children can learn through play situations, in the case of Julia Streit-Lehmann and Andrea Peter-Koop using a series of games and activities, the main focus in both chapters is on instruction or how to ensure that children from low-income families are supported by their families to gain the knowledge they may need to begin school well. The initial results from their pilot study suggest that although some students were able to maintain their mathematical achievement, there was no lasting effect of the intervention on children who come from immigrant families. They suggest that this may be because of the paper and pencil test that is used to assess children's mathematical achievement in school.

Whilst Streit-Lehmann and Peter-Koop are based in Germany, Ann Gervasoni and Bob Perry describe an intervention study from Australia in which early childhood educators work with parents on how to support young children's mathematical understandings in everyday situations. Their results show significant differences to that of a control group at the end of preschool, indicating that these children were better prepared in a number of mathematical content areas to start school. Unlike the German intervention which was state funded through universities, the Australian study was funded from a charity set up to support low-income families.

The third part is about mathematical processes and includes chapters which focus on the kinds of thinking that children develop. It comprises several perspectives which examine interactions between children, teachers, and mathematics. Some chapters are empirical studies, whilst others consider theoretical or methodological issues, with empirical examples being provided to illustrate specific points. It is interesting to note that five out of the seven chapters come from Swedish authors. This perhaps reflects the emphasis given in the Swedish preschool curriculum to mathematical processes, with three out of four goals indicating that preschools should provide activities for children that:

- develop their understanding of space, shapes, location and direction, and the basic properties of sets, quantity, order and number concepts, also for measurement, time and change,
- develop their ability to use mathematics to investigate, reflect over and test different solutions to problems raised by themselves and others,
- develop their ability to distinguish, express, examine and use mathematical concepts and their interrelationships,
- develop their mathematical skill in putting forward and following reasoning. (Skolverket 2011, vp. 10)

The first chapter in this part is by Ola Helenius, Maria L. Johansson, Troels Lange, Tamsin Meaney, Eva Riesbeck, and Anna Wernberg and extends Bishop's (1988) mathematical activity, playing, to consider what it might be for young children. To do this, they synthesise the attributes of play as well as the playful mathematics that mathematicians are documented as engaging in. Using empirical material of a group of 6-year-olds engaged in a free-play situation involving buying and selling an imaginary popsicle, the authors illustrate when and when not the interaction could be classified as mathematical play. As the focus is on how the children develop the interaction, we consider it to be more about construction than instruction.

The chapter by Lovisa Sumpter focuses on theoretical frameworks that can be used for analysing young children's reasoning. Her argument is that individual reasoning and collective reasoning are difficult to analyse using the same framework. She therefore proposes the need for two frameworks and uses these to analyse examples of both kinds of reasoning. Her conclusions describe what is highlighted and what becomes invisible when the frameworks are used and the implications of this on trying to understand young children's mathematical reasoning. This chapter can be considered to be about how to make sense of children's construction of mathematical ideas.

The following chapter is by Götz Krummheuer and Marcus Schütte. They used the same framework, the interactional niche in the development of mathematical thinking, as Bayraktar. In this chapter, they look at how a child at two different points in her life, 3 years apart, simultaneously adapts her participation as well as the theme of the interaction. When the theme is turned toward mathematical topics and the child becomes more autonomous in her interactions, then she is likely to develop her mathematical thinking. From the authors' perspective, this investigation is about construction rather than instruction because of its focus on what the child does within the interaction.

As was the case with other researchers from Frankfurt am Main, Melanie Beck also modifies the interactional niche in the development of mathematical thinking in order to use it to understand how young children's creativity might be linked to their attachment patterns. Attachment patterns are the different ways that young children connect to their caregivers. Her hypothesis is that children with secure attachment patterns are likely to be creative in different ways than those children with insecure attachment patterns. In order to test her hypothesis, she analyses the interactions of two children with different attachment patterns when engaging in a purposely designed game around different solids hidden in a bag. As her investigation is about the children's contributions to the interaction, it can be considered to be more about construction than instruction.

Kerstin Bäckman's chapter is about the potential of play as providing mathematical learning opportunities for children. To investigate this, she examines two videos from a wider study, in which children play with shapes and patterns. However, her main focus is not on what the children do but rather on the teachers' contributions to the interactions. Her focus is on the "here-and-now" teachable moments that teachers need to be aware of in order for them to become learning moments for children. Therefore, in this "dance", the focus is more on instructional aspects. However, it is also clear that the instructional aspects are led by the constructional aspects of the children's engagement in play.

In Dorota Lembrér and Tamsin Meaney's chapter, the focus is also on the "dance" in that the affordances of a game for an interactive table are evaluated in regard to how they supported two young children to learn about conjecturing, justifying, and interpreting. The authors also acknowledge the importance of the teacher being prompted by what appeared on the screen to ask the children questions. Consequently, this chapter can be seen as being focused on how instruction, through the game, contributed to children's construction. In their analysis, they compared the affordances in the game with the list of affordances of information and communication technologies (ICT) described in earlier research as being potentially beneficial for young children's learning of mathematics.

The final chapter in this part is about the introduction of problem-solving lessons to 6-year-olds in Sweden. Hanna Palmér in setting up an intervention study concentrates on how to utilise problem-solving as a way of introducing children to mathematical content. In so doing, she focuses on instructional aspects on the lessons. As well, her evaluation of the intervention involved interviewing the children about their experiences of the problem-solving lessons 6–7 weeks after the intervention was completed. The children's responses illustrated that they were reflective about their learning in regard to both the problem-solving lessons and the mathematics lessons that were not part of the intervention. This is the only chapter in this book in which children's opinions about their learning was sought and used as data to be analysed.

In the fourth part, the chapters utilise several perspectives but all have in common the focus on mathematical content: length, place value, algebraic thinking, number sense, geometrical shapes, and patterns. There are, also, similarities with the chapters in the previous parts with empirical data coming from studies done with toddlers to children attending their second year of school. Some chapters examine children's mathematics learning, mostly in playful situations, whilst others focus more explicitly on the teacher's role. When the teacher's role is highlighted, it is generally from the pole of instruction, showcasing how teachers use planned situations to challenge children's mathematical knowledge. Two chapters (Ladel and Kortenkamp, Steinweg) examine how information and communication technologies (ICT) can be used to foster children's mathematical content learning.

The first chapter in this part is about toddlers' playful learning about patterns through interactions both with concrete materials and with their teachers who were involved in an intervention based on variation theory and learning study. Camilla Björklund's focus in this chapter is on how these very young children learn about differences and similarities between objects and as such the chapter is focused on construction. Nevertheless, the materials the teacher makes available and how they interact with the children mean that instructional aspects are also considered in regard to how they support the children's learning.

The next chapter by Silke Ladel and Ulrich Kortenkamp uses artefact-centric activity theory as a theoretical framework for understanding how young children in school develop a flexible understanding of place value. In a diagnostic interview with 52 children in year 2, it was found that the use of base-ten blocks with place value charts was likely to cause some children confusion over what was being recorded on the chart, thus inhibiting understandings about Western number system. A quantitative study with larger numbers of students showed that just understanding that there could be different representations for showing a particular amount in the place value charts was sufficient to support children to develop and use place value in flexible ways. In this "dance", how children were constructing their understanding of place value was used to inform instruction practices.

Nathalie Silvia Anwandter Cuellar, Manon Boily, Geneviève Lessard, and Danielle Mailhot discuss 5-year-old children's understanding of equality and equivalence and how this understanding is connected to algebraic thinking. Their analysis is of the children's reasoning of how to produce equivalent groups. As a consequence, this chapter has connections to the earlier chapters by Sumpter and Lembrér and Meaney. The differences in children's responses are illustrated in the chapter. The tasks that the children engage in are purposely designed to have children describe their understanding of equality and equivalence, but the children's responses can be used to inform the development of tasks to support learning of these. Therefore, the "dance" of construction with instruction moves seamlessly between the poles.

Similarly in the chapter by Pessia Tsamir, Dina Tirosh, Esther Levenson, Ruthi Barkai, and Michal Tabach, attention moves between the knowledge needed by preschool caregivers in order to support the development of early geometrical understandings and how children make sense of the mathematical activities. As teacher educators, the authors worked with both the caregivers and the children and analysed their interactions with them in regard to predetermine geometrical knowledge. Part of their analysis was about the impact that their work with the caregivers had on the possibilities for children to develop their understandings. This suggests that although the main focus was on instruction, their ultimate aim was to improve the possibilities for children to construct their own knowledge. They raise several issues to do with providing professional development to caregivers, suggesting that there are links between this chapter and those in the final part.

In the chapter by Anna Susanne Steinweg, the focus is on the development of a mathematical app that can be used on a tablet. In this app, children are expected to work through different levels of games in a sequential order in order to gain specific number understandings. Although some data from a child trialling the app is included, the results are analysed in regard to practical issues to do with how it was used. The focus of the chapter is thus about instruction as it is anticipated that children will learn number understandings in the same way from engaging in the games provided.

Like the previous chapter, Johanna Zöllner and Christiane Benz focus on instruction by outlining the measurement knowledge that teachers need in order to support young children learn about length. Their chapter reviews the research literature on this issue. However, they consider the knowledge to be connected as a net. For them, when working with children in informal learning situations, teachers need to be aware that different aspects can arise at different times, and it is for the teachers to make connections between them.

The final chapter in this part is about subitising as an important component of what Judy Sayers, Paul Andrews, and Lisa Björklund Boistrup call foundational number sense. Their aim is to document the complexity of subitising and how it is taught from case studies from Hungary and Sweden, indicating that their focus was on instruction. This is connected to their aim of producing a framework about number sense which can be used to analyse lessons from different cultural contexts as well as indicating how best to support teachers to develop teaching practices about these ideas.

The final part is about professional learning and has only two chapters, although as noted, some of the earlier chapters had links to this part, particularly the chapter by Pessia et al. It is perhaps not so surprising in a conference that deals specifically with issues to do with construction and instruction that there is not so much about the professional learning of teachers in early childhood settings. On the other hand, given the rapid changes in policy and knowledge from mathematics education research, it is surprising that so little attention is paid to how teachers are responding to changes in the field in regard to shifting between construction and instruction.

Laurence Delacour's research is about how four teachers in two preschools in Sweden make sense of the revisions in the preschool curriculum in regard to mathematics goals. She is able to show that the different philosophies of the preschools in which they work and their own professional life histories affect how they introduce children to mathematical understandings. In one preschool which draws its inspiration from Reggio Emilio, the teachers provided opportunities that built on children's own interests and thus could be considered as being connected to construction. At the other preschool, one of the teachers had previously taught in school and both teachers' approach was more about preparing children for school, indicating a closer connection to instruction. The advantages and disadvantages of both approaches are discussed by the author.

In the final chapter of the book, Christiane Benz considers the role of reflection in the professional learning of kindergarten teachers and student teachers. In an innovative study where both groups were able to work together, she was able to draw out how reflection contributed to increased opportunities for children to engage with mathematical ideas. Reflection on actual interactions with children, rather than just on theoretical understandings supported both groups of teachers to go deeper with their reflections. In this way, as was the case with the previous chapter, the author manoeuvres between instruction and construction as the centre of attention.

Conclusion

This chapter introduces the book by both providing a background of the current issues facing mathematics education in the early years and describing how the chapters in the book consider Presmeg's (2014) "dance of construction with instruction". As noted in the previous part, most chapters show that research being done in this area oscillates between concerns for children's construction as active human agents and the need to be instructed in socially valued mathematics knowledge, generally so that they can be better prepared for school. However, the dance looks very different in each of the chapters, depending on the early years of education environment and the experiences and interests of the participants in the study, as well as the interests of the researchers. Consequently, we see this book as providing important insights to the questions which dominate POEM conferences: In which way—and how much—should children be "educated" in mathematics before entering primary school?

References

Bishop, A. J. (1988). *Mathematical enculturation: A cultural perspective on mathematics education*. Dordrecht: Kluwer.

Björklund, C. (2008). Toddlers' opportunities to learn mathematics. *International Journal of Early Childhood, 40*(1), 81–95. doi:10.1007/BF03168365.

Broman, I. T. (2010). Svensk förskola—ett kvalitetsbegrepp. In B. Riddерspore & S. Persson (Eds.), *Utbildningsvetenskap för förskolan* (pp. 21–38). Stockholm: Natur & Kultur.
Carr, M., & May, H. (1996). The politics and processes of the implementation of Te Whaariki, the New Zealand national early childhood curriculum 1993-6. In M. Carr & H. May (Eds.), *Implementing Te Whaariki* (pp. 1–13). Wellington: Institute for Early Childhood Studies.
Clarke, B., Clarke, D. M., & Cheeseman, J. (2006). The mathematical knowledge and understanding young children bring to school. *Mathematics Education Research Journal, 18*(1), 78–102.
Clements, D. H., & Sarama, J. (2007). Early childhood mathematics learning. In F. K. Lester (Ed.), *Second handbook of research in mathematics teaching and learning* (pp. 461–555). Charlotte, NC: Information Age.
Clements, D. H., Sarama, J., Farran, D., Lipsey, M., Hofer, K. G., & Bilbrey, C. (2011, March 3–5). *An examination of the building blocks math curriculum: Results of a longitudinal scale-up study*. Paper presented at Society for Research on Educational Effectiveness Spring 2011 Conference, Washington DC. Available from http://www.eric.ed.gov/PDFS/ED518182.pdf.
Doverborg, E. (2006). Svensk förskola. In E. Doverborg & G. Emanuelsson (Eds.), *Små barns matematik* (pp. 1–10). Göteborg: NCM Göteborgs Universitet.
Doverborg, E., & Samuelsson, I. P. (2011). Early mathematics in the preschool context. In N. Pramling & I. P. Samuelsson (Eds.), *Educational encounters: Nordic studies in early childhood didactics* (pp. 37–64). Dordrecht: Springer.
Duncan, G. J., Dowsett, C. J., Claessens, A., Magnuson, K., Huston, A. C., Klebanow, P., et al. (2007). School readiness and later achievement. *Developmental Psychology, 43*(6), 1428–1446. doi:10.1037/0012-1649.43.6.1428.
Farquhar, S.-E. (2003). *Quality teaching early foundations: Best evidence synthesis iteration*. Wellington: New Zealand Ministry of Education.
Gasteiger, H. (2014). Professionalization of early childhood educators with a focus on natural learning situations and individual development of mathematical competencies: Results from an evaluation study. In U. Kortenkamp, B. Brandt, C. Benz, G. Krummheuer, S. Ladel, & R. Vogel (Eds.), *Early mathematics learning: Selected papers of the POEM 2012 conference* (pp. 223–236). New York: Springer.
Greenes, C., Ginsburg, H. P., & Balfanz, R. (2004). Big math for little kids. *Early Childhood Research Quarterly, 19*(1), 159–166. doi:10.1016/j.ecresq.2004.01.010.
Haynes, M. (2000). Mathematics education for early childhood: A partnership of two curriculums. *Mathematics Teacher Education and Development, 2*, 93–104. Available from http://www.merga.net.au/node/43?volume=2.
Helenius, O., Johansson, M. L., Lange, T., Meaney, T., Riesbeck, E., & Wernberg, A. (2014, February 4–5). Preschool teachers' awareness of mathematics. In O. Helenius, A. Engström, T. Meaney, P. Nilsson, E. Norén, J. Sayers, et al. (Eds.), *Development of mathematics teaching: Design, scale, effects: Dimensions and perspectives: Proceedings from Madif9: Nionde forskningsseminariet med Svensk Förening för Matematikdidaktisk Forskning, Umeå*. Forthcoming.
Kortenkamp, U., Brandt, B., Benz, C., Krummheuer, G., Ladel, S., & Vogel, R. (Eds.). (2014). *Early mathematics learning: Selected papers of the POEM 2012 conference*. New York: Springer.
Lange, T. (2008). A child's perspective on being in difficulty in mathematics. *The Philosophy of Mathematics Education Journal, 23*. Available from http://people.exeter.ac.uk/PErnest/pome23/index.htm.
Lange, T., Meaney, T., Riesbeck, E., & Wernberg, A. (2014). Mathematical teaching moments: Between instruction and construction. In U. Kortenkamp, B. Brandt, C. Benz, G. Krummheuer, S. Ladel, & R. Vogel (Eds.), *Early mathematics learning: Selected papers of the POEM 2012 conference* (pp. 37–54). New York: Springer.
Lee, J. S., & Ginsburg, H. P. (2009). Early childhood teachers' misconceptions about mathematics education for young children in the United States. *Australasian Journal of Early Childhood, 34*(4), 37–45. Available from http://www.earlychildhoodaustralia.org.au/australian_journal_of_early_childhood/ajec_index_abstracts/ajec_vol_34_no_4_december_2009.html.

Lembrér, D., & Meaney, T. (2014). Socialisation tensions in the Swedish preschool curriculum: The case of mathematics. *Educare, 2014*(2), 82–98.

Marcon, R. A. (2002). Moving up the grades: Relationship between preschool model and later school success. *Early Childhood Research and Practice, 4*(1). Available from http://ecrp.uiuc.edu/v4n1/index.html.

Meaney, T. (2013, April 2–7). The privileging of English in mathematics education research, just a necessary evil? Keynote presentation. In M. Berger, K. Brodie, V. Frith, & K. le Roux (Eds.), *Proceedings of the seventh international mathematics education and society, Cape Town, South Africa* (pp. 65–84). Available from http://mescommunity.info/.

Meaney, T. (2014). Back to the future? Children living in poverty, early childhood centres and mathematics education. *ZDM Mathematics Education,* (46), 999–1011. doi:10.1007/s11858-014-0578-y.

Nordahl, M. (2011). Små barns matematik—på små barns vis. In T. Wedege (Ed.), *Vardagsmatematik: Från förskola over grundskolan till gymnasiet [Everyday mathematics: From preschool over compulsory school to senior highschool]* (pp. 11–18). Malmö: FoU Malmö-utbildning, Avdeling barn och ungdom, Malmö Stad.

Papic, M. M., Mulligan, J. T., & Mitchelmore, M. C. (2011). Assessing the development of preschoolers' mathematical patterning. *Journal for Research in Mathematics Education, 42*(3), 237–268.

Presmeg, N. C. (2014). A dance of instruction with construction in mathematics education. In U. Kortenkamp, B. Brandt, C. Benz, G. Krummheuer, S. Ladel, & R. Vogel (Eds.), *Early mathematics learning: Selected papers of the POEM 2012 conference* (pp. 9–17). New York: Springer.

Sarama, J., & Clements, D. H. (2004). Building blocks for early childhood mathematics. *Early Childhood Research Quarterly, 19*(1), 181–189. doi:10.1016/j.ecresq.2004.01.014.

Skolverket. (1998). *Läroplan för förskolan Lpfö 98*. Stockholm: Skolverket.

Skolverket. (2010). *Läroplan för förskolan Lpfö 98: Reviderad 2010 [Lpfö 98]*. Stockholm: Skolverket.

Skolverket. (2011). *Curriculum for the preschool Lpfö 98: Revised 2010*. Stockholm: Skolverket.

Wernberg, A., Larsson, K., & Riesbeck, E. (2010). Matematik i förskolan [Mathematics in preschool]. In B. Ridderspore & S. Persson (Eds.), *Utbildningsvetenskap för förskolan [Educational science for preschool]* (pp. 157–171). Stockholm: Natur & Kultur.

Wylie, C. (2001). *Ten years old & competent: The fourth stage of the competent children's project—A summary of the main findings*. Wellington: New Zealand Council for Educational Research. Available from http://www.nzcer.org.nz/research/publications/ten-years-old-and-competent-fourth-stage-competent-children-project-summary-ma.

A Historical Overview of Early Education Policy and Pedagogy: Global Perspectives and Particular Examples

Helen May

Abstract The history of early children institutions is littered with debates about how best to educate young children, including in regard to mathematics. In this chapter, definitions of instruction and construction are provided as a way of describing aspects of these debates. Some of those debates are discussed in regard to the filling of "socially empty spaces" by pedagogies and practices that drew on the ideologies of proposers of different projects. Drawings and photos are provided to illustrate visually some of the ways that the empty spaces were filled. The most current debates centre on the pulling down of school knowledge and practices into before-school institutions such as preschools.

The "poles of instruction and construction" as a theme for a symposium on early mathematics learning is also a broad metaphor illustrative of the debates and divides concerning the role and nature of early childhood institutions and curricula in relation to school. The "poles" are examined in this chapter as a historical narrative charting the mainly Western European landscape of early years education for young children through centuries and across continents in institutions established to complement (and even replace) the child rearing of family life. Through a mix of experiment and expediency, the "socially empty spaces" of early childhood education [ECE], firstly within a few homes and later a few institutions outside of the home, were gradually filled with pedagogies and practices, shaped by the people, politics and ideologies of the respective times and places, to become the almost global foundation of modern education systems (Singer 1992). This often-revolutionary story is usually framed around ideals of social progress and individual betterment. The story is also littered with political, ideological and pedagogical disputes concerning the appropriate elements that variously define the role of institutions for early years education. Situating such institutions within the social and educational spectrum has been (and still is) buffeted by divides concerning the class, culture and age of young children (Meaney 2014), underpinned too by shifting political understandings and interest in public investment in the infrastructure of early education

H. May (✉)
University of Otago, PO Box 56, Dunedin, New Zealand
e-mail: helen.may@otago.ac.nz

T. Meaney et al. (eds.), *Mathematics Education in the Early Years*
DOI 10.1007/978-3-319-23935-4_2

(May 2005, 2009, 2011, 2013). Before charting some key developments of previous centuries, it is useful to situate this narrative in current times illustrative of the centuries-old "poles" of politics, practice and pedagogy within early years education, including both preschool and school variations. Broadly, the "poles" characterise contesting paradigms of childhood:

- The child as nature whose holistic development is a natural process and who learns through play and discovery—construction
- The child as a reproducer of knowledge, who as an empty vessel is filled with agreed knowledge, skills and cultural values—instruction

Political (Un)Interest in ECE Pedagogy (1945–1990s)

In post-war Western education systems, the institutions of ECE, after more than a century of experimental and/or charitable endeavour, were, in various ways, positioned more formally on the political landscape. The state had a clearer rationale in the aftermath of troubled times for intervention and investments in support of pedagogical practices distinctive from school in so-called broadly understood "preschool" institutions. While the policy infrastructure around these institutions was diverse across countries and cultures, there was a shared acceptance that ECE provided a beneficial social preparation for children prior to school. Psychological rationales around the notion of a sane society significantly underpinned the interest of the state: a view summarised by the influential Swedish educator Alva Myrdal (1948, p. 4):

> The world is sick and troubled…We know that if children were given the right opportunities and handled with the right educational care, so much of the mental ill-health which is now crippling individuals should be prevented, and so many of the conflicts and tensions which are harassing the world could be turned into productive forces.

Most Western countries set about building an infrastructure for provision and funding of preschool education, although there were differences in the levels of participation and access. More cohesive in this burgeoning state interest was an acceptance that ECE pedagogy was primarily a matter for the profession or the organisations providing ECE. State intervention in pedagogy was "light touch", an exception being the Eastern European Soviet bloc countries. By the mid-twentieth century, Western approaches to ECE pedagogy were broadly positioned towards the pole of "construction" stemming from progressive new education ideals of child-centred learning (Howlett 2013). These roots will be elaborated in later sections. From the 1960s, as governments, such as in the USA, embarked on a "war on poverty" with preschool education for the so-called disadvantaged in the vanguard (Pines 1966), there was a tipping towards a more "instructional" curriculum for such children to compensate for their poor home background and provide a "head start" to catch up prior to school entry (Ritchie 1978). But again, the state mainly refrained from formal intervention in ECE pedagogy. This was about to change.

Global Rationales for National ECE Curricula (1990s–2015)

By the 1990s, the policy and political interest in ECE was more inclusive of the institutions of childcare linked to the workforce participation of parents. This integration is still incomplete in many countries, but globally shared understandings of the role early childhood education and care (ECEC) have emerged. Political interest has been shaped by the economic agendas of nation states enforcing more political control of school curricula. This has spilled into the early years education with the development of parallel national early childhood curriculum documents. Such initiatives are linked to wider education reform and each government's interest in ensuring that its investment in education delivers the appropriate outcomes for successful participation of its citizen children in a competitive global economy (Oberheumer 2005). Recognising the links between participation in ECE, success at school and economic development, the education directorate of the OECD (Organisation for Economic Co-operation and Development) launched a wave of cross-country reviews of ECEC policy (OECD 2001, 2006). Peter Moss (2014, p. 19) sums up this new phenomenon:

> Interest [in ECEC] has spread far and wide, attracting the attention of international organisations and nation states, the political classes and policy wonks, and a range of academics from disciplines that have not previously shown interest in the subject—in particular economists. Countries that have previously neglected early childhood education are now putting money into developing services.

Moss is a critic of the recent discourse of "high returns" and describes:

> a story of control and calculation, technology and measurement that, in a nutshell, goes like this. Find, invest in an apply the correct human technologies—aka 'quality'—during early childhood and you will get high returns on investment including education, employment and earnings and reduced social problems. (p. 3)

Early childhood pedagogy has become a key indicator of quality, exemplified in *Starting Strong 111: A Quality Toolbox for Early Childhood* (OECD 2012). This "toolbox" simplifies, for government policy usage, the necessary balance of "technologies" required as measures of success.

The economic environment underpinning ECEC policy has led towards policies for almost universal provision of a preschool experience with a particular focus on disadvantaged communities that pose the most economic and social risk to society (Gambaro et al. 2014). The question of the most appropriate pedagogy for ECEC towards realising these outcomes has become political, with a swing in the pendulum towards the "pole of instruction" particularly in relation to the mantras of literacy and numeracy. Leaving the acquisition of these early skills and understandings to the playful chance of self-discovery is contradictory to the political intent to "invest early and invest smartly" (Moss 2014, p. 3). This tightrope balance between professional and political interests and understandings of ECEC has played out differently across countries in their respective national curricula with the "poles" illustrated in the more "constructive" focus of the New Zealand curriculum *Te Wh riki* (NZ Ministry of Education 1996) and the more "instructional" approach of the recently updated English *Statutory Framework for the Early Years Foundations Stage* (UK Department of Education 2014). The prime objective of the latter is "school readiness" and its strong focus on building the stepping stone skills of literacy and numeracy. By contrast, *Te Whāriki* (translated from Māori as a woven map for all to stand on) is aspirational and framed around the principle of empowerment with literacy and numeracy embedded within "socially and culturally mediated learning and of responsive and reciprocal relationships for children with people, places and things" (NZ Ministry of Education 1996, p. 9).

Despite these "poles", early childhood curriculum documents have drawn a line between school curriculum and early childhood curriculum, and it is the nature of the relationship between the compulsory school sector and the still-voluntary early childhood sector that shapes the curriculum emphasis. Moss (2013, p. 9) suggests that:

> In the relationship of 'readying for school' ECE assumes a subordinate role of preparing young children to perform well in CSE [school], by governing the child effectively to ensure that he or she acquires the knowledge, skills and dispositions to be a successful learner in compulsory education, for example, ready for the rapid acquisition of literacy and numeracy.

Conversely, in a relationship between school and ECEC, identified by OECD in *Starting Strong 1* (OECD 2001) ideally as a "strong and equal partnership", early childhood is a distinctive period "where children live out their own lives" and "the specific character and traditions and quality of early childhood practice are preserved" (OECD 2001, p. 129). Tracing the roots of these ideas requires a step backwards.

An Enlightened Emphasis on Home Instruction of the Very Young (1700s–1800s)

Prior to the emergence of experimental institutions for early education in the nineteenth century, it is useful to signpost key enlightenment individuals, whose collective philosophical and practical ideas on the rearing and education of young children laid the foundation of later infant school and kindergarten pedagogies. The new wisdom was intended to build upon a child's propensity to play and to please and avoid the harsh dogmas and methods of meaningless rote learning that characterised European schooling. Czech-born John Amos Comenius (1592–1670) was a teacher and pastor who spent his life in exile because of his dangerous ideas on religious tolerance and education. He promoted more kindly and interesting approaches to schooling in the vernacular language, instead of Latin. *The School of Infancy* (1631) was intended mainly for mothers and concerning children from birth to 6 years. Comenius described activities, games, plays, tasks and tales that the "mother school" at home could provide as the foundation for later schooling and education. These were the years in which Comenius suggested, "we learn to KNOW some things, to DO some things, and to SAY some things" (Comenius 1831/1956, p. 73). He told parents that "too much sitting still ... is not a good sign" and young children should be able to "play freely". Comenius outlined a curriculum framed around knowledge and experience of the natural world, optics, astronomy, geography, chronology, history, household affairs and politics. In each area, he provided ideas of what might be relevant for the young child. The examples he describes for teaching mathematics are illustrative of the mix of informal learning and more formal content:

> About the second year the principles of *geometry* may be perceived, when we say of anything that it is large or small: they will afterwards easily know what is short, long wide or narrow. (p. 41)

> The elements of *Arithmetic*...for children in their third year; as soon as they begin to count to five and afterwards to ten, or at least pronounce the numbers correctly, although they may not at first understand what those numbers really are, for then they will observe the use to which this enumeration is used. (p. 41)

The philosophical and political writings of John Locke (1632–1704) in England and Jean Jacques Rousseau (1712–1778) from Switzerland fuelled new thinking on the nature of education, childhood and family in an enlightened society (Locke 1693; Rousseau 1762/1911). Unlike Comenius, neither Locke nor Rousseau was a teacher, either by inclination or experience. In *Some Thoughts Concerning Education* (1693), Locke set out a view of human development in which the child came into the world with a mind as a "blank tablet". He saw the minds of children as fertile "garden plots" which in the right environment could be cultivated and moulded through early education experiences, with the implication that if parents treated their children as rational beings, they could be guided rather than punished towards good action. Locke, like Comenius, signposted the direction for child-centred approaches to education by suggesting that "Learning must be a Play and Recreation to Children, and they must be brought to desire to be taught" (Locke 1693/1989, p. 208). Locke combined abstract understandings of arithmetic, geometry, astronomy and geography with the view that "children may be taught anything that falls within their senses" (p. 181). "When he has the natural parts of the globe well fixed in his memory, it may then be time to begin arithmetic...Arithmetic is the easiest, and consequently the first sort of abstract reasoning which the mind commonly bears, or accustoms itself to" (p. 179).

Rousseau's ideas on education brought him fame and persecution. *Emile* (1762) was the fictional story of the upbringing of a boy and a blueprint to save children from vice and corruption: "All things are good as their creator makes them; but degenerate in the hands of man" (Rousseau 1762/1911, p.1). In opposing the biblical doctrine of original sin, Rousseau believed that children came to learn the limits and possibilities of their actions through freedom. Rousseau's infants were to be unswaddled and allowed freedom of movement to explore their environment. Curiosity was encouraged. Rousseau's prescription of freedom from swaddling and constraint was synonymous with political freedom. Neil Postman (1999, p. 120) claims:

> Rousseau's writings aroused a curiosity about the nature of childhood that persists to the present day. We might fairly say that Friedrich Froebel, Johann Pestalozzi, Marie Montessori, Jean Piaget, Arnold Gesell, and A. S. Neill are all Rousseau's intellectual heirs … Certainly their work proceeded from the assumption that the psychology of children is fundamentally different from that of adults, and is to be valued for itself.

Rousseau laid down no mantras on the subject content of Emile's early education, which was to be shaped by Emile's interests and explorations under the loving eye of his mother.

On a more practical level are the writings of Anglo-Irish father and daughter, Richard Edgeworth (1744–1817) and Maria Edgeworth (1767–1849), best remembered for their popular book *Practical Education* (Edgeworth and Edgeworth 1798). In their travels, they met Rousseau and the Swiss teacher Johann Pestalozzi (1746–1827) who was applying some of Rousseau's principles of education into actual school settings. Pestalozzi, like Comenius, did not believe in schools for very young children. In his fictional book, *How Gertrude Teaches Her Children* (1801/1894), the mother is the teacher who demonstrates how an ordinary home environment can be used for education. She shows how including children in everyday work tasks develops senses and guides observations. *Practical Education* was based on the Edgeworths' family experience with Richard's 17 children. It captured the spirit of enlightened thinking with the child being encouraged to learn through play, discovery and invention rather than by rote and discipline: "Children work hard at play, therefore we should let them play at work" (p. 55). Adults were admonished to follow the child's pace and interests:

> An infant should never be interrupted in its operations; whilst it wishes to use its hands, we should not be impatient to make it walk …When children are busily trying experiments upon objects within their reach, we should not … break the course of their ideas, and totally prevent them from acquiring knowledge by their own experience. (p. 910)

In the Edgeworth's large family, children would have card, pasteboard, scissors, wood, wire, gum and wax, balls and pulleys, and they were encouraged to invent, construct, discuss and find out for themselves (pp. 5–6). This was an approach to learning based on the scientific experiment rather than formal teaching, where fostering curiosity and the interests of the child was the primary method. With their interests in science, the Edgeworths detailed the necessary subject content that should be encouraged in discovery and playful tasks. For example, in a chapter on "arithmetick", they write:

> Many children who have thought to be slow in learning arithmetick, have after their escape from the hands of pedagogues, become remarkable for their quickness….(p. 425)

> We recommend the use of plain regular solids, cubes, globes, made of wood as playthings… For teaching arithmetick half inch cubes, which can be easily grasped by infant fingers… they can be arranged in various combinations; the eye can sufficiently take in a sufficient number at once…[and] consider assemblages as they relate to quantity or shape. (p. 425)

There are chapters on geometry, mechanics and chemistry. The Edgeworths' approach to early education was that it be both playful and purposeful, albeit the experiment was set amid the resources and spaces of a wealthy home environment.

By the end of the eighteenth century, there were sufficient clues in the writings of enlightened educators and thinkers to inform an alternative pedagogy of early education practice to the formal rote 3Rs (reading, writing, arithmetic) instruction evident in burgeoning school institutions that were forerunners of later public school systems.

"Socially Empty Space" and the Apparatus of Early Education (1800–1900)

Elly Singer (1992, p. 34) suggests the concept of "socially empty space" in which "There were no pedagogic traditions for working with young children in groups... The entire learning experience had to be specially designed for this children's world". These traditions were established during the nineteenth century mainly in the context of the British infant school and the German kindergarten. Both institutions created a space outside the family and separate from school traditions of learning and teaching. Both institutions underwent further transformation and standardisation that softened their early radical beginnings. The ideas of both institutions were exported across continents and cultures. Broadly, the later infant school pedagogic tradition was towards the pole of instruction; its intent is to provide a disciplined haven for children otherwise playing "wild" on the streets (May 2013). The early kindergarten trended towards the pole of construction, although never with the extreme of Rousseau's freedoms. With the advent of the later urban *volks-kindergartens* (free kindergartens), for inner-city poor, instruction was to the fore. In both pedagogic traditions, there was an emphasis on removing children from external ills, promoting the notion of happy childhood, kindly teachers, interesting visual environments, teaching apparatus to stimulate senses and conversation, outdoor play, movement and music. Both traditions led to the creation of specially trained professionals to work with the very young.

In 1816, Robert Owen (1721–1858) established an infant school at his cotton mill factory in New Lanark, Scotland, for the young children of women mill workers. James Buchanan was the first teacher. Owen's instructions were simple:

> They were on no account ever to beat any of the children or threaten them with any word of action ... but were to speak to them with a pleasant voice and a kindly manner. They should tell the infants and children ... that they must do all they could to make their playfellows happy. (Donnachie 2000, pp. 166–167)

Owen also instructed that young children should not be "annoyed" with books or religion. The detail of the curriculum for the infant school owed much to Buchanan's inventiveness who devised a programme of games, singing, outdoor play, dancing and storytelling and included Pestalozzi's methods of observing natural objects, which, Owen reported, "always excited their curiosity and created an animated conversation between the children and their instructor" (cited in Morton 1962, p. 103). John Griscom describes the "baby school":

> One apartment of the school afforded a novel and pleasing spectacle. It consisted of a great number of children, from one to three or four years of age. They are assembled in a large room, under the care of a judicious female, who allows them to amuse themselves with various selected toys, and occasionally collects the oldest into a class and teaches them their letters. (Griscom 1823, pp. 385–386)

It can be assumed that the children were also learning their numbers. The infant school museum at New Lanark has a standard counting frame for infant school classrooms. Between 1816 and 1825, around 20,000 curious visitors, educators and reformers are reported to have arrived at New Lanark (Donnachie 2004). The performances of the children for the benefit of visitors were a practical demonstration that an enlightenment education could benefit the children of the poorer working classes. Ian Donnachie claims that (Fig. 1):

> [Owen's] basic assumption that character could be transformed under favourable conditions seemed to work … he succeeded in creating a system [of education] which was able to produce conforming and apparently happy (or docile) children equipped with basic literacy and numeracy. (Donnachie 2000, pp. 170–171)

Samuel Wilderspin (1791–1866) can be credited with transforming Owen's experiment into a blueprint more acceptable to church backers (Wilderspin 1829). From the 1820s, the idea spread across Britain and its colonies (May et al. 2014) and was exported to Europe and the USA. Wilderspin invented the infant gallery for instruction and, by borrowing methods from monitorial schools for older children, standardised a pedagogy for teaching the basics of the 3Rs to large groups of young children (Figs. 2 and 3).

Fig. 1 "View of an infant school" from *The Infant School Manual, or Teacher's Assistant*, by Mrs. Howland (1831)

Fig. 2 "The three Rs" from *A Manual Detailing the System of Instruction Pursued at the Infant School*, Bristol, by David Goyder (1826)

Fig. 3 "Gallery lessons" from*Infant Education*, by Samuel Wilderspin (1829)

Friedrich Froebel (1782–1852) established the first kindergarten at Blankenburg, in the Prussian state of Thuringia, in 1837. A small-scale innovation by a visionary but elderly educationist became, in the space of 50 years, a successful blueprint for the early education in many countries. Froebel did not want young children to be "schooled". Instead, this new institution recreated a homelike atmosphere that extended the possibilities for learning with a programme structured around play and self-activity. Froebel (1826/1892, p. 54) earlier argued that:

> Play is the highest expression of human development … The child that plays thoroughly, with self active determination … will surely be a thoroughly determined man … The plays of childhood are the germinal seeds of all later life.

He also outlined methods for teaching young children about numbers and geometry.

The kindergarten signified both a "garden for children" where they could observe and interact with nature and a "garden of children" who could develop freely, under the guidance of the "gardeners of children" the kindergarten teachers. Froebel is most remembered for the kindergarten *spielgaben*, a progression of geometric toys called the "gifts" and craft "occupations". The first "gift" was a soft knitted ball for the young infant. Froebel saw the gifts as playthings designed to stimulate and develop the child's senses. Subsequent gifts were sets of blocks and tablets of increasing complexity for self-activity and construction. Each gift functioned as a means to explore the forms of life though block building, the forms of beauty through pattern making and the forms of knowledge for understanding mathematics (Figs. 4, 5 and 6).

Whereas surviving infant school timetables have daily timeslots for "arithmetic", "multiplication", "pence table" and "tables" (Goyder 1826), kindergarten methods were integrated into timetabled tasks of playful activities such as block building,

Fig. 4 A selection of Froebel gifts and blocks

pattern making, stick laying and pricking (Pösche 1862). The mantra of "learning through play" at kindergarten was its distinctive pedagogical tradition. The formal teaching of the 3Rs was the task of the school, although this did not prevent the kindergarten introducing the foundations of literacy and numeracy.

By the end of the century, the pedagogic traditions of the infant school and kindergarten, as preschool education systems, had filled what had been a "socially empty space". The mass expansion and export of these approaches had mediated their radical origins, and there was growing criticism of the rigidity of their programmes, although kindergarten methods were cautiously being adapted for the large infant school classrooms in public school systems. In the USA, the kindergarten itself was attached as the first rung of public school. In the twentieth century, new pedagogical approaches and ECEC institutions were poised to emerge, although both infant school and kindergarten traditions adapted and survived in a variety of ways.

Refilling the Space and "New Education" Ideals (1900–1940s)

With a new century, notions of new education for a new era became the vogue. Medicine, the physical sciences and the emerging fields of sociology and psychology provided new tools and rationales not only for so-called progressive pedagogies of ECE but also for schooling, although in the case of the latter it took half a century before experimental ventures became mainstream (May 2011). At a conference of

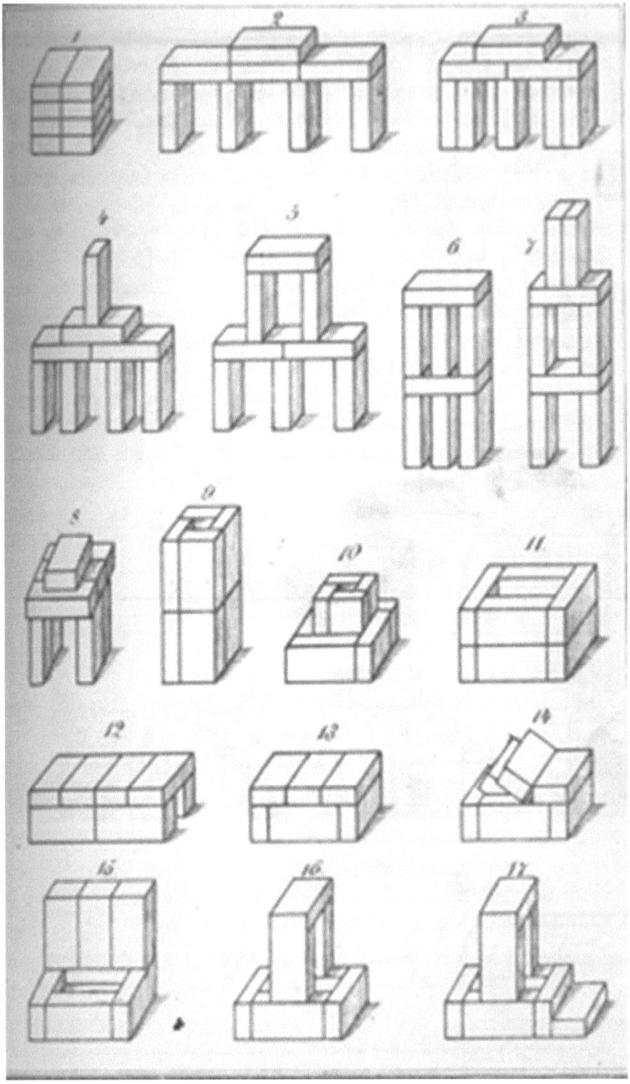

Fig. 5 The third gift, in *The Gifts of the Kindergarten*, by H. Goldammer (1882)

inspectors in 1904 to launch a new primary school syllabus, the priorities of George Hogben, the New Zealand inspector general, were clear:

> The important thing … is not the amount of things that are taught, but the spirit, character, and method of teaching in relation to its purpose of developing the child's powers … We must believe with Froebel and others of the most enlightened of the world's educators, that the child will learn best, not so much by reading about things in books as by doing: that is exercising his natural activities by making things, by observing and testing things for himself; and then afterwards by reasoning about them and expressing thoughts about them. (Hogben 1904, p. 2)

Fig. 6 Early kindergarten classroom, Tavistock Place, London, *The Lady's Newspaper* 13 October 1855

Froebelian activities and in particular their timetabled application in schools were, however, poised for change. In the USA, the reappraisal was spearheaded in laboratory-style kindergartens and nurseries. The reformulation was a combination of new philosophical insight from people such as John Dewey and the drive of reformist kindergarten practitioners such as Patty Smith-Hill (Weber 1984). From Italy, the work of Maria Montessori came to the attention of the international community (Montessori 1915), and for a few years, her methods were heralded as the "cure-all" curriculum blueprint for early childhood. In England, Margaret McMillan's child health campaigns and nursery school initiatives brought a sharper political perspective to early education (McMillan 1919). From a broader theoretical perspective, the radical insights of Sigmund Freud were transforming psychology. Anna Freud and Melanie Klein were creating the new field of child analysis (Young-Bruehl 1991), and Susan Isaacs was demonstrating how self-expression in early childhood was a foundation for psychological well-being and intellectual growth (Isaacs 1930). At the same time, Jean Piaget was formulating his theories on the development of rational thinking in young children that, he postulated, grew out of the child's spontaneous play and interaction with objects. As Piaget's ideas trickled into practice, teachers were urged to provide more hands-on experiences and to observe rather than to intervene (Piaget 1926).

Cumulatively, the impact was dramatic, and the "new education", with the possibilities of both individual (psychological, intellectual and behavioural) and collective (sociological) transformation, was promising a pathway to various new social orders. That this pathway began in the early childhood years was the crux of new education, although there were broad aims across the spectrum of education, based on the belief that education has to do with the whole child, the child's individual personality was of primary importance, the child's needs and interests were more important than predetermined subject matter and individual motivation for learning rather than external pressure should be the basis of schooling (Boyd and Rawson 1965). Some examples concerning new pedagogical approaches to early mathematics usefully illustrate the diverse interpretations of new education in early childhood settings.

It was under the influence of John Dewey (1859–1952) that reformist kindergarteners began to abandon the standardised kindergarten equipment and programme. Dewey was primarily a philosopher, and his philosophical conceptions of the relationship between education and society became a core strand of progressivism. For most practitioners, however, maxims such as "learning through experience" or "learning by doing" became the guiding mantra. For Dewey, educational theory was about adjustment and adaptation to the social environment. The school would form a miniature community of learners where there was an emphasis on co-operative and problem-solving activities that were part of everyday life. Children as participants in a classroom community were engaging in the "reconstruction of experience" in the play and activity that Dewey believed would be the basis of training for living in a democratic. Schools therefore must:

> reproduce on the child's plane the typical doings and occupations of the larger, and maturer society into which he is finally to go forth … it is through production and creative use that valuable knowledge is secured and clinched. (Dewey 1900, pp. 143–144)

At the laboratory school kindergarten established in 1896 at the University of Chicago, activities based around family life came to be the basis of the new curriculum, with a view to bridging the gap between home and school and tapping into the child's natural interest in everyday activities. In this context, mathematics education for young children was mainly a product of their play: in the shop, at the post office, cooking, sewing or building houses with blocks or collage.

In 1906, in Italy, Maria Montessori (1870–1952) established the first *casa dei bambini* for children on a housing estate in Rome with a programme intended to first "educate the senses" including teaching the 3Rs to young children. There was immediate worldwide interest from educators interested in new education methods useful in schools:

> Once the education of the senses is underway, along with the arousal of interest, we can begin real instruction. We can introduce the alphabet, not in a book, but on a little table on which are raised letters, painted different colours, that can be touched and traced with the fingers. (quoted in Kramer 1976, pp. 76–77)

Montessori developed a range of "didactic apparatus" such as gradated rods, cylinders, colour tablets, number shapes and boards and geometric insets. Each piece was designed for different stages of development and to stimulate specific sensory feelings (Fig. 7).

The apparatus was designed for children to work with individually; the aim was for the child to gain a sense of autonomy over the object and the environment. The role of the teacher was to observe and guide where necessary, but not to interfere.

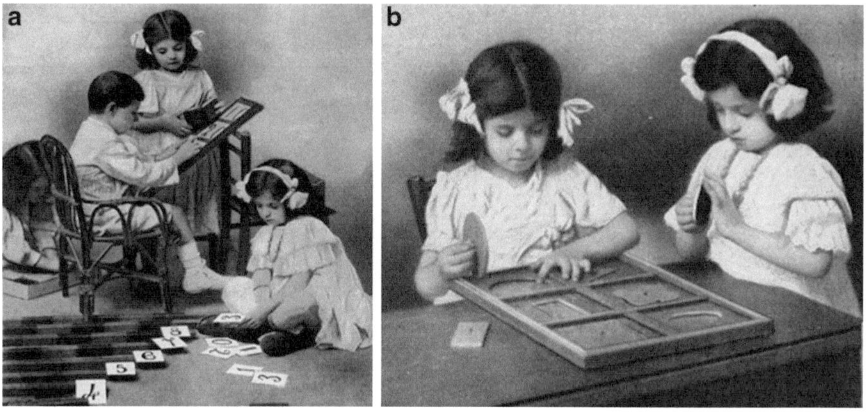

Fig. 7 (**a**) "A lesson in arithmetic"; (**b**) "fitting geometric insets" from *McClure Magazine*, vol. 37, 1911

The Montessori equipment, if not the programme, was adapted into progressive school and kindergarten settings. Similarly, new styles of blocks saw the phasing out of the Froebel gifts in kindergartens. In the USA, the large Patty Smith-Hill blocks and the Caroline Pratt multiple unit blocks became standard fare in new nursery schools. While play with blocks retained mathematical purposes, its functions were increasingly social (Fig. 8).

Shifting the emphasis further from the formal teaching of early mathematics was the influence of psychoanalytic pedagogical approaches in experimental nursery schools espousing a free play programme. They are cited because their influence on the ECE pedagogy is still strong. They created a mantra of no direct teaching and a view that children would develop the necessary understandings through play and in contexts that were relevant. The Freudian focus on the psychic conflicts of childhood caused the meaning of children's play to take on a new dimension. Free play became the medium for self-expression. There was minimal adult intervention, but by observing the child, the adult would get clues about understanding the child's behaviour and emotional needs. Observation (i.e. analysis) became the key role for the teacher. Play materials also took on new dimensions, and "blocks could be a medium for the expression of feelings; painting and drawing revealed unconscious urges" (Weber 1984, p. 118). Creative expression became a cornerstone of the progressive education movement.

At the Bank Street nursery school in New York, considerable attention was given to selecting equipment and activities that fostered opportunities for enabling children to develop and grow naturally and for staff to study that growth:

> The equipment provides ample opportunity for the vigorous climbing, swinging, balancing, jumping running, which are so important at this age: there are also blocks, trains, dolls, crayons, paints, clay and other plastic materials to invite constructive activity: and of course there is a sandbox for the younger children and a workshop for the older ones.[1]

[1] "The Harriet Johnson Nursery School", in *Children yes*, The Bank Street Schools, New York, *c*. 1939, p. 4, Bank Street School Archive

a

b

Fig. 8 (**a**) Patty Smith-Hill blocks, Columbia Kindergarten c. 1920s; (**b**) Caroline Pratt multiple unit blocks, Bank Street School, New York, c. 1920s

Between 1924 and 1927, Malting House School near Cambridge in England became an influential beacon of new education ideas in Britain, Australia and New Zealand. Led by Susan Isaacs (1885–1948), Malting House was an experiment conceived as a scientific laboratory intended to provide an educational environment for young children, based on the best psychological thinking then available. Isaacs' "infant scientists" encouraged (with some safety restrictions) to "find out" and "discover" for themselves (Cameron 2006, p. 851) (Fig. 9).

The first data from these observations was published in *Intellectual Growth in Young Children* (1930). The view was that the school would be a "point of vantage" rather than a "screen" for the outside world (Isaacs 1930, p. 21). Isaacs saw herself applying some of Dewey's work to very young children:

> We have been content to apply our new psychological knowledge of *how* the child learns ... We have not used it to enrich our understanding of *what* he needs to learn, nor what experiences the school should bring him [emphasis in original]. (Isaacs 1930, p. 21)

Isaacs regarded Montessori's activities as too narrow, although both the Malting House children and teachers made frequent use of Montessori's number cards and rods, illustrating how numbers, counting and geometry could be integrated into a free play discovery programme. Through the observations of children in their play, Isaacs saw how very young children showed a capacity for logical thinking.

New education approaches shaping ECE curriculum were also moving into school classrooms. At a 1943 conference in New Zealand, infant advisors to schools considered progress made towards introducing "activity period work", "play-way aids" and "manipulative toys" into infant classrooms. An hour of "developmental value of free play" first thing in the morning was recommended. Teachers were told

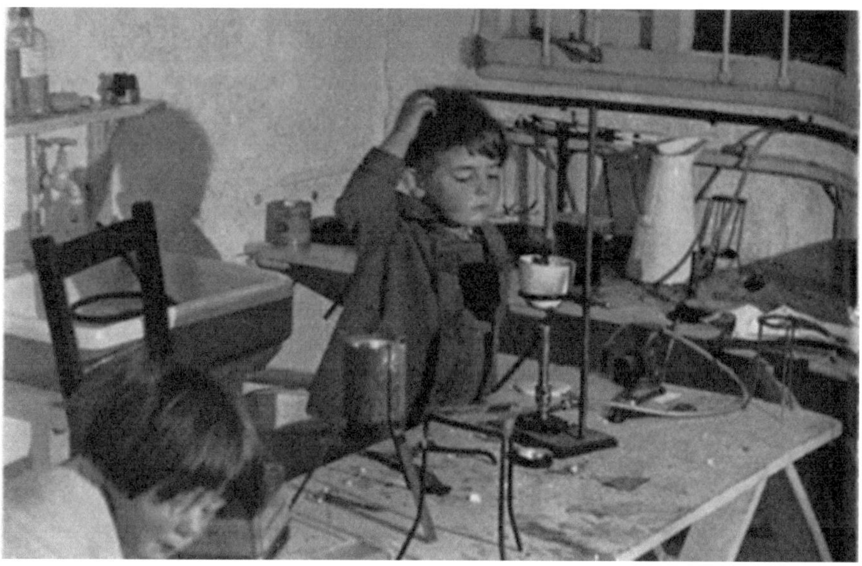

Fig. 9 Children at work in the science laboratory

Fig. 10 (**a**) "Matching, placing and counting", (**b**) "ancient game of darts" and (**c**) "spatial relationships" from *Number Work in the InfantRoom* (1944)

a

MATCHING, PLACING, AND COUNTING
Primer One Children using Counters, Blocks, and Coloured Pegs, reproduce Designs of Boats, Flags, &c., and count the Colour Groups.

b

TION GAZETTE 2 AUGUST 1943

The Ancient Game of Darts serves a Purpose in Infant Number Work

c

This Skilfully-contrived piece of Infant-room Apparatus introduces Children to the Study of Spatial Relationships

that, during this time, they should "restrain all impulses to teach and direct—in other words … develop a capacity for passive watchfulness … advice is given only when asked" (Lowrie 1943, p. 271). A new syllabus, *Number Work in the Infant Room* (NZ Department of Education 1944), rejected the memorising of facts and tables "without any knowledge of practical work"; the syllabus made "discovery work" the official method. On paper, the pole of construction was to the fore; in practice, most teachers preferred instruction (Fig. 10).

Summary

This circular historical overview of the pedagogical landscape of ECEC concludes at the mid-twentieth century. New education ideas and approaches had refilled and/or overlaid nineteenth-century initiatives to create a pedagogy for the institutions of early childhood outside of the family, but separate from school. Alignments with school curricula and the ages at which young children moved into school institutions were variable and contested, as they still are. By the mid-twentieth century, the early years school curricula in subject areas such as mathematics were increasingly shaped by new education approaches. This too became contested and continues to be reshaped by curricula demands in the higher levels of schooling and global dictates linking 3Rs' performance with economic prosperity. With the current trend towards the almost universal participation in ECEC within OECD countries at least, governments are no longer so tolerant of separate preschool and school curricula paradigms. The rhetoric of "transition to school" and "continuity of learning" has often resulted in a pulling down of school agendas into ECEC. It has become increasingly hard to shift ECE agenda upwards into school. For example, many countries have shifted the school entry age downwards, albeit with some safeguards about appropriate curricula, or in the case of the UK, the government has preferred to fund schools to provide 1 year free "ECE" for 4-year-olds rather than the traditional nursery schools which cost more. This contest is most acutely played out in Ireland where 4-year-old children, who all attend school, are subject to the dictates of both the holistic national early childhood curriculum framework, *Aistear* (National Council for Curriculum and Assessment 2009), and the formal subject-based national school curriculum. Two different government departments, with different positions on the paradigm "poles of instruction and construction", are at war for the same group of children.

References

Boyd, W., & Rawson, W. (1965). *The story of the new education*. London, UK: Heinemann.

Cameron, L. (2006). Science, nature, and hatred: 'Finding out' at the Malting House School, 1924–29. *Environment and Planning D: Society and Space, 24*(6), 851–872.

Comenius, J. A. (1831/1956). *The school of infancy.* E. M. Miller (Ed.). Chapel Hill, NC: University of North Carolina Press.
Dewey, J. (1900). Froebel's educational principles. *Elementary School Record,* June, 143–151.
Donnachie, I. (2000). *Robert Owen: Owen of New Lanark and New Harmony.* East Linton, Scotland, UK: Tuckwell Press.
Donnachie, I. (2004). Historic tourism to New Lanark and the Falls of Clyde 1795–1839: The evidence of contemporary visiting books and related sources. *Journal of Tourism and Cultural Change, 2*(3), 145–162.
Edgeworth, M., & Edgeworth, R. L. (1798). *Practical education* (Vol. 1). London: J. Johnson.
Froebel, F. (1826/1892). *The education of man.* W. N. Hailmann (Ed.). New York: D. Appleton.
Gambaro, L. F., Stewart, K., & Waldfogel, J. (Eds.). (2014). *An equal start? Providing quality early childhood education and care for disadvantaged children.* London: Policy Press.
Goyder, D. G. (1826). *A treatise on the management of infant schools.* London: Simpkin & Marshall.
Griscom, J. (1823). *A year in Europe* (Vol. 2). New York: Collins/E. Bliss & E. White.
Hogben, G. H. (1904). Report on conference of primary inspectors. *Appendices to Journals of the House of Representatives*, E-1C. Wellington: Government Printer.
Howlett, J. (2013). *Progressive education: A critical introduction.* London: Bloomsbury Press.
Isaacs, S. (1930). *Intellectual growth in young children.* London: Routledge.
Kramer, R. (1976). *Maria Montessori: A biography.* Oxford, UK: Basil Blackwell.
Locke, J. (1693/1989). *Some thoughts concerning education.* J. W. Yolton & J. S. Yolton (Eds.). Oxford, UK: Oxford University Press.
Lowrie, F. E. (1943). Play activity in the infant room. *The Education Gazette,* November, 271–272.
May, H. (2005). *School beginnings: A nineteenth century colonial story.* Wellington: NZCER Press.
May, H. (2009). *Politics in the playground: The world of early childhood in New Zealand* (2nd ed.). Wellington: NZCER Press.
May, H. (2011). *'I am five and I go to school': Early years schooling in New Zealand 1900–2010.* Dunedin: University of Otago Press.
May, H. (2013). *The discovery of early childhood* (2nd ed.). Wellington: NZCER Press.
May, H., Kaur, B., & Prochner, L. (2014). *Empire education and indigenous childhood: Missionary infant schools in three British colonies.* Farnham, Surrey, UK: Ashgate.
McMillan, M. (1919). *The nursery school.* London: J.M. Dent.
Meaney, T. (2014, April). Back to the future? Children living in poverty, early childhood centres and mathematics education. *ZDM The International Journal of Mathematics Education.* The final publication is available at the link Springer.com (online).
Montessori, M. (1915). *The Montessori method.* London: Heinemann.
Morton, A. L. (1962). *The life and ideas of Robert Owen.* London: Lawrence & Wishart.
Moss, P. (Ed.). (2013). *Early childhood and compulsory education: Reconceptualising the relationship.* London: Routledge.
Moss, P. (2014). *Transformative change in real utopias in early childhood education: A story of democracy, experimentation and potentiality.* London: Routledge.
Myrdal, A. (1948, August). *The first conference of the World Council for Early Childhood Education, Conseil Mondial pour l'Education Prescholaire, Svetova Rada Pro Presdskolni Vychoevu, Charles University, Prague.*
National Council for Curriculum and Assessment. (2009). *Aistear, The early childhood curriculum framework.* Dublin: NCCA. http://www.ncca.biz/Aistear/pdfs/UserGuide_ENG.pdf.
New Zealand Ministry of Education. (1996). *Te Whāriki: He Whāriki matauranga mo nga mokopuna o Aotearoa. Early childhood curriculum.* Wellington: Learning Media. http://www.educate.ece.govt.nz/learning/curriculumAndLearning/TeWhariki.aspx.
New Zealand Department of Education. (1944). *Number work in the infant room> Some suggestions for teachers.* Wellington: Government Printer.

Oberheumer, P. (2005). International perspectives on early childhood curricula. *International Journal of Early Childhood, 37*(1), 27–37.
OECD. (2001). *Starting strong: Early childhood education and care*. Paris: OECD. http://www.oecd.org/edu/school/37423778.pdf.
OECD. (2006). *Starting strong II: Early childhood education and care*. Paris: OECD. http://www.oecd.org/newsroom/37425999.pdf.
OECD. (2012). *Starting strong III: A quality toolbox for early childhood*. Paris: OECD. http://www.oecd.org/edu/school/49325825.pdf.
Pestalozzi, J. H. (1801/1894). *How Gertrude teaches her children: An attempt to help mothers to teach their own children*. L. E. Holland & F. C. Turner (Eds.). London: Sonnenschein.
Piaget, J. (1926). *The language and thought of the child*. London: Kegan Paul.
Pines, M. (1966). *Revolution in learning: The years from birth to five*. London: Allen Lane, Penguin.
Pösche, von H. (1862). *Friedrich Fröbel's entwickelnderziehende menschenbildung kindergarten pädagogikal system*. Hamburg: Hoffmann, Campe.
Postman, N. (1999). *Building a bridge to the 18th century: How the past can improve the future*. New York: Alfred A. Knopf.
Ritchie, J. (1978). *Chance to be equal*. Whatamonga Bay, Queen Charlotte Sound, New Zealand: Cape Catley.
Rousseau, J. J. (1762/1911). *Émile*. B. Foxley (Ed.). London: Dent Everyman Library.
Singer, E. (1992). *Childcare and the psychology of development*. London: Routledge.
UK Department of Education. (2014). *Statutory framework for the early years foundations*. London: UK Department of Education. https://www.gov.uk/government/uploads/system/uploads/attachment_data/file/335504/EYFS_framework_from_1_September_2014.
Weber, E. (1984). *Ideas influencing early childhood education: A theoretical analysis*. New York: Teachers College Press.
Wilderspin, S. (1829). *Infant education; or, practical remarks on the importance of educating the infant poor*. London: James Hodson.
Young-Bruehl, E. (1991). *Anna Freud*. London: Papermac.

Part II
Family and Transitions

Can Values Awareness Help Teachers and Parents Transition Preschool Learners into Mathematics Learning?

Alan J. Bishop

Abstract This chapter focuses on different kinds of transitioning that young children do. These are transitioning from home to preschool, from home to school and from preschool to school. With each of these transitions, children may come into contact with different perspectives on mathematics. Six mathematical activities and six mathematical values are described as a way of conceptualising the possible differences young children might experience as they transition across different contexts. A call for more research in this area is made at the end of this chapter.

Introduction

I congratulate those who research the issues of preschool mathematics learning as that is the level at which young learners either switch on to mathematics or switch off from it. To paraphrase the old Jesuit adage, 'Give me the child before they are 7 and I will give you the person'. While teachers and parents have known about the influence of the home background on learning for a long time, it is only relatively recently that we have developed the theoretical tools and research practices to help us study the influences of home/preschool/school relations and values.

My chapter is framed within the general construct of mathematics as a form of cultural knowledge. I outline some of the trends in research in this area, and in particular I present my ideas about the relatively new concept of values in mathematical education. I will also focus on ideas from research, as that is the area I know best. I am not a preschool teacher, and my parental experiences are in the distant past! However, I believe it is the theoretical content of research which is giving us the most useful practical ideas concerning the influences of preschool, school and home mathematical practices.

I am also an enthusiastic promoter of the work of researchers in the area of mathematics education. These researchers around the world, along with most workers in education, are now experiencing hard financial times, thanks to increasingly

A.J. Bishop (✉)
Faculty of Education, Monash University, Melbourne, VIC, Australia
e-mail: alan.bishop@monash.edu

T. Meaney et al. (eds.), *Mathematics Education in the Early Years*,
DOI 10.1007/978-3-319-23935-4_3

right-wing governments who think they know what is 'best practice' in this area. However, my emphasis on research should not be read as ignoring the excellent work done by many preschool and school teachers in real situations. Indeed, the best research in my opinion usually involves both practising teachers and practising researchers.

My talk is structured around three basic contributions from research involving cultural issues:

- The contribution to *curriculum* thinking of culturally based mathematical knowledge
- The contribution of research on culturally situated *learning*
- The contribution of mathematical values to mathematical *pedagogy*

In each case I will describe the implications for research and practice concerning mathematical transitioning of preschoolers. I will concentrate mainly on the values area since this work is relatively new and increasingly influential.

The Contribution of Ethnomathematics to a Culturally Based Mathematics Curriculum

One of the most significant areas of research development in the last three decades has been that of ethnomathematics. It has not only generated a great deal of interesting evidence, but it has fundamentally changed many of our ideas and constructs (see, e.g. Ascher and Ascher 1997; D'Ambrosio 1985; Gerdes 1997; Joseph 1991).

For me ethnomathematics is not a branch of mathematics but is the study of relationships between mathematics and culture, and the most significant influences from this research have been in relation to:

- Human interactions. Ethnomathematics concerns mathematical activities and practices in society, which mostly take place outside school, and it thereby draws attention to the roles which people other than teachers play in mathematics education, especially family members.
- Values and beliefs. Ethnomathematics makes us realise that any mathematical activity involves values, beliefs and personal choices.
- Interactions between mathematics and languages. Languages act as the principal carriers of mathematical ideas and values in different cultures.

In general, these points have provoked mathematics educators into giving more consideration to the overall structure of the mathematics curriculum and to how it responds, or sadly more usually how it does not respond, to the challenge of culturally based knowledge.

From the perspective of preschool education, my early work in this area (Bishop 1988) identified six general activities which every studied culture performs, in different ways and to different levels of sophistication. As well as researching mathematical practices outside the classroom, they also have been developed into firstly the commonly recognisable curriculum topics we know today, but also the complex academic activities we know about in universities.

The six universal activities are the following together with examples from non-mainstream mathematics cultures:

Counting This is to do with answering the question 'How many?', with inventing ways to describe numbers, recording them and calculating with them. Fingers, parts of the body, stones, sticks and string are just some of the objects which are used as 'counters'.

Locating This concerns finding your way around, navigating, orienting yourself and describing where things are in relation to one another. Compass direction, stars, the sun, wind and maps are used by people all over the world to find their ways and position themselves. Many spatial geometric ideas come from this activity.

Measuring 'How much?' is a question asked and answered everywhere. Whether it is amounts of cloth, food, land or money which are valued, measuring is a skill all people develop. Parts of the body, pots, baskets, string, beads and coins have all been used as units, as have written and drawn amounts on paper or cloth.

Designing Shapes are very important in geometry, and these come originally from designing objects to serve different purposes. The objects can be small and mundane like a spoon or symbolically important like a temple. Mathematically, we are interested in the shapes and the designs which are used, together with their different properties.

Playing Everyone plays and everyone takes playing ,very seriously! Not all play is important from a mathematical viewpoint, but puzzles, logical paradoxes, rules of games, strategies for winning, guessing, chance and gambling all demonstrate how playing contributes to the development of mathematical thinking.

Explaining Understanding why things happen the way they do is a universal human quest. In mathematics, we are interested in why number patterns happen, why geometric shapes go together, why one result leads to another, why some of the natural world seems to follow mathematical laws and in the process of trying to symbolise answers to these kinds of 'why' questions. A proof is one kind of symbolic answer, but there are many others, depending on what else you believe to be true.

Certainly every family group will engage in these activities to more or less an extent, and at the micro-level of culture, we can understand the need to assist teachers with making the links between the curriculum content in-school and in-preschool and the mathematical practices of the families outside school (Civil 1998). It also is an excellent basis for building on the constructivist approach to education at this level. As I understand it, Sweden has used this six-activity structure to develop a preschool mathematics curriculum. I hope it is proving to be a useful structure to help all your homeschoolers and preschoolers make the transition into formal mathematics successfully. But for this to happen, it is necessary to embed these activities within the value-laden frameworks of the classroom or home contexts.

The Contribution of Research on Culturally Situated Learning

Compared with Australia, Sweden appears to the outside world as a monocultural society with few multicultural schools, although this situation has changed during the last few decades. This is a markedly different educational context from that in Australia where we have the second largest number of different migrant groups in the world attending our multicultural schools. We can easily see that the culture experienced by learners in their homes is rarely the same as that represented by the school through its curriculum, its pedagogy and its values.

This kind of disjunction can easily lead to what I have called for many years 'cultural conflicts' (Bishop 1994). For many children around the world, the mathematics experience in schools is in cultural conflict with their home experience. Their situation is one of cultural dissonance, and the educational process is for them one of acculturation, rather than enculturation. The social groupings in which learners exist and learn mathematics inside and outside school have their own cultures, customs, languages and values.

This argument has been the basis for the development of the research on 'situated cognition' (Lave and Wenger 1991; Kirshner and Whitson 1997). When you learn something, you learn it in a context. For example, studies of the 'failures' of bilingual learners in a monolingual classroom, or of farmers' children studying in totally urban-centred curriculum, or of handicapped learners in mainstream classrooms all help to shed light on other explanations of failure and success besides the attributes of the learners themselves.

More research needs to focus on the transitions in learning as experienced by learners in cultural conflict situations. A good start has been made with the text by Abreu et al. (2002a) which is focused around five empirical studies of learners in cultural transition, with important theoretical perspectives added. These studies illustrate the range of contexts being researched now in mathematics learning from a cultural perspective.

The first empirical study is by Gorgorio et al. (2002). Their work was with immigrant students in Catalonia, an area of Spain, in which there are many migrants, and their study deals with the classroom complexities of cultural conflict and the enormous challenges facing the teachers in those schools and preschools.

The second study, reported by Bishop (2002a), also involved immigrant students but focused on the classroom challenges facing them, as they come to terms with different expectations, norms of teacher and student behaviour and the significant values.

The third study was carried out in Cabo Verde by Santos and Matos (2002). It was carried out in the city and focused on the mathematical activities learnt and practised by the *ardinas* who are the local newspaper sellers.

In the fourth study, Abreu et al. (2002b) describe the ways that parents participate in their children's mathematical progress at their multi-ethnic primary schools. They analyse the ways parents try to support their children's transitions from home to preschool and school, and they point to the conflicts experienced by both parents and their children as they try to come to terms with the unfamiliar, unwritten and unspoken values and 'rules' in the schools.

The fifth study, by Civil and Andrade (2002), also is concerned with the home/school relationship and features Mexican-American families and their children. As well as exploring the challenges the parents, their children and the teachers face, the study focuses on an experimental structured programme whereby teachers visit the homes of their students' families.

In all these situations and studies, the learners are clearly faced with negotiating transitions in knowledge and knowing, but they must also make transitions in values, language, customs and behaviours. These studies and perspectives enable us to see that learners are not just learning the cultural knowledge that they are being taught (as well as other knowledge that they are not taught, of course). They are in fact co-constructing that knowledge. This is the importance of the ideas behind the term 'transition' in this chapter, where it is used both as a noun and as a verb. More usually it is the noun that is emphasised, but the verb emphasises the complexity of the transition process. In particular it focuses our research attention on the contribution of the learners themselves.

This is probably also the most important contribution of constructivism—that it is not the individual who is constructing her/his own personal knowledge. Of course from a psychological point of view, that is important, but it is also rather obvious. What is much more important is the quality of the *social* situation that enables the learners to socially co-construct their new cultural knowledge. Knowledge changes with every generation, and it is mediated in that change by teachers and by learners of all cultural persuasions.

The Contribution of Research on Values to Mathematical Transitions

If we now consider research on mathematics pedagogy from the cultural perspective, one aspect that seems to be the most fundamental is, strangely, one of the most ignored, namely, that of values in mathematics teaching. It is ignored in both theory and practice. It seems that in keeping with a common idea that many people still believe, that mathematics education, like mathematics itself, is universal and therefore culture-free, it is also perceived by them to be value-free. This does not mean that they think mathematics has no value, but rather that they do not think that mathematics education conveys, or is the result of, any values over and above those values a particular society or culture is promoting through its educational, political and social institutions.

What should be of great concern to educators and parents alike is that values teaching and learning does of course occur in mathematics classrooms all the time, for example, whenever teachers or parents make decisions that affect their children's learning. Moreover, because most of this decision-making appears from our research to be done implicitly, there is only a limited understanding at present of what and how values are being developed.

So why should we study values and mathematics teaching? Surely teachers have enough to worry about teaching numbers, fractions, etc. without more abstract ideas. Here are some answers to that question:

1. Values emphasise the emotional and affective side of mathematics education, which despite its well-recognised significance is still not well researched.
2. Values are often ignored by researchers and teachers in their work context, yet they have a profound influence on the quality of mathematics learning in our preschools and schools.
3. Teaching mathematics without considering values is a nonsense—indeed I believe that the main reason many promising curriculum and teaching developments are not sustained is precisely because they do not take into account the value changes which are often implied.
4. Indeed, from a research perspective, trying to develop new curriculum and pedagogy practices without understanding the changing values involved is a futile exercise. For example, teaching the six universal mathematical activities above without considering values is to devalue their pedagogical usefulness. These activities have developed over the centuries and in diverse socio-historical mathematical contexts. The values involved are culturally hugely significant.

My research on values in mathematics education started in the 1980s as part of the concerns above. My particular interest was in the cultural dimension of mathematics education, and in the book on mathematical enculturation (Bishop 1988), I proposed six sets of values that I argued to be the main values adopted by mathematicians as they developed historically what has come to be called 'Western' mathematics. The analysis relied on various historical interpretations of the development of the activities of mathematics.

My educational argument is that these values are then 'carried' by the mathematics that is currently taught in schools and universities all over the world. They are mathematical values, as distinct from mathematics **educational** values which are imbued with values associated with each educational situation. Furthermore, these mathematical values will always be mediated by teachers and by every education system and will receive relative emphases in their teaching. Nevertheless, I argue that wherever Western mathematics is being taught, it is reasonable to assume that these values will always be portrayed by teachers and parents to some degree.

Some values will appear recognisable, while others may be rather more obtuse. They have been further discussed in the literature, explored in research studies and recognised as being important for educational purposes. In relation to this conference, I want to see how they could contribute to the transitioning of preschool learners into school. So far, the research has been mainly with primary and secondary teachers and their students, but it will be interesting to see what the ideas could offer to learners who are in the home/school transition process of preschool learning.

My particular research on values has its conceptual basis in a seminal work by White (1959). Briefly White, a cultural anthropologist who was neither mathematician nor educator, proposed four drivers of any culture, namely, technology, ideology, sentiment and sociology, the last three of which I have argued are the value drivers of the culture of mathematics, viewed itself as a symbolic technology (Bishop 1988, 1991). Their structure is connected with the six value clusters as follows, in three sets of complementary pairs:

Ideology: rationalism and objectism
Sentiment: control and progress
Sociology: openness and mystery

In providing details of these value clusters below, note that I am talking in general educational terms, not addressing preschool mathematics education specifically. I have also included various pedagogical activities, which enable the values to be seen in the learning context. Perhaps the best way to contextualise them in this chapter is to consider them in relation to your own knowledge and values.

Ideology: Rationalism

Valuing rationalism means appreciating argument, reasoning, logical analysis and explanation. It concerns theory and hypothetical and abstract situations. It includes appreciation of the aesthetics and beauty of mathematical proofs and is the main value cluster that people think about with mathematics. Pedagogical questions might be: Do you encourage your students to argue in your classes? Do you have debates? Do you emphasise mathematical proving? Could you show the students examples of proofs from history (e.g. different proofs of Pythagoras' theorem) and discuss their beauty and elegance?

Ideology: Objectism

Valuing objectism means appreciating and creating mathematical objects, the objectifying process and application ideas in mathematics. This cluster favours analogical thinking, symbolising and the presentation and use of data. Mathematicians throughout history have created symbols and other forms of representation for their ideas and have then treated those symbols as the object source for the next level of abstraction and theorising. Encouraging students to search for different ways to symbolise and represent ideas, and then to compare their symbols for conciseness and efficiency, is a good way to encourage appreciation of this value. Do you use geometric diagrams to illustrate algebraic relationships? Could you show the students different numerals used by different cultural groups in history? Could you discuss the need for simplicity and conciseness in choosing symbols? And why that helps with further abstractions?

Sentiment: Control

Valuing control means appreciating the power of mathematical knowledge through mastery of rules, facts, procedures and established criteria. It also promotes security in knowledge and the ability to predict. The value of 'control' is another one of which most people are very conscious. It involves aspects such as having rules and being able to predict, and it is one of the main reasons that people like mathematics. It has right answers that can always be checked. Do you emphasise not just 'right' answers but also the checking of answers and the reasons for other answers not being 'right'? Do you encourage the analysis and understanding of why routine calculations and algorithms 'work'? Could you emphasise more the bases of these algorithms? Do you always show examples of how the mathematical ideas you are teaching are used in society?

Sentiment: Progress

Valuing progress means appreciating the ways that mathematical ideas grow and develop, through alternative theories, development of new methods and the questioning of existing ideas. This cluster is also about the values of individual liberty and creativity. Because mathematics can feel like such secure knowledge, mathematicians feel able to explore and progress their ideas. This value cluster is involved in ideas such as abstracting and generalising, which is how mathematics grows. Do you emphasise alternative, and non-routine, solution strategies together with their reasons? Do you encourage students to extend and generalise ideas from particular examples? Could you stimulate them with stories of mathematical developments in history?

Sociology: Openness

Valuing openness means appreciating the democratisation of knowledge, through demonstrations, proofs and individual explanations. Verification of hypotheses, clear articulation and critical thinking are also significant in this cluster, as is the transparency of procedures and assumptions. Mathematicians believe in the public verification of their ideas by proofs and demonstrations. Asking students to explain their ideas to the whole class is good practice for developing the openness value. Do you encourage your students to defend and justify their answers publicly to the class? Do you encourage the creation of posters, for example, so that the students can display their ideas?

Sociology: Mystery

Valuing mystery means appreciating the wonder, fascination and mystique of mathematical ideas. It promotes thinking about the origins and nature of knowledge and of the creative process, as well as the abstractness and dehumanised nature of mathematical knowledge. Do you tell the students any stories about mathematical puzzles in the past, about, for example, the search for negative numbers or for zero? Do you stimulate their mathematical imagination with pictures, artworks, images of infinity, etc.?

These then are what I consider to be the fundamental values underlying so-called Western mathematics. But do they accord with others' ideas? I have preferred an approach (White's categories) which gives a good theoretical basis for the clusters and categories. However, the real issue is: Can those descriptions give a complete value curriculum picture, and framework, such as already exists in the traditional conceptual and technique curriculum? Is this values listing both necessary and sufficient? And how can that question best be answered?

A way of testing the ideas is to consider them in relation to various past curriculum projects. Changing curricula means changing values, or rather changing the balance between the six values. In Table 1, I have used a simple reference device, and I have assumed that each curriculum project emphasises the three clusters but with differences within each cluster. For example, if we assume that current/traditional curricula emphasise Ob, C and M , then other projects have emphasised other balances (note: Rat = Rationalism, Obj = Objectism, etc.).

So in Table 1 we can see that:

'New Math' emphasised R, C and M.
'Realistic Math' emphasised Ob, C and Op.
'Critical Math' emphasised R, C and Op.
'Investigations approach' emphasised R, P and Op (the complete opposite to the current curriculum emphasis.

In relation to the preschool (home?) phase, I would suggest that there is a need to emphasise Ob, C and M, while the in-school phase should transition the learners into R, P and Op. The research work goes on!

In another research context, concerning teachers and their values, several research projects based at Monash University from 1999 have been specifically concerned with mathematics teachers' values in primary and secondary schools. These were under the umbrella of the Values and Mathematics Project (VAMP) and included Bishop et al. (1999), Clarkson et al. (2000) and FitzSimons et al. (2000). In these studies we rarely found any explicit values teaching going on in mathematics classrooms.

Table 1 Mathematics curricula projects and the six values

	Rat	Obj	Con	Pro	Ope	Mys
New math	*		*			*
Realistic math		*	*		*	
Critical math	*		*		*	
Investigate	*			*	*	
Preschool		*	*			*
School	*			*	*	

Additionally, few mathematics teachers admitted to any explicit values teaching. Here are some other general conclusions that have come from these studies:

1. Teachers find it difficult to discuss values and mathematics, because they are not used to doing so. Often they do not have the words. But primary teachers and teachers in schools which incorporate values programmes are, generally, more able to do so than most secondary mathematics teachers.
2. Mathematics teachers do hold values about mathematics and about mathematics education; some of these are explicitly recognised and able to be articulated; others are tacitly or implicitly held, or perhaps not even recognised.
3. Teachers have many goals in planning for lessons, all of which involve value judgements, yet the values behind the goals and plans are rarely articulated.
4. Teachers may choose to make explicit certain mathematics or mathematics education values, or they may deal with them implicitly, although it is unclear just how consciously they make this choice.
5. In the actual classroom situation, teachers face a constantly evolving and unpredictable situation as each lesson unfolds. They are often faced with value conflicts which have to be resolved immediately and pragmatically, revealing the obvious need for some coherent structuring of teachers' values frameworks.
6. It is easier for teachers to think about and recognise the values they are teaching than to implement new values. This conclusion is, of course, not really surprising, but it carries huge implications for any curriculum development process. No teacher is autonomous and any developmental process must bear in mind the influences that affect the values that teachers can impart and develop in their classrooms.

We currently are lacking comparable studies which explore family's values relating to mathematics, although Abreu et al. (2002a, b) could be considered an example. Also Civil and Andrade's (2002) research indicates that teachers can be encouraged to visit the homes of their students and to find ways to bridge the gaps between the values taught in school and in the home. In some other research, reference is made to the 'elders' of the family and the community, and perhaps some of the research on community values may be helpful here, rather than focusing solely on the family unit. The 'community of practice' research may also inform this part of our research (see, e.g. César and Santos 2006).

The Adequacy or Otherwise of Our Research Methods

The essential goal of research in mathematics education is to help us understand phenomena in richer ways so that we can improve the mathematics teaching and learning situation for as many students as possible. But as we embrace fully the implications of a cultural perspective on mathematics curriculum, teaching and learning, are our research methods and procedures themselves adequate for the task?

There are several researchers who argue 'no' and that we need to change how research is carried out and conceptualised if we are to address these cultural aspects in the thorough way that they need to be addressed. As an example of this, at the PME conference in 1998, Valero and Vithal (1998) criticised the mathematics education research community for its imposition of research methods from the relatively developed 'north' onto researchers and students from the relatively underdeveloped 'south' part of the world. They argued that methods developed in one cultural context are not necessarily appropriate or helpful in another cultural context, in terms of what is considered 'normal'.

It is possible to use, for example, questionnaire approaches (see Appendix for an example of two questions which are based on the six value clusters). However, to develop more sensitive research on preschool mathematical transition, we clearly need to take on board the procedures and practices of anthropological and social psychological research. In general, our research approaches must move to a more collaborative teamwork style, involving not only practitioners and researchers, but also to include the learners and their families as partners in the research process (see Flecha and Gómez 2004; Goos et al. 2005; Stathopoulou 2007).

Just as we have found it necessary and beneficial to do research 'with' rather than 'on' teachers, so there is a need to develop ways of researching 'with' rather than 'on' learners and their families. Already qualitative methodologies have moved us closer to that goal, and if we are really serious about trying to improve our understanding of family mathematics education, then we have little choice but to engage them fully in the inquiry process.

This implies, as well as taking into account the cultural contexts of the learners and their teachers, we must also take into account our contexts as researchers. Just as we recognise the influences that their cultural contexts have on their situations, so we need to recognise the influences that our cultural contexts and values have on us and on our research. This is particularly the case with values which are often hidden. Values are often the 'hidden persuaders' of mathematics education (Bishop et al. 2003). Fortunately thanks to the kinds of research illustrated here, values are hidden no longer. What is needed now is for these ideas to be accepted and to be explored throughout the mathematics education community. Values awareness can certainly help parents and teachers transition learners into formal mathematics education.

Appendix: Two Items from the VAMP Teacher Questionnaire

3. "For me, Mathematics is valued in the school curriculum because…"	
……………………………………*Ranking*	
It develops creativity, basing alternative and new ideas on established ones	
It develops rational thinking and logical argument	

It develops articulation, explanation and criticism of ideas	
It provides an understanding of the world around us	
It is a secure subject, dealing with routine procedures and established rules	
It emphasises the wonder, fascination and mystique of surprising ideas	

4. "For me, Mathematics is valuable knowledge because…"	
…………………………………… *Ranking*	
It emphasises argument, reasoning and logical analysis	
It deals with situations and ideas that come from the real world	
It emphasises the control of situations through its applications	
New knowledge is created from already established structures	
Its ideas and methods are testable and verifiable	
It is full of fascinating ideas which seem to exist independently of human actions	

References

Abreu, G. de (2002). Towards a cultural psychology perspective on transitions between contexts of mathematical practices. In G. de Abreu, A. J. Bishop & N. C. Presmeg (Eds.), *Transitions between contexts for mathematics learning* (pp. 173–192). Dordrecht: Kluwer.

Abreu, G. de, Cline, T., & Shamsi, T. (2002). Exploring ways parents participate in their children's school mathematics learning: Case studies in multi-ethnic primary schools. In G. de Abreu, A. J. Bishop & N. C. Presmeg (Eds.), *Transitions between contexts for mathematics learning* (pp.123–147). Dordrecht: Kluwer.

Ascher, M., & Ascher, R. (1997). Ethnomathematics. In A. B. Powell & M. Frankenstein (Eds.), *Ethnomathematics: Challenging Eurocentrism in mathematics education* (pp. 25–50). New York: State University of New York Press.

Bishop, A. J. (1988). *Mathematical enculturation: A cultural perspective on mathematics education*. Dordrecht: Kluwer.

Bishop, A. J. (1990). Western mathematics: The secret weapon of cultural imperialism. *Race and Class, 32*(2), 51–65.

Bishop, A. J. (1991). Teaching mathematics to ethnic minority pupils in secondary school. In D. Pimm & E. Love (Eds.), *Teaching and learning school mathematics* (pp. 26–43). London: Hodder and Stoughton.

Bishop, A. J. (1994). Cultural conflicts in mathematics education: Developing a research agenda. *For the Learning of Mathematics, 14*(2), 15–18.

Bishop, A. J. (1999). Mathematics teaching and values education—An intersection in need of research. *Zentralblatt fur Didaktik der Mathematik, 31*(1), 1–4.

Bishop, A. J. (2001). What values do you teach when you teach mathematics? *Teaching Children Mathematics, 7*(6), 346–349.

Bishop, A. J. (2002a). The transition experience of immigrant secondary school students: Dilemmas and decisions. In G. de Abreu, A. J. Bishop, & N. C. Presmeg (Eds.), *Transitions between contexts for mathematics learning* (pp. 53–79). Dordrecht: Kluwer.

Bishop, A. J. (2002b, April). *Research policy and practice: The case of values*. Paper presented to the third conference of the Mathematics Education and Society Group, Helsingor, Denmark.

Bishop, A. J. (2008). Teachers' mathematical values for developing mathematical thinking in classrooms: Theory, research and policy. *The Mathematics Educator, 11*(1/2), 79–88.

Bishop, A. J., Clarkson, P. C., FitzSimons, G. E., & Seah, W. T. (1999, December). *Values in mathematics education: Making values teaching explicit in the mathematics classroom*. Paper presented at the 25th Australian Association for Research in Education conference.

Bishop, A. J., Seah, W. T., & Chin, C. (2003). Values in mathematics teaching: The hidden persuaders? In A. J. Bishop, K. Clements, C. Keitel, J. Kilpatrick, & F. Leong (Eds.), *International handbook of mathematics education* (2nd ed., pp. 715–763). Dordrecht: Kluwer.

César, M., & Santos, N. (2006). From exclusion into inclusion: Collaborative work contributions to more inclusive learning settings. *European Journal of Psychology of Education, XXI*(3), 333–346.

Civil, M. (1998). *Bridging in-school mathematics and out-of-school mathematics*. Paper presented at the annual meeting of the American Educational Research Association.

Civil, M., & Andrade, R. (2002). Transitions between home and school mathematics: Rays of hope amidst the passing clouds. In G. de Abreu, A. J. Bishop, & N. C. Presmeg (Eds.), *Transitions between contexts for mathematics learning* (pp. 149–169). Dordrecht: Kluwer.

Clarkson, P. C., Bishop, A. J., FitzSimons, G. E., & Seah, W. T. (2000). Challenges and constraints in researching values. In J. Bana & A. Chapman (Eds.), *Mathematics education beyond 2000. Proceedings of the 23rd annual conference of the Mathematics Education Research Group of Australasia* (pp. 188–195). Perth: Mathematics Education Research Group of Australasia.

Clarkson, P. C., Bishop, A., & Seah, W. T. (2010). Mathematics education and student values: The cultivation of mathematical wellbeing. In T. Lovat, R. Toomey, & N. Clement (Eds.), *International research handbook on values education and student wellbeing* (pp. 111–136). Dordrecht: Springer.

D'Ambrosio, U. (1985). Ethnomathematics and its place in the history and pedagogy of mathematics. *For the Learning of Mathematics, 5*(1), 44–48.

FitzSimons, G. E., Seah, W. T., Bishop, A. J., & Clarkson, P. C. (2000). Conceptions of values and mathematics education held by Australian primary teachers: Preliminary findings from VAMP (Values And Maths Project). In W. -S. Hong & F. -L. Lin (Eds.), *Proceedings of the HPM 2000 conference, History in mathematics education: Challenges for the new millennium* (Vol. II, pp. 163–171). Taipei: National Taiwan Normal University.

Flecha, R., & Gómez, J. (2004). Participatory paradigms: Researching "with" rather than "on". In B. Crossan, J. Gallacher, & M. Osborne (Eds.), *Researching widening access: Issues and approaches in an international context* (pp. 129–140). London: Routledge.

Gerdes, P. (1997). Survey of current work on ethnomathematics. In A. B. Powell & M. Frankenstein (Eds.), *Ethnomathematics: Challenging Eurocentrism in mathematics education* (pp. 331–372). Albany, NY: State University of New York Press.

Goos, M., Jolly, L., & Kostogriz, A. (2005). Home, school and community partnerships for numeracy education in a remote Indigenous community. In M. Goos, C. Kanes, & R. Brown (Eds.), *Proceedings of the 4th International Mathematics Education and Society Conference* (pp. 176–186). Gold Coast: Griffith University.

Gorgorio, N., Planas, N., & Vilella, X. (2002). Immigrant children learning mathematics in mainstream schools. In G. de Abreu, A. J. Bishop, & N. C. Presmeg (Eds.), *Transitions between contexts for mathematics learning* (pp. 23–52). Dordrecht: Kluwer.

Joseph, G. G. (1991). *The crest of the peacock: Non-European roots of mathematics*. London: I. B. Tauris.

Kirshner, D., & Whitson, J. A. (1997). *Situated cognition: Social, semiotic, and psychological perspectives*. Mahwah, NJ: Lawrence Erlbaum.

Lave, J., & Wenger, E. (1991). *Situated learning: Legitimate peripheral participation*. Cambridge: Cambridge University Press.

Santos, M., & Matos, J. P. (2002). Thinking about mathematical learning with Cabo Verde ardinas. In G. de Abreu, A. J. Bishop, & N. C. Presmeg (Eds.), *Transitions between contexts for mathematics learning* (pp. 81–122). Dordrecht: Kluwer.

Seah, W. T. & Bishop, A. J. (2000). Researching values with migrant/expatriate teachers of mathematics. In J. Wakefield (Ed.), *Shaping the future. Proceedings of the 37th annual conference of the Mathematical Association of Victoria* (pp. 348–355). Melbourne: Mathematical Association of Victoria.

Stathopoulou, C. (2007). The way a Romany (Gipsy) community perceives space notions and some educational implications for Romany students. *LIBEC Line—Revista em Literacia e Bem-Estar da Criança, 1*(1), 32–41.

Valero, P., & Vithal, R. (1998). Research methods from the "north" revisited from the "south". In A. Olivier & K. Newstead (Eds.), *Proceedings of the 22nd conference of the International Study Group for the Psychology of Mathematics Education* (Vol 4, pp. 153–160). Stellenbosch, South Africa: University of Stellenbosch.

White, L. A. (1959). *The Evolution of Culture*. New York: McGraw-Hill.

Negotiating with Family Members in a Block Play

Ergi Acar Bayraktar

Abstract In this chapter, a study on the impact of the familial socialisation on mathematics learning is described. The aim of the study is the development of theoretical insights in the functioning of familial interactions for the formation of children's mathematical thinking. The concept of the 'interactional niche in the development of mathematical thinking' is adapted to the special needs of familial interaction processes. It is integrated with the idea of Mathematical Learning Support System in order to shed light on how an elder sibling and a grandmother can be supportive or helpful for the mathematics learning process of a child. In this sense, the negotiation of meaning during the block play is observed and identified using interaction analysis. The result demonstrates that a block play with an elder sibling and a grandmother takes place as a social act for a child and an elder sibling and a grandmother provides different learning opportunities to the child, who is exposed to learning about giving, receiving, sharing and expressing his ideas and feelings. On the basis of this result, it can be concluded that through a block play with family members, a child gets an opportunity to think, to talk, to learn and to be 'educated' in mathematics and in cognitive, social-emotional competences as well.

Introduction

> The play, for the child and for the adult alike, is a way of using mind, or better yet, an attitude toward the use of mind. (Bruner 1983, p. 69)

Family is a group of people affiliated by consanguinity, affinity or co-residence. This principal institution is a micro unit of the social system and the cornerstone for socialisation of children (see Bayraktar 2014a, b; Acar 2011). In the context of everyday practices, it provides different resources, which are the basis for configuring options for individual education paths and offering more or less favourable conditions for the school success (Hawighorst 2000). In this regard, the education

E.A. Bayraktar (✉)
Goethe University Frankfurt, Frankfurt am Main, Germany
e-mail: acar@math.uni-frankfurt.de

T. Meaney et al. (eds.), *Mathematics Education in the Early Years*,
DOI 10.1007/978-3-319-23935-4_4

process of children begins first in the family and then continues in the school. A family enables children to grow up socially, while they experience school or the academic life individually. In this regard, family is the first institution for children, to prepare themselves for the wider world. The family makes up a child's community and lets them grow up within a closely linked network of people (Hughes 1986). The German Federal Ministry of Family Affairs, Senior Citizens, Women and Youth suggests that children test, discover and come to know their world with its forms, colours, sounds, surprises and regulations through their own family (Bundesministerium für Familie, Senioren, Frauen und Jugend 2002). Thus, the childcare provided by a family is a kind of 'strategy' (Kim and Fram 2009, p. 78), which enables children to live fruitfully. While the social status and educational level of family members appear to have significant influence on childcare, the relations of family members also have a significant influence on the formative values and beliefs about how children should develop (Kim and Fram 2009). Thus, the family provides plenty of opportunity for children to play, explore and make positive contributions to each other's life. However, how families combine their values and beliefs about how children should develop into support for learning of academic subject knowledge, such as mathematics, is not well understood.

Therefore, in this chapter, the participation of family members in block play is examined, in order to answer the question 'in which way and how much children should be "educated" in mathematics before entering primary school?'. The empirical material comes from the early Steps in Mathematics Learning-Family Study (erStMaL-FaSt). To undertake this investigation, the theoretical concept of *interactional niche in the development of mathematical thinking in the familial context* is used (see Bayraktar and Krummheuer 2011). It is combined with Mathematics Learning Support System (MLSS) (see also Bayraktar 2014c) in order to respond to the sub-question 'How and in what way does a support system in the familial context operate in order to enable children to be "educated" in mathematics before entering primary school?'

erStMaL Family Study

The family study investigates the impact of the familial socialisation on young children's mathematics learning. It is designed as a longitudinal study. In the wider project erStMaL, emphasis has been placed on understanding the experiences of immigrant children in Frankfurt, Germany, and their mathematical development from the age of three until the third year of primary school has been investigated (Bayraktar et al. 2011).

Eight children were chosen from the project erStMaL to participate in the family study. They were chosen based on the ethnic background of children (German or Turkish), the duration of the formal education of the parents and grandparents (more or less than 10 years) and the sibling situation within the families (see Bayraktar and Krummheuer 2011; Bayraktar 2012). The data collection included recorded videos and their transcripts. From 2010 to 2013, once a year, an appointment was arranged with each family, in which a total of three observation phases for each child were produced (see Bayraktar 2014a, b). At each appointment, the erStMaL child was video recorded together with members of the family while they are playing in different mathematical settings designed by the author (see Bayraktar 2012, 2014a, b).

For the family study, two mathematical domains were chosen: geometry and measurement. For each mathematical domain, two play situations were designed. 'Building' and 'geometric bodies' were connected to the mathematical domain 'geometry', whereas 'weight' and 'towers' were connected to the mathematical domain 'measurement' (see Bayraktar 2012, 2014b). The members of the family are asked to choose at least two games out of four presented and to play them. In order to assist families in the play situation, instruction manuals both in German and Turkish for each play situation are made available. Family members can speak either or both languages during the play situations. Additionally, all game materials are made available to the family in the recording room at the university, at home or in the kindergarten so that they can feel free to play with their children. Each play situation is arranged for limited play rounds: 'Towers' and 'building' are scheduled for five rounds. 'Geometric bodies' and 'weight' are scheduled for three rounds.

Theoretical Basis of Interactional Niche and MLSS

During each play situation with the participation of a family member such as father, mother, sibling and grandmother, the child explores something about the mathematical topic. The attendance of family members provides children with different learning opportunities that arise from the interactive process in regard to the negotiation of meaning about the mathematical play. In this way, different forms of participation and support emerge during the interaction processes.

From a socio-constructivist perspective, the cognitive development of an individual is bound to their participation in a variety of social interactions. Moreover, individuals can support their own and others' development while they are negotiating with each other interactively.

The analysis of these concepts of the 'interactional niche in the development of mathematical thinking' (NMT) consists of the aspects of allocation, situation and the child's contribution (Krummheuer 2011a–c, 2012, 2014; Krummheuer and Schütte 2014). Allocation refers to the provided learning offerings of a group or a society, which specifically highlight cultural representations. The aspect of situation consists of the emerging performance occurring within the process of negotiating meaning. Lastly, the aspect of the child's contribution is the situational and individual contribution of the focused child.

NMT-Family is constructed as a *subconcept* of NMT. It offers the possibility for closer analyses and comparisons between familial mathematical learning opportunities for children in early childhood and primary school. In Table 1, the three components of NMT-Family are described in relationship to content, the cooperation and the pedagogy and education perspective.

Table 1 The structure of NMT-Family (Bayraktar 2014c)

NMT-Family	Component: content	Component: cooperation	Component: pedagogy and education
Aspect: allocation	Mathematical domains: geometry and measurement	Play as a familial arrangement for cooperation	Developmental theories of mathematics education and proposals of active participation of family members on this theoretical basis
Aspect: situation	Interactive negotiation of the rules of play and the content	Leeway of participation	Folk theories of mathematics education, everyday routines in mathematics education
Aspect: child's contribution	Individual actions	Individual participation profile	Competence theories

Each cell of Table 1 is described as follows:

Content

Content × Allocation In the practice of the family study, children and their families are confronted with mathematical play situations.

Content × Situation The play situations are designed to offer families different opportunities to negotiate meaning interactively. The rules of the play situation and/or mathematical topics are the focus of the negotiation processes between family members.

Content × Contribution The focused child might *contribute* more or less actively to the negotiation processes in the play situations. In such processes, either different forms of efficient and original ideas can be expressed and realised or the activities of other participants can be pursued (see Krummheuer and Brandt 2001).

Cooperation

Cooperation × Allocation In the play situations, each family member can cooperate with each other. This process of cooperation between family members and the child provides different opportunities in order to refine their thinking and make their performance more effective.

Cooperation × Situation Depending on the cooperation process, different leeways of participation emerge. 'Leeway' is perceived as a 'room for freedom of action'

(Krummheuer 2012, p. 322). This type of participation in the interaction encourages the child to explore, by 'co-construct aspects of the cultural environment' during play (Brandt 2004, pp. 32–43).

Cooperation × Contribution Brandt (2004) explains that the participants interactively accomplish different versions of leeways of participation that are conducive or restrictive to the mathematical development of a child (see also Krummheuer 2011c, 2012). The children are involved in the social settings in the play situations, which are variously structured as in child–parent interactions and/or child–sibling interactions. These social settings need to fulfil the process of interaction. In this way, different 'leeways of participation' for the child emerge (see Brandt 2004, 2006), in which individually different participation profiles of the child are generated in the joint interaction.

Pedagogy and Education

Pedagogy and Education × Allocation Developmental theories and theories of mathematics education describe and delineate learning paths in the familial context for the children's mathematical growth (see also Bayraktar and Krummheuer 2012).

Pedagogy and Education × Situation Bruner emphasises that folk pedagogy reflects a variety of assumptions about children: they may be seen as willful and needing correction, as innocent and to be protected from a vulgar society, as needing skills to be developed only through practice, as empty vessels to be filled with knowledge that only adults can provide and as egocentric and in need of socialisation (1996, p. 49). Brandt highlights that the different concepts of folk pedagogy of Bruner are linked to different concepts or ideas of 'mind' and 'knowledge' in general (Bruner 1996, p. 50 f.), which offer the opportunity for content-related deliberations of different instruction practices (Brandt 2014, p. 57). With respect to these ideas of folk pedagogy (Bruner 1996, see also Brandt 2013, 2014), the participating adults and children become *situationally* active and operant in the concrete interaction process.

Pedagogy and Education × Contribution In the manner of the child's participation profile (see Brandt 2004, 2006), the learning process of the focused child can be characterised. The child is intuitively able to describe the change and/or progress of own participation in development of mathematical thinking. In this sense, appropriate theories like 'self-regulation' (see Nader-Grosbois et al. 2008) or 'situated learning' (see Lave and Wenger 1991) can be grouped under the title 'competence theories'.

Mathematics Learning Support System Bruner emphasises that play is a way of using the mind or an attitude towards the use of the mind for children and their families (Bruner 1983, p. 69). In this sense, a child and each family member interact with each other regularly, socially and unintentionally during play, in which different learning opportunities are provided. Consequently, a Mathematics Learning Support System (MLSS) is generated:

> The interaction system adapts itself to the possibilities of participation of the involved children in that it generates a kind of conversation that enables at least some children to contribute actively to this interaction. This adaptation can result in a pattern of interaction that one could characterize as a format. In general, the result of this adaption is the MLSS (see Krummheuer and Schütte in this book, p. 171).

An MLSS can occur in different ways through patterns and routines of interactions between the child and family members during play.

Block Play as an Intervention

Block play enables children to learn a diverse range of valuable competencies and knowledge, from social skills to the foundations for later mathematics achievement (Kersh et al. 2008; Hewitt 2001). In the block play, the child can experience taking turns, sharing and respecting the rights of others and learning to cooperate and have knowledge of various roles and skills while exploring, matching and classifying the sizes, shapes, distances and proportions (Bullock 1992; Cartwright 1988; Bayraktar 2014b). Through building, stacking and balancing blocks, the child gets an opportunity to learn the sense of balance and symmetry. Moreover, as Bullock (1992) emphasised, such a block play situation enables the child to learn patience, increase independence and contribute to a sense of accomplishment. Therefore, it was chosen in order to observe the learning opportunities available in the interactions between siblings and a grandmother. Bullock (1992) emphasises that the block play contributes to children gaining physical, social, emotional and cognitive growth (see also Cartwright 1988; Bayraktar 2014b). The negotiation of meaning during the block play was identified using interaction analysis (for more details, see Krummheuer 2011d). Furthermore, different participation models of family members during the block play will be considered.

The Block Play of Family Gül

Family Gül is a German-Turkish family who lives in a major German city. Berk is the focus child who is aged 7 years and 1 month old. He speaks German and rudimentary Turkish. He has an elder brother, who is about 13 years old and goes to a secondary school in Germany. The elder brother can speak fluently German and Turkish. The child's grandmother, the mother of the child's father, has a close relationship with Family Gül and cares for Berk and his elder brother when they return from school. She can speak Turkish and rudimentary German. Her education consisted of 5 years in the Turkish school system.

The first analysis consists of the play 'Building 02', about geometry and spatial thinking, which occurred in the second observation phase. The family is supposed to build a three-dimensional version of a provided picture with wooden blocks, all having the same size and weight. In doing so, they are supposed to relate two-dimensional representations to three-dimensional representations. The player chooses one card from the deck and replicates the building that they see in the image on the card. During the play, the cards are placed on the table face down.

In the first extract from the video recording to be discussed, Berk is playing with his elder brother and grandmother, who is called granny. They play ten rounds by taking turns. The extract comes from the second round when Berk and the granny pick up the same card from the deck. It begins with Berk's turn. The beginning of the extract is briefly explained, and then two key points of Berk's turn are highlighted and analysed. Afterwards, the turn of the elder brother and the beginning of granny's turn are briefly described. Then the final part of granny's turn is analysed.

During the whole interaction process, this group speak both in German and Turkish by code switching between them. In the transcripts, speech which was originally stated in German is written in normal font, and speech which was originally written in Turkish is underlined. Sometimes, the granny spoke in English. These comments are underlined twice.

The video extract began with Berk picking up a card from the deck and showing it to his brother and granny (see Fig. 1). He then started building immediately. He put the first three blocks (K1, K2, K3) vertically next and parallel to the each other (see Fig. 1). Then he put two blocks (K4, K5) next to each other horizontally on the initial three blocks. After this, he placed a block (K6) vertically where blocks K4 and K5 meet and directly above K1. The final block (K7) is put on K6 horizontally in a parallel direction to K4 and K5. During the block-building process, granny acknowledged Berk's effort, whereas his elder brother stated that Berk did not make the building identical to the figure on the chosen card. The elder brother smiled and told granny in Turkish that Berk built it wrong. Granny confirmed his statement and agreed that Berk did not replicated correctly what was on the card.

In the figure on the card, there are three horizontal blocks (d, e, h) which can be seen in Fig. 4. In Berk's building, there are only two horizontal blocks (K5, K4) (see Figs. 2 and 3). Thus, the building resembles but is not identical to the figure on the card.

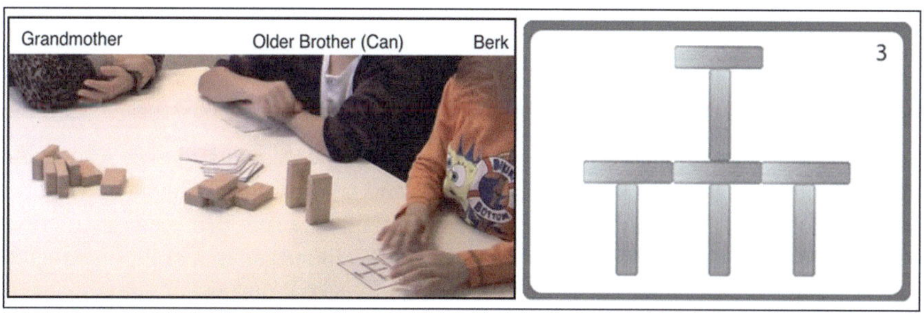

Fig. 1 Recording position and the chosen card

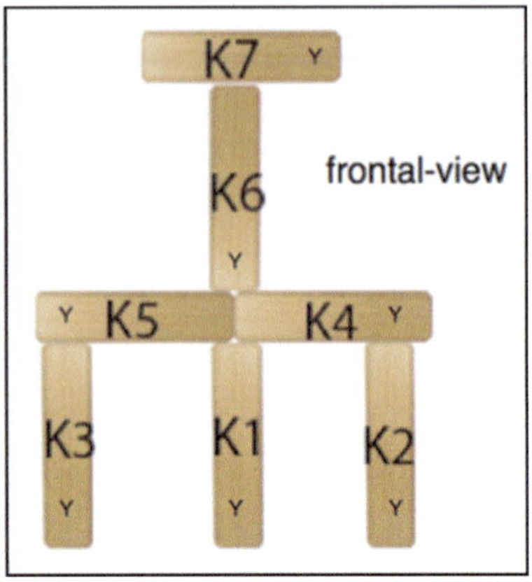

Fig. 2 The built construction by Berk

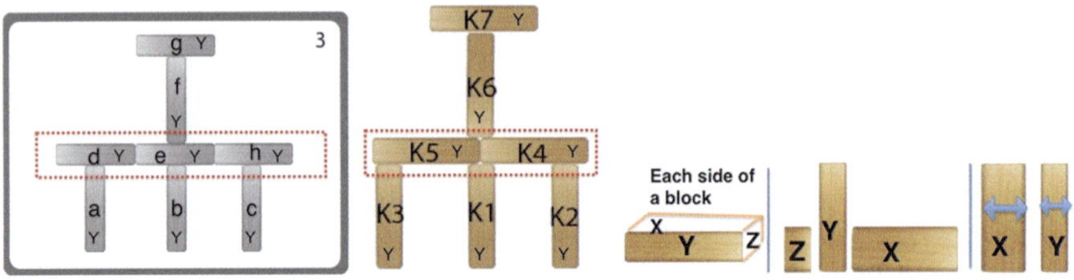

Fig. 3 The chosen card and the built construction with the named blocks and sides from the front view, and the wide side (X) and narrow side (Y) of building blocks

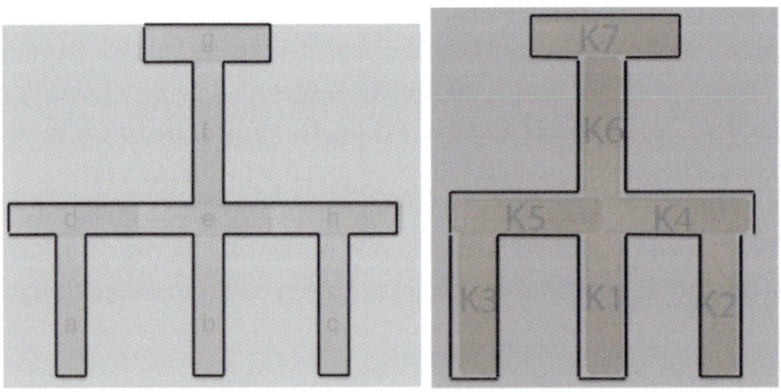

Fig. 4 'Gestalt' of the chosen card and the built construction

To build a nonidentical corpus to the figure on the card can be either a 'gestalt' problem (1) in which Berk sees only the whole and not the individual parts or 'static balance' problem (2) in which Berk focused only on ensuring that the construction was stable (see Fig. 4). Clements and Sarama (2007) identified that children balance blocks intuitively and so often they place blocks off centre. Instead of situating each block on one block, Berk might prefer to situate K5 on two blocks (K3 and K1) and K4 on two blocks (K1 and K2) (see Fig. 3). In such a way, he makes the construction more stable.

It seems that for Berk, the construction resembled the figure on the card. From a developmental perspective, he was able to represent block sides at the detailed level. However, topologically he might have either an adaptation problem by decomposing and then composing shapes or a spatial problem by regulating spatial relations of the objects and figures.

Berk looked at the chosen card and asked why the construction was wrong. His older brother gave him a feedback immediately by rubbing the surface of blocks K4, K5 and K6 from the construction while saying in German, 'this stands in the middle and there are three of them'. His answer can be interpreted in different ways:

- 'This is in the middle': 'In the middle' could be referring to blocks d, e and h in the figure or K4 and K5 in the construction, which are shown with the red lines in Fig. 3. He might be comparing block e or h in the picture with block K4 in the construction.
- When the picture and the construction are compared, it can be seen that one block is missing in the construction. Consequently, the older brother's explanation 'This is in the middle' could be interpreted as:

 1. 'This block (K4) has to be between the blocks d and h, so that K4 could be in the middle of construction'. If a block was not missing in the construction, then block K4 would have been in the middle. In this way, block e in Fig. 3 could represent block K4 in the construction, and the missing block in the construction would be block h in the picture.
 2. 'This block (e) is in the middle of the construction. Here it has to be one more block in the middle'. If this block were not missing in the construction, then between the blocks K4 and K5, there would have been one more block in the middle. Thus, block e (see Fig. 3) is the missing block, while K5 represents block d and K4 represents block h in the construction.

- 'There are three of it': This explanation can be interpreted as follows:

 3. 'There are three blocks (d, e, h) in the middle of the figure (see Fig. 3), so there have to be three blocks in the construction'.
 4. 'The block (K4) has to be in the middle of the construction; thus in total, there have to be three blocks. But here there are two.'
 5. 'There have to be three blocks and K4 has to be in the middle. But Berk used only two blocks'.
 6. 'The block (e) (see Fig. 3) is in the middle of the figure. In the picture, there are three blocks, which have red lines around them in Fig. 3. But there are two in the construction. There have to be three blocks.'

- 'This is in the middle and there are three of it': the brother might mean that the block K6 stands in the middle of the construction and there have to be two more blocks which are the same as K6. Then together there would be three blocks.

From the participatory point of view, the elder brother took the role of an autonomous person, who expressed his ideas and so had the freedom to act independently (see Krummheuer and Brandt 2001). He stated the reason why he considered the construction to be incorrect, so taking on the standard role of an adult mentor.

It could have been expected that this role might have been reserved for granny. From the socio-constructivist perspectives, the adult mentors the child's learning process, by being an experienced and trusted adviser. Moreover, the adult mentor provides aid for the child that guides inherently.

The elder brother responds to his younger brother by acting as an adviser. However, although his response contains sufficient knowledge on geometry, it does not utilise sufficient knowledge on pedagogy. By conveying his thoughts, the elder brother only gave Berk a geometrical hint but took no notice of whether Berk understood it. His mentoring was technical but pedagogically insufficient as Berk showed a lack of comprehension about the inadequacies of his construction.

Consequently, the elder brother removed the chosen card away and placed it on the deck on the table. In this way, he indicated that Berk's turn had expired and it was his turn. The older brother chose a card from the deck and built an identical construction to the figure on the chosen card. Then granny began her turn as the last player in the second round. She chooses the same card from the deck as Berk and started to build up a construction similar to Berk's (see Fig. 5).

During granny's block building, Berk's brother was laughing. The grandmother asked whether the building construction was wrong. Berk's brother confirmed this and granny asked how it was wrong. The brother gave the same explanation as he had for Berk's turn: 'there are three pieces just look. In the middle there are three pieces'. This time, granny asked for an exact definition of what he meant by 'three pieces'. Berk's brother picked one block (K1) from the building construction and set it in front of him. He put on it another block (K6), which lay on the table. As he said 'like that', he took two more blocks (K7, K8) from the pile of blocks and set them next to K1 and K6 (see Fig. 6).

In this way, the elder brother showed granny how to build the construction correctly, using action rather than words to inform her what to do. He modelled behaviour for granny so she could observe and make sense of what is required. Through the modelling process, the elder brother made his reasoning publicly accountable and visible. From a participatory point of view, the elder brother again adopted the role of adult mentor and in a technical and operational way that he mentors granny and Berk who was also present and watching. He appeared to be an *adviser* for his younger

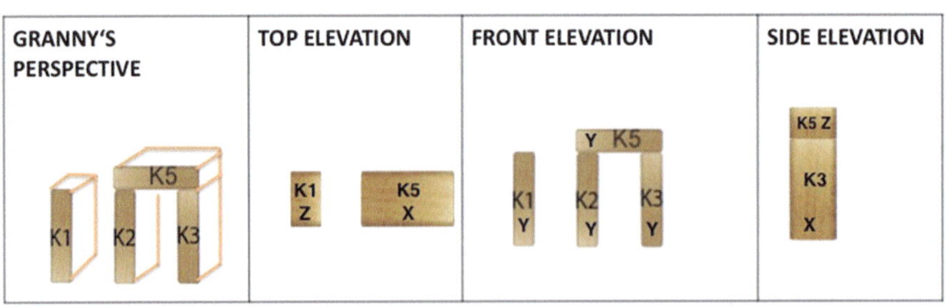

Fig. 5 Granny's built construction over time

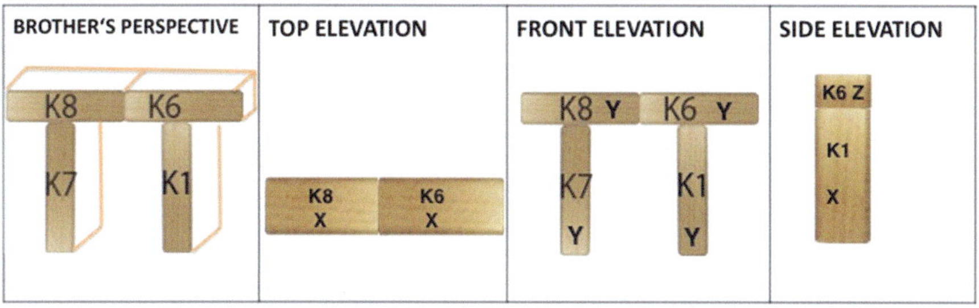

Fig. 6 The modelled blocks by the elder brother

brother and granny. From a developmental perspective, the ongoing action of the elder brother supported Berk to explore and to examine situating each block right.

This can be seen in the following transcript:

Transcript[1]

527	Granny	*(Took K4 from her construction and set it on the right side of K7. Then she took K5 from her construction and centred it horizontally on K4)* (see Figs. 7 and 8). One more isn't it?
530	Brother	*(Nodded).* Too far..yes.
531	Granny	Yes. *(Took K2 from her construction and placed it vertically and centred on K8 on the older brother's construction)* (see Figs. 9 and 10)
533	Berk	Granny says yes
534	Granny	*(Laughed, took K3 from her construction and put it horizontally on K2)* O.K.? (see Fig. 11)
536	Brother	*(Nodded)*

By posing the question 'one more isn't it?' (line 527), granny appeared to request a clue from the elder brother, about whether she was building the construction correctly. Her reaction might also have shown that she had internalised the elder brother's action in the modelling and his knowledge about geometry. In this way, she may have tried to provide a supportive learning situation (Bruner 1978) indirectly for Berk, showing how to obtain more detailed information from the elder brother. Following the instructions, she was better able to complete the task successfully, and in doing so, she modelled for Berk how each block should be placed to build the construction correctly. In this way, she continued the actions of the elder brother. Granny's reaction showed that she was in agreement with her elder grandson that the construction should be built correctly. The actions of granny and the older brother positioned Berk as the 'legitimate peripheral participant' (Lave and Wenger 1991). From the developmental perspective, granny's actions support Berk to

[1] Rules of Transcription

Column 1	Column 2	Column 3
Serially numbered lines	Abbreviations for the names of the interacting people	Verbal (regular font) and non-verbal (italic font) actions

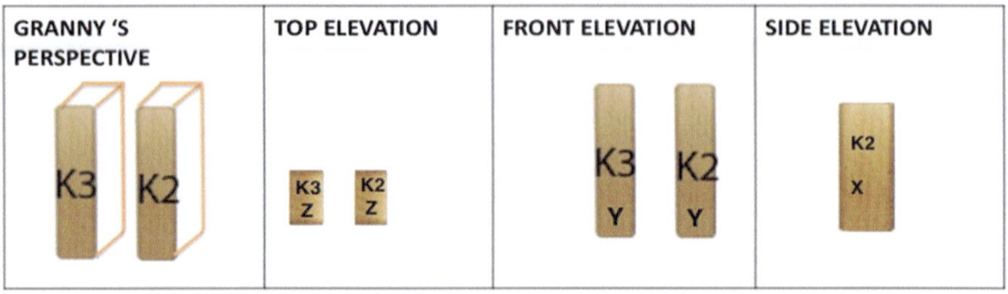

Fig. 7 Set blocks in front of granny and their elevations according to granny's perspective

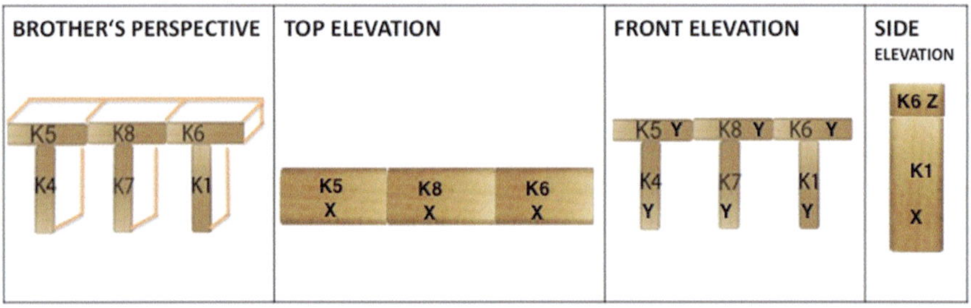

Fig. 8 Set blocks in front of Berk's brother and their elevations according to brother's perspective

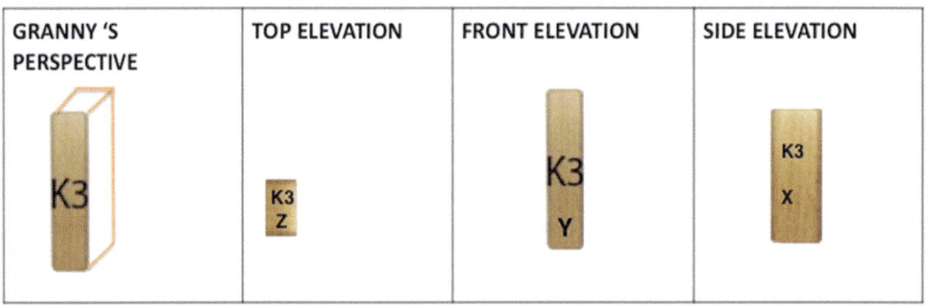

Fig. 9 Remained block in front of granny and its elevations according to granny's perspective

continue exploring and examining how to place each block in order to build an exact replica. In this way, Berk's relatives encourage him to engage in the building and learning process (Fig. 12).

When the brother nodded and then said to granny in Turkish '<u>too far..yes</u>' (line 530), he probably was directing her on how she should set the blocks. By saying '<u>too far</u>', he could have meant that she placed the blocks K4 and K5 too far apart or from the construction or to the place, where he built his construction. After this, he seemed to approve of granny's action to place the blocks. In this way, he continued to provide information also to Berk. From the developmental perspective, the elder brother provides Berk with a learning opportunity on how to position the blocks and

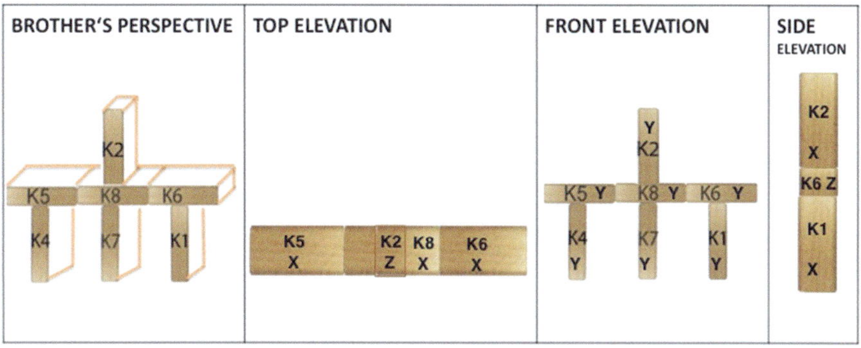

Fig. 10 The building construction and its elevations according to brother's perspective

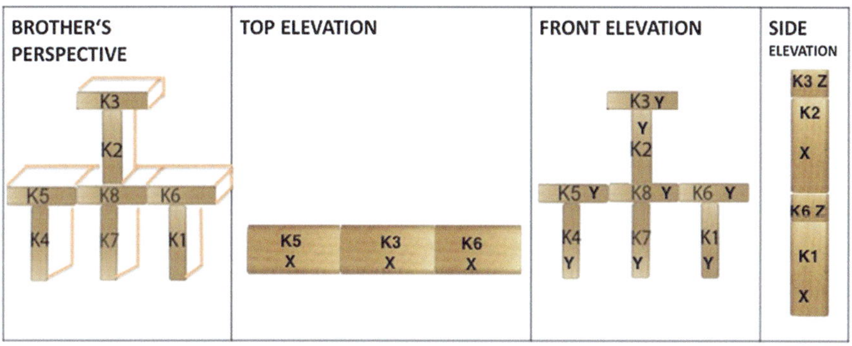

Fig. 11 The built construction and its elevations according to brother's perspective

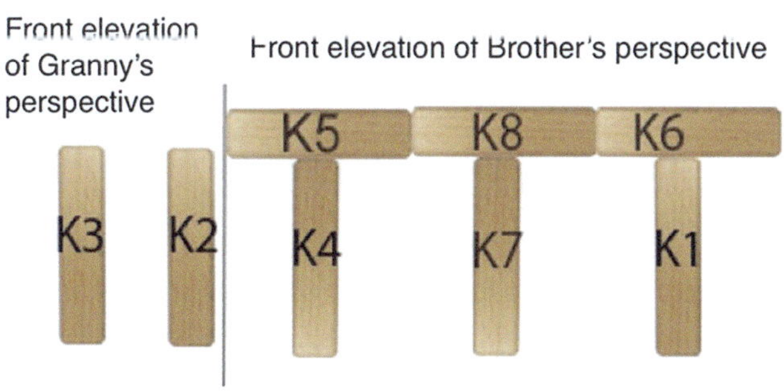

Fig. 12 The built constructions by granny and elder brother at the beginning of the chosen transcript

how spatial relations and kinaesthetic imagery can be configured. From the participatory point of view, the elder brother continued with the role of an adviser, while Berk continued as a legitimate peripheral participator.

When Granny said 'yes' in English (line 531) as she placed K2 vertically centred on K8, it may be that switching languages indicated a greediness for success, which

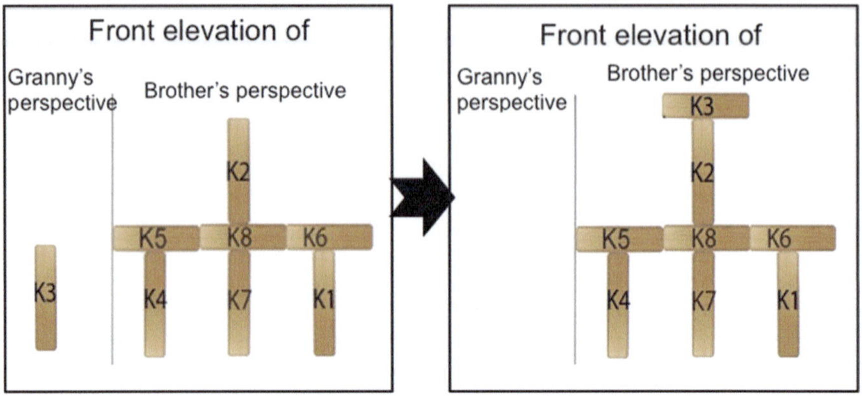

Fig. 13 Granny goes on building and completes the construction

can be considered an emotional expression. Maybe granny praised herself in this way because of the elder brother's confirmation of her answer. The use of English by granny is unusual and remarked on by Berk (line 533). From the participatory point of view, Berk gave the impression of having two roles, an observer and an overhearer.

Following Berk's surprised comment, granny laughed and placed K3 horizontally on K2 (see Fig. 13). Her laugh reinforces the assumption that she had used English as a way of congratulating herself on her perseverance and achievement.

Then she asked elder brother if her construction is OK in line 534 (see Fig. 13). It may be that she hoped that the elder brother would give her detailed information about building the correct construction as he has done before. In so doing, she provided Berk with a learning opportunity about building an identical construction to the figure on the card. From participatory point of view, granny appeared to ascribe the role adviser to the elder brother and observer and overhearer to Berk.

Delineation of the Interactional Niche of Family Gül

In the extract from the video, the elder brother took on the role of an adult mentor, which is *unexpected* for the standard social-constructivist approaches (see Brandt 2004, 2006, 2013, 2014; Bruner 1978, 1983,1986, 1990, 1996; Cobb 2000a–c; Cobb and Bauersfeld 1995; Cobb et al. 1993, 2000; Cobb and McClain 2001; Ernest 1991, 1998; Palincsar 1998; Rogoff 1990; Rogoff et al. 1984; Sfard 2001, 2008; Tiedemann 2010, 2012). Generally, it is an adult who takes on the role of a 'mentor' in the child's learning process. In the extract, older sibling tended to guide and to encourage his younger sibling's play (for more details, see Vandermaas-Peeler 2008; Doron 2009), although he is not an adult person. He acted like a wise adult rather than the granny. He paid attention to the building activities of Berk and granny and gave them instructions, feedback and clues. As his reactions look like guiding and encouraging, three components of the interactional developmental niche (NMT) can be suggested for Berk in this familial context:

Content

Content × Allocation The play is in the mathematical domain 'geometry', so each family member works on their own spatial skills, through spatial structuring and operating with shapes during the block play.

Content × Situation Berk negotiated with his elder brother and granny about building the replica of the picture. There was a consensus that Berk had built the construction incorrectly, whereas in granny's turn, a modelling process for Berk emerged when granny and the elder brother negotiated how to build the construction correctly. The elder brother provided more information about his reasoning through his actions and with granny participated critically but cooperatively.

Content × Contribution During his turn, Berk built a construction actively and autonomously and did not comment on his construction. In granny's turn, he observed the activities of granny and the elder brother and overhears their arguments.

Cooperation

Cooperation × Allocation The video-recorded extract focused on the two turns of Berk and granny from one play round. With the interventions of Berk's older brother, a polyadic interaction process emerged.

Cooperation × Situation During Berk's turn, he was an active participant, who negotiated with his older brother about his construction. The elder brother mostly directed the play situation and supported Berk briefly to explore how the construction had to be built. However, the interaction developed in such a way that Berk's leeway of participation was limited to discussing what was wrong with his construction, first with his elder brother and then with his granny. During granny's turn, negotiation of the building activities occurred between her and the elder brother. This provided Berk with different learning possibilities, from observing and overhearing. In this way, granny and the elder brother ascribed to him the role of a 'legitimate peripheral participant'. Consequently, it can be said that Berk was involved in different leeways of participation.

Cooperation × Contribution Berk presented his own idea during the block-building process and so acted as an autonomous person who expressed his own ideas and had the freedom to act independently (see Krummheuer and Brandt 2001). As well, he tried to negotiate with his elder brother about building the construction, which situated him as an active participant. During granny's turn, she tried to provide a position for Berk, in which the elder brother gave more detailed information about how to build a correct construction. In this way, granny ascribed the role of 'adviser' to the elder brother, while she reserved for Berk the role of an observer and overhearer. In the interaction process during granny's turn, Berk gained experience of a situated learning process. By observing the situated activities of his elder brother and granny, Berk was able to mentally compare his construction with the final construction and so took the role of legitimate peripheral participant. By recognising the differences between the two built constructions, he could understand what was wrong with his construction.

Pedagogy and Education

Pedagogy and Education × Allocation In the play, four goals are pursued: spatial structuring, manipulating shapes and figures, static balance between blocks and identifying the faces of 3-D shapes to 2-D shapes. The US National Research Council (2009) reports that 5-year-old children can understand and replicate the perspectives of different viewers. These competencies reflect an initial development of thinking about relating parts and wholes (National Research Council 2009, p. 191). The analysed extract from the video recording was therefore about exploring and examining the spatial structuring, visualising and kinaesthetic imagery.

Pedagogy and Education × Situation In the extract, Berk and his family members are involved in a spatial play situation. In terms of folk pedagogy (Bruner 1996), granny participated in the mathematical play situation by using her pedagogical knowledge, which integrally supports the mathematical development of Berk. Using her pedagogical knowledge, she put forward her suggestion on how to build the construction in order to provide Berk with vicarious learning (Bruner 1978). To do this, she encouraged the elder brother to give more detailed information about building a replica construction of the figure on the card. In this way, they make visible their reasoning and reached eventual agreements of how to achieve the correct solution. Granny's accumulated set of beliefs, conceptions and assumptions (Bruner 1996) enabled her to direct the elder brother pedagogically and provide Berk with an indirect learning situation. The elder brother directed the play situation by his technical and operational mentoring. In terms of folk pedagogy (Bruner 1996), granny used her pedagogical knowledge to let her elder grandson experience the practice of teaching but with the purpose of benefitting Berk. The elder brother modelled the correct way of building the construction and provided Berk with a way of recognising his error and discovering how to build the construction correctly. Thus the graphical and spatial development of Berk are indirectly assisted.

Pedagogy and Education × Contribution For learning to occur, the child did not have to operate with shapes and figures physically in the leeway of play. During Berk's turn, he examined the static balance of the construction through his active and autonomous participation. Thus, he could explore the idea of building the stable construction. He also experienced the spatial relations between 2-D and 3-D objects so that he could relate some parts with the whole. Furthermore, he grasped the whole of the figure on the chosen card and his construction.

During granny's turn, Berk indirectly examined the process of placing each block. Through the modelling process, the negotiation between the elder brother and granny was made public, making their reasoning more visible by orienting the play to critical, cooperative and situated reasoning. Berk gained an opportunity to observe their activities and structure shapes, figures in his mind. Paradise and Rogoff (2009) point out that learning can be 'observational' (p. 107), while children are participating by paying close attention to ongoing events. The context of a practical engagement provided Berk with different learning opportunities about how the position of the blocks could be represented and how spatial relations and kinaesthetic imagery could be configured. Therefore, granny's turn was informative for Berk that he got a chance to compare his construction with the correct construction in his mind and to realise his mistake.

Summarising the discussion of the results, NMT is represented in Table 2.

Table 2 The structure of NMT-Family Gül

NMT-Family Gül Building 02	Component: content	Component: cooperation	Component: pedagogy and education
Aspect of allocation	Mathematical domain: 'Geometry and spatial thinking', using spatial skills at the building activity	Playing with elder brother and granny	Theory of the development of spatial skills and spatial structuring: identifying the faces of 3-D shapes to 2-D shapes, relating parts and wholes, replicating the perspectives of different viewers and directly and indirectly operating shapes and figures
Aspect of situation	Negotiation between brother, granny and Berk about the building construction; *consensus*	Different leeways of participation	Modelling, by which granny and elder brother directly act with the play materials Enabling him to examine his construction with the last built construction and to realise his mistake Providing Berk with an opportunity to confront his mistake *Technical* and *operational* mentoring by elder brother
Aspect of contribution	Examining the resembling gestalt of the figure and the construction Overhearing all the situated activities of elder brother and granny	Active participant in his own turn Observer, overhearer and legitimate peripheral participant in granny's turn	Exploring static balance to build the robust construction Witnessing of all the situated activities of elder brother and granny Learning opportunities, how the position of the blocks can be represented and how spatial relations and kinaesthetic imagery can be configured

Berk's interactional developmental niche came from being mentored by an elder brother and being confronted with his own mistake. In this way, Berk had an opportunity to 'see' what was wrong with his construction. To be mentored by his elder brother provided Berk with the opportunity to improve his spatial abilities. The spatial processing in young children is not qualitatively different from that of older children or adults, but children produce progressively more elaborate constructions, as they get older (Clements and Sarama 2007). In this sense, spatial processing by Berk could be enhanced so that he could produce more elaborate constructions with the help of his brother's mentoring.

Mathematical Learning Support System for Berk

In the way he operated, the elder brother could be considered geometrically wise but pedagogically insufficient. He gave Berk geometrical hints but not pedagogically in that he did not take note of Berk's understanding. He mentored Berk in technical and operational ways. On the other hand, granny, who has limited education and only rudimentary German, set up an MLSS for Berk by utilising the participation of the elder brother. Granny is knowledgeable about her grandchildren's rearing and their custodial caring (Goodfellow and Laverty 2003). She was able to transmit her beliefs and behaviours across the generations in terms of folk pedagogy (Bruner 1996), by directing Berk's elder brother pedagogically so that Berk was provided with an indirect learning situation. She directly influenced the children's acts through her face-to-face interactions (Smith 2005). Berk seemed to accept the directions from his older sibling (Abramovitch et al. 2014) and to *elicit* many more explanations from him than from an adult (Smith and Drew 2002). A favoured teaching status is enabled in which granny provided for both children. The elder brother came to be treated as an expert and Berk a novice (see Abramovitch et al. 2014). Granny's folk-pedagogical knowledge contributed to the elder brother becoming responsible to his younger brother as an adviser. In the negotiation process, granny took an indirect pedagogic role towards her grandchildren. In the learning situation, Berk has the status of a legitimate peripheral participant. In this sense for Berk, a mathematical learning support system is realised through granny and her elder grandson. Summarising the discussion of the results, the structure of MLSS is represented in the NMT Table of Berk and showed with the bold-labelled area in Table 3.

Epilogue

Mathematical play situations conducted in the familial context seemed to be a possible contribution to the child's mathematical development. Berk experienced different learning opportunities during block play with his brother and granny.

Table 3 The structure of MLSS-Family Gül

NMT-Family Gül Building 02	Component: content	Component: cooperation	Component: pedagogy and education
Aspect of allocation	Mathematical domain: 'Geometry and spatial thinking', using spatial skills at the building activity	Playing with elder brother and granny	Theory of the development of spatial skills and spatial structuring: identifying the faces of 3-D shapes to 2-D shapes, relating parts and wholes, replicating the perspectives of different viewers, directly and indirectly operating shapes and figures
Aspect of situation	Negotiation between brother, granny and Berk about the building construction; *consensus*	**Different leeways of participation**	**Modelling, by which granny and elder brother directly act with the play materials** **Enabling him to examine his construction with the last built construction and to realise the mistake** **Providing Berk with an opportunity to confront his mistake** ***Technical* and *operational* mentoring by elder brother**
Aspect of contribution	**Examining the resembling gestalt of the figure and the construction** **Overhearing all the situated activities of elder brother and granny**	**Active participant in his own turn** **Observer, overhearer and legitimate peripheral participant in granny's turn**	**Exploring static balance to build the robust construction** **Witnessing of all the situated activities of elder brother and granny** **Learning opportunities, how the position of the blocks can be represented and how spatial relations and kinaesthetic imagery can be configured**

He was exposed to learning about giving, receiving, sharing and expressing his ideas and feelings. From his granny and elder brother, he observed their choice making. The play was a social act for Berk and he had an opportunity to think, to talk, to learn and perhaps, as Bruner said (1983), to be himself. In this way, he becomes 'educated' in mathematics and in cognitive, social-emotional competences as well. Consequently there occur a mathematical learning support system and an interactional niche in the development of spatial thinking for Berk.

Mostly, the child grows up in a closely linked network of family members, where early learning occurs within play activities (Pound 2006; Hughes 1986). With the participation of each family member, a block play situation can be productive and fruitful for the child. Regardless if the family member has adequate knowledge about geometrical activities, the interaction process can lead the child to learn something. In the extract, the grandmother who may not have had a lot of geometrical knowledge coordinates the activities of the elder brother so that he modelled being an adviser for his younger brother. Family members can impart knowledge to each other and provide new interpretations, which highly and constructively support the child's development in the block play. In this way, a child is exposed to being 'educated' in mathematics before entering primary school or kindergarten and can reach the relatively high levels of achievement and learning.

Considering the question 'in which way and how much children should be "educated" in mathematics before entering primary school?', it seems that a block play with family members might have a positive effect on children's development. Their participation, negotiation and interaction processes can emerge differently, but such play situations enable the children to get different learning opportunities. In this analysis, a block play situation with family members facilitated the child's exploration and way of using his mind (Bruner 1983) through linking their own ideas with others. Different family members are likely to provide many learning opportunities about mathematical ideas. Thus, the more children are exposed to a block play with family members, the better they can be 'educated' in mathematics before entering primary school.

References

Abramovitch, R., Pepler, D., & Corter, C. (2014). Patterns of sibling interaction among preschool age children. In M. E. Lamb & B. Sutton-Smith (Eds.), *Sibling relationships: Their nature and significance across the lifespan* (pp. 61–86). Hillsdale, NJ: Erlbaum.

Acar, E. (2011). Mathematics learning in a familial context (Mathematiklernen in einer familialen Spielsituation). In R. Haug & L. Holzäpfel (Eds.), *Beiträge zum Mathematikunterricht 2011* (pp. 43–46). Münster: WTM.

Bayraktar, E. A. (2012). The first discernment into the interactional niche in the development of mathematical thinking in the familial context. In *Proceedings of the first congress of a mathematics education perspective on early mathematics learning between the poles of instruction and construction.* http://cermat.org/poem2012/main/proceedings_files/Acar-POEM2012.pdf. Accessed 12 December 2012.

Bayraktar, E. A. (2014a). The second discernment into the interactional niche in the development of mathematical thinking in the familial context. In B. Ubuz, Ç. Haser, & M. A. Mariotti (Eds.), *Proceedings of the 8th congress of the European Society for Research in Mathematics Education* (pp. 2078–2088). Ankara: Middle East Technical University. ISBN:978-975-429-315-9.

Bayraktar, E. A. (2014b). The reflection of spatial thinking on the interactional niche in the family. In C. Benz, B. Brandt, U. Kortenkamp, G. Krummheuer, S. Ladel, & R. Vogel (Eds.), *Early mathematics learning selected papers of the POEM 2012 conference* (pp. 85–107). New York: Springer.

Bayraktar, E. A. (2014c). Interactional niche of spatial thinking of children in the familial context (Interaktionale Nische der mathematischen Raumvorstellung den Vorschulkindern im familialen Kontext). In E. Niehaus, R. Rasch, J. Roth, H.-S. Siller, & W. Zillmer (Eds.), *Beiträge zum Mathematikunterricht 2014* (pp. 93–96). Münster: WTM.

Bayraktar, E. A., Hümmer, A.-M., Huth, M., Münz, M., & Reimann, M. (2011). Research methods and settings of erStMaL and MaKreKi projects (Forschungsmethodischer Rahmen der Projekte erStMaL und MaKreKi). In B. Brandt, R. Vogel, & G. Krummheuer (Eds.), *Die Projekte erStMaL und MaKreKi. Mathematikdidaktische Forschung am Center for Individual Development and Adaptive Education* (pp. 11–24). Berlin: Waxmann.

Bayraktar, E. A., & Krummheuer, G. (2011). Thematisation of spatial relationships and perspectives in play situations of two families. The first discernment into the interactional niche in the development of mathematical thinking in the familial context. (Die Thematisierung von Lagebeziehungen und Perspektiven in zwei familialen Spielsituationen. Erste Einsichten in die Struktur "interaktionaler Nischen mathematischer Denkentwicklung" im familialen Kontext). In B. Brandt, R. Vogel, & G. Krummheuer (Eds.), *Die Projekte erStMaL und MaKreKi. Mathematikdidaktische Forschung am "Center for Individual Development and Adaptive Education"* (pp. 11–24). Berlin: Waxmann.

Brandt, B. (2004). *Children as learners. Leeway of participation and participation profiles (Kinder als Lernende. Partizipationsspielräume und -profile im Klassenzimmer).* Frankfurt am Main: Peter Lang.

Brandt, B. (2006). Children as learners in mathematics classrooms in primary school (Kinder als Lernende im Mathematikunterricht der Grundschule). In H. Jungwirth & G. Krummheuer (Eds.), *Der Blick nach innen. Aspekte der alltäglichen Lebenswelt Mathematikunterricht* (pp. 19–51). Münster: Waxmann.

Brandt, B. (2013). Everyday pedagogical practices in mathematical play situations in German "Kindergarten". *Educational Studies in Mathematics, 84*(2), 227–248.

Brandt, B. (2014). I have little job for you. In C. Benz, B. Brandt, U. Kortenkamp, G. Krummheuer, S. Ladel, & R. Vogel (Eds.), *Early mathematics learning selected papers of the POEM 2012 conference* (pp. 55–70). New York: Springer.

Bruner, J. (1978). The role of dialogue in language acquisition. In A. Sinclair, R. Jarvella, & W. J. M. Levelt (Eds.), *The child's conception of language* (pp. 241–256). New York: Springer.

Bruner, J. S. (1983). Play, thought, and language. *Peabody Journal of Education, 60*(3), 60–69.

Bruner, J. S. (1986). *Actual minds, possible worlds*. Cambridge, MA: Harvard University Press.

Bruner, J. S. (1990). *Acts of meaning*. Cambridge, MA: Harvard University Press.

Bruner, J. S. (1996). *The culture of education*. Cambridge, MA: Harvard University Press.

Bullock, J. R. (1992). Learning through block play. *Early Childhood Education Journal, 19*(3), 16–18. doi:10.1007/BF01617077. Netherlands: Springer.

Bundesministerium für Familie, Senioren, Frauen und Jugend (BMFuS). (2002). The importance of education-policy of families—Conclusions of PISA study (Die bildungspolitische Bedeutung der Familie—Folgerungen aus der PISA-Studie). *Wissenschaftlicher Beirat für Familienfragen, Band 224.* Stuttgart: W. Kohlhammer.

Cartwright, S. (1988). Play can be the building blocks of learning. *Young Children, 43*, 44–47.

Clements, D. H., & Sarama, J. (2007). Early childhood mathematics learning. In F. K. Lester (Ed.), *Second handbook of research on mathematics teaching and learning* (pp. 461–555). New York: Information Age.

Cobb, P. (2000a). Constructivism. In A. E. Kazdin (Ed.), *Encyclopedia of psychology* (Vol. 2, pp. 277–279). Washington, DC: American Psychological Association and Oxford University Press.

Cobb, P. (2000b). Constructivism in social context. In L. P. Steffe & P. W. Thompson (Eds.), *Radical constructivism in action: Building on the pioneering work of Ernst van Glasersfeld* (pp. 152–178). London: Falmer.

Cobb, P. (2000c). The importance of a situated view of learning to the design of research and instruction. In J. Boaler (Ed.), *Multiple perspectives on mathematical teaching and learning* (pp. 45–82). Stamford, CT: Ablex.

Cobb, P., & Bauersfeld, H. (1995). *Emergence of mathematical meaning: Instruction in classroom cultures*. Hillsdale, NJ: Erlbaum.

Cobb, P., & McClain, K. (2001). An approach for supporting teachers' learning in social content. In F. L. Lin & T. Cooney (Eds.), *Making sense of mathematics teacher education* (pp. 207–231). Dordrecht: Kluwer.

Cobb, P., Wood, T., & Yackel, E. (1993). Discourse, mathematical thinking, and classroom practice. In E. A. Forman, N. Minick, & C. A. Stone (Eds.), *Contexts for learning: Sociocultural dynamics in children's development* (pp. 91–119). New York: Oxford University Press.

Cobb, P., Yackel, E., & McClain, K. (2000). *Symbolizing and communicating in mathematics classrooms: Perspectives on discourse, tools, and instructional design*. Mahwah, NJ: Erlbaum.

Doron, H. (2009). Birth order, traits and emotions in the sibling system as predictive factors of couple relationships. *The Open Family Studies Journal, 2*, 23–30.

Ernest, P. (1991). *The philosophy of mathematics education*. London: Routledge Falmer.

Ernest, P. (1998). *Social constructivism as a philosophy of mathematics*. Albany, NY: State University of New York Press.

Goodfellow, J., & Laverty, J. (2003). Grandparents supporting working families. Satisfaction and choice in the provision of child care. *Family Matters, 66*, 14–19.

Hawighorst, B. (2000). Mathematics education in the context of the family. An intercultural comparison of parents' education aspirations (Mathematische Bildung im Kontext der Familie. Über einen interkulturellen Vergleich elterlicher Bildungsorientierungen). *Zeitschrift für Erziehungswissenschaft, 10.Jahrg.Heft 1/2007*, 31–48.

Hewitt, K. (2001). Blocks as a tool for learning: Historical and contemporary perspectives. *The Journal of the National Association of Young Children, 56*, 6–13.

Hughes, M. (1986). *Young children learning in the community, in involving parents in the primary curriculum*. Exeter: Exeter University.

Kersh, J. E., Casey, B., & Young, J. M. (2008). Research on spatial skills and block building in girls and boys. In O. N. Saracho & B. Spodek (Eds.), *Contemporary perspectives on mathematics in early childhood education* (pp. 233–251). New York: Information Age.

Kim, J., & Fram, M. S. (2009). Profiles of choice: Parents' patterns of priority in child care decision-making. *Early Childhood Research Quarterly, 24*(1), 77–91.

Krummheuer, G. (2011a). The interactional niche in the development of mathematical thinking (Die Interaktionale Nische mathematischer Denkentwicklung). *Beiträge zum Mathematikunterricht 2011* (pp. 495–498). Münster: WTM.

Krummheuer, G. (2011b). The empirical founded derivation of the term the "interactional niche in the development of mathematical thinking" (Die empirisch begründete Herleitung des Begriffs der "Interaktionalen Nische mathematischer Denkentwicklung"). In B. Brandt, R. Vogel, & G. Krummheuer (Eds.), *Die Projekte erStMaL und MaKreKi. Mathematikdidaktische Forschung am "Center for Individual Development and Adaptive Education" (IDeA)* (Vol. 1, pp. 25–90). Münster: Waxmann.

Krummheuer, G. (2011c). Representation of the notion "learning-as-participation" in everyday situations of mathematics classes. *Zentralblatt für Didaktik der Mathematik (The International Journal on Mathematics Education), 43*(1), 81–90. doi:10.1007/s11858-010-0294-1.

Krummheuer, G. (2011d, November 8). The interaction analysis. Arithmetical problem on the shirt numbers (Die Interaktionsanalyse. Rechenaufgabe mit Trikot-Rückennummern), 1–8. Retrieved from http://www.fallarchiv.uni-kassel.de/wp-content/uploads/2011/01/krumm_trikot_ofas.pdf.

Krummheuer, G. (2012). The "non-canonical" solution and the "improvisation" as conditions for early years mathematics learning processes: The concept of the "interactional niche in the development of mathematical thinking" (NMT). *Journal für Mathematik-Didaktik, 33*(2), 317–338.

Krummheuer, G. (2014). Adaptability as a developmental aspect of mathematical thinking in the early years. In *Proceedings of the second congress of a mathematics education perspective on early mathematics learning between the poles of instruction and construction*. Retrieved from http://www.mah.se/english/faculties/Faculty-of-education-and-society/A-Mathematics-Education-Perspective-on-early-Mathematics-Learning-between-the-Poles-of-Instruction-and-Construction-POEM/Online-Proceedings/.

Krummheuer, G., & Brandt, B. (2001). *Paraphrase and traduction. Participation theoretical elements of interaction theory of mathematics learning in the primary school (Paraphrase und Traduktion. Partizipationstheoretische Elemente einer Interaktionstheorie des Mathematiklernens in der Grundschule)*. Weinheim: Beltz Wissenschaft Deutsche Studien.

Krummheuer, G., & Schütte, M. (2014). The change between mathematical domains—A competence, which doesn't exist in educational standards (Das Wechseln zwischen mathematischen Inhaltsbereichen—Eine Kompetenz, die nicht in den Bildungsstandards steht). *Zeitschrift für Grundschulforschung, 7*(1), 126–128. ISSN:1865-3553.

Krummheuer, G., & Schütte, M. (in this book). Adaptability as a developmental aspect of mathematical thinking in the early years. In T. Meaney, T. Lange, A. Wernberg, O. Helenius, & M. L. Johansson (Eds.), *Mathematics education in the early years—Results from the POEM conference 2014*. Cham: Springer.
Lave, J., & Wenger, E. (1991). *Situated learning: Legitimate peripheral participation*. Cambridge: Cambridge University Press. ISBN 0-521-42374-0.
Nader-Grosbois, N., Normandeau, S., Ricard-Cossette, M., & Quintal, G. (2008). Mother's, father's regulation and child's self-regulation in a computer-mediated learning situation. *European Journal of Psychology of Education, XXIII*(1), 95–115.
Palincsar, A. S. (1998). Social constructivist perspectives on teaching and learning. *Annual Review of Psychology, 49*, 345–375.
Paradise, R., & Rogoff, B. (2009). Side by side: Learning by observing and pitching in. *Ethos, 37*(1), 102–138.
Pound, L. (2006). *Supporting mathematical development in the early years* (2nd ed.). Buckingham: Open University Press.
Rogoff, B. (1990). *Apprenticeship in thinking*. New York: Oxford University Press.
Rogoff, B., Ellis, S., & Gardner, W. (1984). Adjustment of adult-child instruction according to child's age and task. *Developmental Psychology, 20*(2), 193–199.
Sfard, A. (2001). Learning discourse: Sociocultural approaches to research in mathematics education. *Educational Studies in Mathematics, 46*(1/3), 1–12.
Sfard, A. (2008). *Thinking as communicating: Human development, the growth of discourses, and mathematizing*. Cambridge: Cambridge University Press.
Smith, P. K. (2005). Grandparents and Grandchildren. *The Psychologist, 18*(11), 684–687.
Smith, P. K., & Drew, L. M. (2002). Grandparenthood. In M. H. Bornstein (Ed.), *Handbook of parenting: Volume 3. Being and becoming a parent* (pp. 141–172). Mahwah, NJ: Lawrence Erlbaum.
Tiedemann, K. (2010). Support in mother-child discourses: Functional observation of an interaction routine (Support in mathematischen Mutter-Kind-Diskursen: Funktionale Betrachtung einer Interaktionsroutine). In B. Brandt, M. Fetzer, & M. Schütte (Eds.), *Auf den Spuren interpretativer Unterrichtsforschung in der Mathematikdidaktik: Götz Krummheuer zum 60. Geburtstag* (pp. 149–175). Münster: Waxmann.
Tiedemann, K. (2012). *Mathematics in the family. The familial support of early mathematics learning in the storybook reading situations and playsituations (Mathematik in der Familie. Zur familialen Unterstützung früher mathematischer Lernprozesse in Vorlese- und Spielsituationen)*. Münster: Waxmann.
Vandermaas-Peeler, M. (2008). Parental guidance of numeracy development in early childhood. In O. N. Saracho & B. Spodek (Eds.), *Contemporary perspectives on mathematics in early childhood education* (pp. 277–290). Charlotte, NC: Information Age.

Mathematical Understanding in Transition from Kindergarten to Primary School: Play as Bridge Between Two Educational Institutions

Dorothea Tubach and Marcus Nührenbörger

Abstract German kindergarten and Grundschule ("primary school") are characterized by different conditions concerning the organization of learning processes. This situation places particular demands on the arrangement of linked and coherent mathematical learning environments in transition from more informal to formal learning situations. Particularly the relational understanding of numbers is one important objective for mathematical learning during the transition. In our qualitative study, we developed three "complementary playing and learning environments" (CLEs) and observed how 20 children explored and discussed relationships between numbers in the last year of kindergarten and the first year of school. We focus on the institutional similarities and differences and discuss them with regard to the arrangement of playing and learning environments in kindergarten and primary school.

Learning Mathematics Before and At the Beginning of Formal Schooling

It is established that mathematical learning processes of children start long before formal schooling in informal ways and take place in other learning locations, such as kindergarten. In Germany, children usually start compulsory school, "Grundschule" (primary school), at the age of 5–6. However, most children voluntarily attend kindergarten for some years before they enter school. Thus, the arrangement of linked and coherent learning processes gains importance, and questions arise as to how to cope with the changing learning situations from kindergarten to primary school. In kindergarten, mathematical learning situations arise in play and daily life in the context of both free and guided activities. In primary school, however, learning situations are more formally arranged in the context of substantial learning environments and geared to curriculum standards. These differences demand that the discontinuities be dealt

D. Tubach (✉) • M. Nührenbörger
TU Dortmund University, Dortmund, Germany
e-mail: dorothea.tubach@tu-dortmund.de; Marcus.Nuehrenboerger@math.tu-dortmund.de

T. Meaney et al. (eds.), *Mathematics Education in the Early Years*,
DOI 10.1007/978-3-319-23935-4_5

with in a productive way, as well as ensuring continuities concerning the learning processes (Roßbach 2010). Cooperation activities between teachers from kindergarten and primary school concerning the coordination and arrangement of academic learning processes seem to have a positive impact on children's learning. However, cooperative activities with a special focus on becoming familiar with the new learning location have no impact on children's learning (Ahtola et al. 2011).

From the mathematics didactics perspective, the coherence of learning arrangements with respect to content, objectives, and requirements is important (e.g., Clements 2004). On the one hand, there are discussions about how to build on individual mathematical learning experiences in primary school, and on the other hand, recommendations for the arrangement of mathematical learning situations in kindergarten have been developed (e.g., Clements et al. 2004; Gasteiger 2012).

However, so far, explicit development studies are rare concerning the design of mathematical learning environments which focus on mathematical relations in kindergarten *and* primary school in due consideration of the particular and different institutional conditions. This chapter responds to this research need and focuses both on how to design linked, complementary learning environments for kindergarten and primary school and on how children explore, identify, and discuss mathematical relations within these learning environments. Hence, the research interest is to gain insights into children's constructions in transition, in order to derive information on how best to arrange learning situations in this context. It is discussed how and in what way complementary learning environments contribute to link mathematical experience from rather informal playing activities with more formal learning in school. Moreover, opportunities and limitations of these learning environments are considered.

Mathematics Learning in Kindergarten and Primary School

With regard to the development of mathematical understanding, the acquisition of the concept of numbers is of central importance. Early learning processes concerning quantity–number competencies seem to have a lasting effect on mathematical learning at school (Krajewski and Schneider 2009). Krajewski and Schneider (2009) characterize the acquisition of early quantity–number competencies over three levels. According to their model, number–word sequence and an understanding of quantity develop in isolation from each other (Level I). Only by linking these "basic numerical skills" do number words gain quantitative meaning (Level II). In Level II, children become aware that number words can be used to describe discrete quantities and that quantities can be determined by counting ("quantity–number concept"). An understanding of "number relationships" develops when children realize that numbers relate to other numbers and that numbers can be used to describe these relations (Level III). Two main relations are distinguished:

1. Relation of composition and decomposition: numbers can be described by *compositions* of other numbers (parts) ($5=3+2$) or can be *decomposed* into numbers.

2. Relation of difference: numbers can describe *differences* between two numbers (the difference between 3 and 5 is 2, $5-3=2$).

From the developmental psychology point of view, this relational understanding has a meaningful role for children's further mathematical learning (Krajewski and Schneider 2009; Langhorst et al. 2012; Resnick 1983). From the mathematics didactics perspective, the ability to identify and use relations between single numbers by counting and calculating is highlighted as a central objective for learning processes in kindergarten, which should be continued in primary school (Wittmann and Müller 2009).

Moreover, from an epistemological point of view, relational understanding is important. Mathematical concepts are not concrete, but characterized by relationships between concrete and abstract objects (Nührenbörger and Steinbring 2009; Steinbring 2005). Consequently, mathematical concepts acquire meaning if children deal with the concrete and later abstract objects in an active way and construe relationships between the objects. In the transition from kindergarten to primary school, the connection of experiences with concrete objects and systematic abstract examinations is challenging (e.g., Hasemann 2005). This connection succeeds if children move beyond the concrete specific situation: "They are requested to see, interpret or discover 'something else', another structure, in the situation" (Steinbring 2005, p. 82). Accordingly, children have to identify general structures in the specific situation and at the same moment consider the specific nature of the situation.

For construing, identifying, and using relations between numbers, interaction and negotiation processes are relevant. When children are encouraged to express and discuss their ways of thinking and acting, they get the chance to reason about mathematical meaning and construct mathematical knowledge (e.g., Nührenbörger and Steinbring 2009).

Playing and Mathematics Learning

Play-based academic learning is one approach to mathematics in kindergarten. Playing can be seen as a social interactive activity, which is characterized by certain rules, some degree of freedom, and high involvement of the actors (van Oers 2014).

Several authors have stressed the importance of play and playfulness for early mathematics learning. Ginsburg (2006) has observed children in kindergarten and categorized different types of mathematical play: Two of these types are "mathematics embedded in play" and "play centering on mathematics."[1] *Mathematics*

[1] Ginsburg (2006, p. 152) mentions a third type "play with the mathematics that has been taught" which is not important here because it seems to concern another aspect: This type describes children's role-play of "teacher." By playing "mathematics lesson," children also "play" with mathematical contents.

embedded in play can arise by playing, for example, mathematically rich games. Playing the game is the focused activity, and mathematical activities happen casually while playing. An example is when young children play "the great race." The focused activity is to roll the die and to set the playing piece to first reach the "end." In playing this game, children casually learn about the number line (Ramani and Siegler 2008). In contrast, *play centering on mathematics* occurs if the objects of play are mathematical patterns and structures. Playing, thus, is characterized by dealing with mathematical relations. If children, for example, "play" with chips and create a "triangle" with them, first one, then two, then three, and so on, the mathematical relation "one more" is used as a creating principle.

Mathematics embedded in play in kindergarten Some studies provide evidence that children achieve mathematical competencies by playing games in kindergarten (e.g., Ramani and Siegler 2008; Stebler et al. 2013). Stebler et al. (2013) demonstrate that games can provide a meaningful context for mathematical activity and support individual mathematical strategies. Schuler (2011) emphasizes the importance of the "conversational management" of educators for the development of the mathematical potential in play situations. The guiding activities of adults, "guided play" in the terms of Hirsh-Pasek et al. (2009), involve not only the organization of playing activities—e.g., selecting appropriate materials and games—but also stimulating discussions and asking for mathematical reflection and reasoning, oriented on children's ability and on the playing process (Ginsburg 2006; Pramling Samuelsson and Asplund Carlsson 2008; van Oers 2010). Van Oers (2010) points out how the role-play of children can be used as meaningful context for further mathematical learning. With the help of game recordings (which means written or drawn documentations of important situations or outcomes of the game), more systematic learning processes can be encouraged.

Mathematics learning in primary school In primary school, mathematical learning processes are prepared and structured systematically by means of challenging tasks or "substantial learning environments" (Wittmann 2001). Substantial learning environments (SLEs), for example, represent "central objectives, contents and principles of teaching mathematics" (Wittmann 2001, p. 2) and promote rich mathematical activities. Thus, content equivalent and integrated learning arrangements, which can be easily adapted to their individual prerequisites, can be provided for all children.

Play centering on mathematics in primary school Formal learning in school can also be playful. "Play centering on mathematics" (Ginsburg 2006) occurs in cooperative discovering, exploring, and inventing patterns and structures. Children play with mathematical objects and explore mathematical relations. They change objects and discover the impact of change and how to react to these changes (Steinweg 2001). To develop "play centering on mathematics," an understanding of mathematics is important: "If mathematics is as much about understanding connections, processes and possibilities as it is about knowing facts, then play and mathematics have much in common" (Dockett and Perry 2010, p. 717). Playful mathematical identification and construction of relations between numbers provide a basis for reflecting and discussing relations among children and guiding adults.

Complementary playing and learning environments (CLEs) The different institutional conditions of kindergarten and primary school have to be considered when designing mathematical learning environments for children transiting between them. Similar to the concept of substantial learning environment (SLE) for formal schooling, a "substantial playing environment" (SPE) in kindergarten can provide rich mathematical experiences and activities by playing games (Stebler et al. 2013). SLE and SPE can be combined to a complementary learning environment (CLE). A CLE is concerned with sustainable mathematical content and materials in kindergarten in the context of an SPE, which are picked up and continued in an SLE in primary school. In the CLE "who has more?" which is presented below, children get acquainted with a mathematically rich game in kindergarten. They gain experience in using the game material and informally explore mathematical relation of difference. These experiences and the game materials are used again in the first class at school. Children create written or drawn recordings, sort, and add them. These activities help them to reflect on their experiences and deepen their insights in relations of differences. The exploration of relations of differences becomes the focused activity. The game "who has more?" loses its meaning as a playing activity but gains in relevance as a meaningful context.

While current studies focus on the importance of playing for mathematics learning in kindergarten, there are fewer findings on how to use mathematical play and playing as a bridge between kindergarten and primary school, as a bridge between informal and formal learning situations, and how mathematical learning processes are performed in kindergarten and primary school. In the following, children's interactive learning processes are studied with regard to these questions:

(I) How do children discover elementary numerical relationships while transitioning from kindergarten to primary school in the interactive context of the learning situations?
(II) Which similarities and differences can be pointed out concerning the mathematical construction process in kindergarten and primary school?
(III) How does complementary learning environment contribute to link informal and formal learning experiences?

Methods

In order to answer these questions, "design experiments" were conducted. The method of "design experiments" draws on Cobb et al.'s (2003) work and includes the (usually iterative) exemplary testing of teaching and learning arrangements or SLEs with learners, e.g., classroom experiments. The purpose of the experimental testing and analyzing of teaching and learning arrangements is to investigate which learning processes are initiated and how they are supported. This gives answers to how the arrangement can be optimized in order to support further or deeper learning processes.

In a cyclic process, a prospective and reflective perspective is adopted on the learning process: "Prototypically, design experiments entail both 'engineering' particular forms of learning and systematically studying those forms of learning within the context defined by the means of supporting them" (Cobb et al. 2003, p. 9).

In the present research, three complementary learning environments (CLEs) were designed to encourage children to reason about mathematical relations as well as to investigate children's learning processes within these learning environments (Nührenbörger and Tubach 2012). Children in the last year of kindergarten were involved in these CLEs in the context of playing environments. At the beginning of the first school year, they were involved again, but now in the context of learning environments, which included three lessons.[2] In total, about 20 children were observed by video over two survey cycles dealing with the playing and learning environments in kindergarten and primary school. In each cycle, about four kindergarten teachers from different kindergartens and two teachers of primary school were introduced to the particular playing and learning environments beforehand. The introduction involved becoming familiar with the learning arrangement and its learning opportunities for the children. Furthermore, it was discussed with them how to introduce it and how to involve children in meaningful mathematical discussions. This was important, since kindergarten teachers in Germany are usually not educated in mathematics and in arranging mathematical learning situations. The video observation of authentic situations is complemented by qualitative interviews of two children after each playing and learning situation in kindergarten and primary school.

The ultimate goal of these design experiments is the development of interactive and linked learning environments for kindergarten and primary school and to investigate its principles, opportunities and limits, as well as the development of local theories concerning the interactive process of understanding and establishing intersubjectivity between children.

Construction of the Learning Environment "Who Has More?"

In this section, one of three designed CLEs is exemplified: "who has more?" As a playing environment, "who has more?" is a game for two children. Each child receives a wooden block of five, a ten frame, and a die (with the numbers 0–5) in the colors blue or red. The game materials are completed with small round gaming pieces, called counters (with a red and an alternate blue side). The structured materials (ten frames and blocks of five) enable the interpretation of numbers in relationship to 5 and 10 (Flexer 1986).

Rules of the game Both players roll their die and put, according to the rolled number, the appropriate number of counters on their respective blocks of five. The player with the higher number of counters is allowed to take the difference of counters (the

[2] School usually starts in August. The experiments in kindergarten were carried out in the period from March to July, in primary school from October to February.

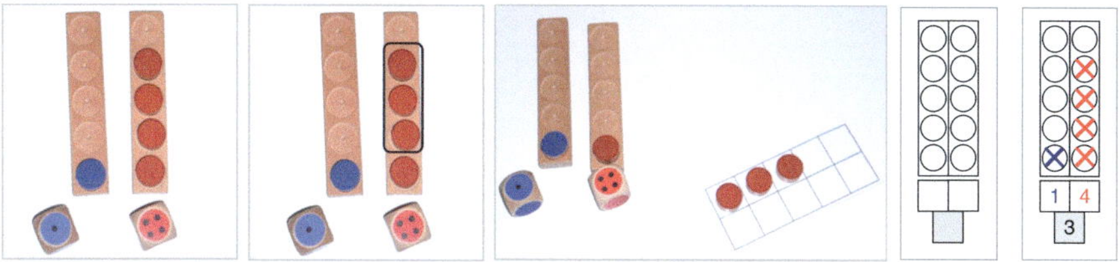

According to the rolled number, counters are put in the block of 5

Comparison: There are 3 red counters more.

The red player is allowed to take the three counters and put them on his or her ten-frame.

Recording sheet to record the rolled numbers.

Fig. 1 Rolling and comparing

ones in excess) and put them on the ten frame (see Fig. 1). Afterwards, the blocks are cleared and the dice are rolled again. The player who fully fills the ten frame first wins the game.

At the heart of the playing environment is the comparison of two numbers and the determination of the difference. This gives children the possibility to gain insights into the relations of differences between two numbers.

(a) There exist different number pairs with the same difference, e.g., 4 and 1 and 5 and 2 which both have the difference of 3.
(b) Number pairs with the same difference are characterized by a compensation relationship: $5-2=(5-1)-(2-1)=4-1$.
(c) The difference increases (decreases) if the minuend is increased (decreased) or the subtrahend is decreased (increased).

Meanwhile, children collect, structure, and determine the number of counters on their ten frame and gain experiences in composing and decomposing of numbers.

The learning environment in primary school reuses the known materials and rules and therefore the previously gained mathematical experiences in kindergarten. An iconic-symbolic form of documentation (recording sheet, see Figs. 1 and 2) complements the learning environment. The children mark the achieved rolled numbers with crosses (every child in his or her color), determine the difference, and write the numerals. After rolling the dice for some times, the created recording sheets (see Fig. 2) can be ordered by equal differences or other criteria, to gain deeper insights into relations of differences. The recordings furthermore offer the possibility to increase or decrease the numbers in the sheet in order to achieve a certain difference, for example.

The learning environment consists of three lessons. In the first lesson, the children play "who has more?," remember the rules and materials, and discuss some controversial and mathematical fruitful play situations. In the second lesson, the recording sheets are introduced. The children record their rolled numbers and the difference. Afterwards, the recording sheets are discussed with respect to the questions: "Which results are good for the red player? Which results are equally good for both players?" Through this discussion, children become aware that there are different number pairs with the same difference (e.g., 5 and 3, 4 and 2, 3 and 1, etc.).

Fig. 2 Game recordings: which results are good for the red player?

In the third lesson, children are asked to create (without rolling a die) number pairs with a given difference "3 more" using a recording sheet that has been increased to ten. In this way, they can explore the effect on the difference by increasing and decreasing the two compared numbers by the same amount and so gain insights into compensation.

Reconstruction of Early Mathematical Understanding Processes

To reconstruct interactive processes of understanding, a qualitative approach is chosen, oriented toward the interpretative classroom research (Krummheuer 2000). Therefore, mathematical understanding processes are interpreted step-by-step. At first, video episodes are selected and transcribed. In a turn-by-turn process, a group of researchers paraphrases the episode and works out plausible interpretations. To advance the validity of interpretations, a consensus is sought between interpretations. For the process of interpretation to be inter subjectively checked, in the following section of this chapter, the relevant episodes are presented, although being aware that even the selection of episodes is an act of interpretation.[3] In a further step, these interpretations are developed and reviewed and theoretical elements are gathered. The theoretical elements are also reviewed and extended by comparing with further episodes. The comparison of diverse interactive episodes increases the chance of determining the specific element of a single episode. Thus, the empirical evidence for the theoretical elements enhances and raises them beyond case studies (Krummheuer 2000). In this way, general findings of the particular case can be provided and local theories can be developed. Hence, mathematical understanding processes can be specified on different levels of mathematical development (e.g., Krajewski and Schneider 2009). Essential for the mathematical analyses is the epistemological triangle, as described by Steinbring (2005). This analysis tool enables identifying specific reference contexts children use by construing and constructing relations. The epistemological analysis focuses in particular on the reconstructions of the interactive process of constructing knowledge on the basis of actions and interactions.

[3] The whole German transcripts are available from the authors.

Early Understanding of Mathematical Relations: Two Exemplary Episodes of "Who Has More?"

With the following two interaction episodes from kindergarten and primary school, different but typical accesses to the understanding of elementary mathematical relationships can be determined. Thus, conclusions can be made about how CLEs link the more informal learning situation in kindergarten and the more formal learning situation in primary school (see footnote 3).

"Who Has More?" in Kindergarten

Mahsum and Dalina play the game "who has more?" for the first time with a guiding adult (GA) in kindergarten. The rules are clear at this stage of the game. At the beginning of the following scene, Mahsum (Ma) has six and Dalina (Da) has seven counters on their respective ten frame (see Fig. 3).

1 Ma I've got only four missing (*takes his die*).
2 GA Well (*pointing with her finger at Mahsum*). You only need four more? What about you? (*Pointing with her finger at Dalina*)
3 Da And I only need three more.
4 GA Only three more. Aha.
5 Ma (*rolls number 4, takes four counters while counting*) One, two, three, four.
6 Da (*rolls number 1, rolls immediately number 1 again, and the third time she rolls number 3, takes three counters, and starts putting them in her ten frame*)
7 GA Stop.
8 Da Oh. (*Removes the counters and puts them in her block of five*)
9 Ma (*meanwhile*) I've got only one more. 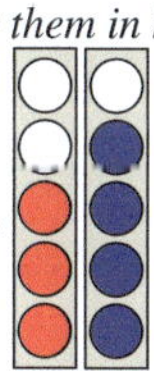
10 Da (*moves her block of five to Mahsum's*) Mahsum has got one more. (…) May I put mine anyway?

In the first lines (1–4), the children point out how many counters are still missing on their ten frames. The term "only" indicates a comparison. Comparing the filled with the unfilled fields or comparing the unfilled before with the unfilled now, there are less unfilled, so "only four." Dalina could also have referred to Mahsum's missing counters. Compared to his four, she only needed three more. In the following,

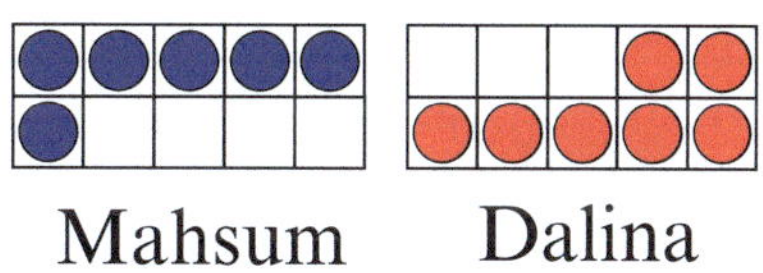

Fig. 3 Ten frames

the children roll the dice again. Mahsum rolls number 4 and transfers it into the corresponding number of counters in his block of five. But Dalina seems not to be satisfied with her number 1. She rolls her dice unnoticed for several times until she has number 3. Two interpretations are possible why Dalina chooses number 3, either she thinks number 3 is better than 1 or she thinks number 3 is best to win. The latter seems most likely since she starts to put three counters on her ten frame (line 6).

After Dalina put her counters in her block of five, it is clear that Mahsum has got one more counter to put on his ten frame (line 9–10). Even before all counters have been put in the block, Mahsum anticipates that it had to be only one more. But the constraint "only" indicates that Mahsum had another expectation after his roll: The number 4 is the second highest possible. Despite the high number, he achieves the smallest gain: $4 - 3 = 1$. However, it is conceivable to have three better results ($4 - 2 = 2$, $4 - 1 = 3$, $4 - 0 = 4$) and only two worse ($4 - 5 = -1$, $4 - 4 = 0$).

In line 10, it becomes clear that Dalina put her counters not only mistakenly on her ten frame (line 6) but considered them to be proper for the unfilled fields of her ten frame ($7 + 3 = 10$). In doing so, she relates the gaps to the rolled number directly although she knows how to determine differences between two numbers of counters in the blocks of five. By comparing these two numbers, she experiences that the rolled number 3 compared to 4 does not help to win three counters and fill the gaps.

The guiding adult denies Dalina's question. The children clear their blocks and roll the dice again.

11 Ma (*rolls number* 2) Two (*takes two counters and puts them in his block of five*)

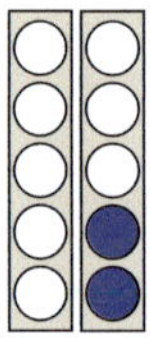

12 Da (*rolls number 4, looks at Mahsum's block of five, turns her dice to number 5, and raises her hands*) Five.

Mahsum again directly transfers the rolled number 2 in the number of counters in the block of five (line 11). Dalina first rolls number 4 but again changes her dice this time without rolling into 5 (line 12). By raising her arms saying "five," she could either distract that she infringed the rules or express that she had won the game.

Due to her endeavor to achieve three counters in the first scene, it can be assumed that here, as well, she tries to achieve three counters. Possibly this time, she first mentally compares her rolled number 4 with the two counters in Mahsum's block and determines the difference 2. She realizes that the difference does not suffice to fill the gaps on her ten frame. Through increasing the rolled number by 1 to 5, the difference increases by the same amount to 3. With the difference 3, Dalina could win the game, but this time it is noticed that she turned her dice and she has to reroll correctly. Compared to the first scene, Dalina increases her rolled number in order to achieve a greater difference instead of achieving the equal number according to the number of gaps. It seems that her view on the three needed counters changed, and she tries to construe them as a difference of two numbers. Nevertheless, it is unlikely that she is completely aware of the relationship: $(4+1)-2=2+1=3$.

Summary Within the playing environment “who has more?,” children are encouraged to determine numbers and relate them to each other especially by comparing. Comparisons are made in a qualitative way (e.g., “only”) and in a quantitative way by determining the difference. Also due to adaptive guiding, children discover a variety of number relations.

1. Determine numbers: Different types of numbers are determined (rolled numbers, number of counters, filled and unfilled fields on the ten frame, and the block of five).
2. Compare numbers: (a) Differences between two linear structured amounts are determined. (b) Rolled numbers are compared directly. (c) The difference is related and compared to the number of unfilled fields. (d) The rolled number is related and compared to the number of unfilled fields. (e) The filled and unfilled fields are compared. (f) Diverse real and desired differences are compared qualitatively.
3. Decompose numbers: The difference between two numbers is seen as a part of the bigger number and is removed.
4. Compose numbers: (a) The number of newly won counters is added to previous counters on the ten frame. (b) The sum of the filled and unfilled fields of the ten frame is always 10.
5. Increase and decrease numbers: greater or specific rolled numbers are desired in order to gain an advantage: (a) The rolled number is increased, so the number of gaps and the rolled number are equal. (b) The rolled number is increased to achieve a greater or a certain difference.

This scene shows—although the children are acquainted with *construing and determining differences* between two linearly represented numbers of counters—that it is challenging to reason about the required counters in terms of a certain relation between two numbers and *create differences* by finding two appropriate number pairs. The first access usually is to desire a rolled number that is equal to the number of gaps. But the experience during the game shows that this only works if the subtrahend is zero. Dalina even changes her rolled number in order to acquire a certain gain. While this operative activity infringes the rules of the game in kindergarten, in primary school, the intention of the lesson is that the students create different “three more results.”

“Three More Results” in Primary School

In the first class of primary school in the third lesson of the learning environment, based on “who has more?,” children are presented a fictitious score (a ten frame with seven blue counters and a ten frame with nine red counters) and are asked to

Fig. 4 Recording sheets

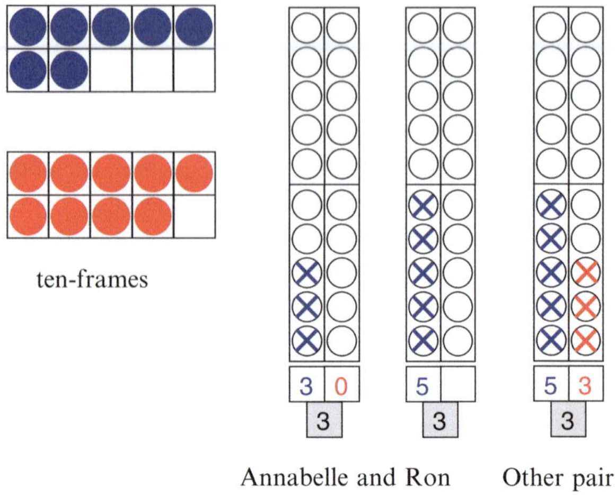

find different possibilities for the blue player to win. The task is to "find three more results." Children are given recording sheets with two rows up to ten instead of blocks to record their findings.

Annabelle and Ron solve this task together; Ron marks the blue and Annabelle the red crosses. On the other side of the table, another pair of students works together (see Fig. 4). Annabelle (Ann) marks the determined number of two crosses with her red color as Ron comments on the solution recording of the pair (Stud) working vis-à-vis:

1 Ron	Oh, you've done something wrong. Blue has to win always three (*shows three fingers and bends over the table*) Oh look, there are only two winning (*pointing on the two blue crosses above*). Oh!
2 Stud	Oh, we need a rubber. Do you have a rubber?
3 Ron	Look here, look here you can make one more here and circle as well (*pointing on the sixth field above the blue crosses and circling with his finger around the 4 to 6 field on the left side*). It's also possible.

This scene exemplifies that the given objective "three more" produces differing solutions, which Ron indicates as wrong (line 1). Ron emphasizes that the difference always has to be 3 and explains that the current difference is "only 2." He creates a relationship between the current and the given difference and distinguishes the current to be smaller. The other pair does not defend their solution, but ask for a rubber. Possibly the process of correction is associated with deleting and remaking (line 2). In the following, Ron formulates another idea, without or with less deleting necessary: If you add a blue cross, the difference increases by 1, $(5+1)-3=2+1$. Instead of deleting, it "is also possible" to add crosses in order to achieve a desired difference (line 3).

The two children are not convinced of this idea. Annabelle lends her rubber, provided that she is allowed to erase by herself. She removes the third red cross and thus benefits from the mathematical relation that the difference can also be increased by removing a red cross: $5-(3-1)=2+1$.

Meanwhile Ron concentrates on the next recording sheet:

4 Ron	(*marks six fields with blue crosses, points on the third field of the left row, and adds four further blue crosses*) I'm rolling ten.
5 Ann	Ten? We aren't allowed here (*pointing with her pen to the upper fields*)
6 Ron	Sure (.) we are.
7 Ann	What am I rolling now?
8 Ron	You're rolling (.) ehm eight (*pointing to the seventh field of the right row*)
9 Ann	Okay (*marks the fields beginning bottom up by counting*) One, two, three, four, five, six, seven
10 Ron	Stop.
11 Ann	Seven (*points at her red crosses and starts to mark a further cross*).
12 Ron	No. No (.) Or else I'll lose. Look, I'll win with this (*pointing to the three upper blue crosses*) and here (*pointing to the right numeral field below the crosses*) you have to write seven.

Ron initially marks six crosses and finds a preliminary solution (the number pair 6 and 3) by pointing on the third field on the left side. This would have been exactly the solution he had offered to his classmates. Maybe because Annabelle is still rubbing or because this solution is not clear, Ron adds four further blue crosses (line 4). With these ten crosses, he exceeds the fifth field for the first time. This possibly causes a short confusion and leads to Annabelle's objection (line 5–6). Ron answers Annabelle's question that she had to roll eight. Possibly he miscounted because at the same time he points on the seventh field of Annabelle's row (line 7–8). But now Annabelle wants to mark the "rolled number" 8 correctly. First she does not comply with the interruption by Ron after seven crosses (line 11). Just at that moment as Annabelle marks the seventh cross, Ron seems to recognize that the field he had pointed before and thus the desired difference of 3 is reached. To conserve this difference of 3 has now priority over marking the predefined number. One reason for avoiding further red crosses in any case might be that he could hardly add blue crosses to repair the difference. Instead, he tells Annabelle to write the numeral 7 in the corresponding field (line 12). Ron argues that in this way he would win, otherwise, he would lose. For the difference 3 not the previous determined number is crucial but that the smaller number of crosses ends three fields below the other: $a-(a-3)=3$.

Summary The analyzed scene shows how children in first class of primary school try to represent the number 3 as a difference between two numbers. In doing so, they explore different solutions (number pairs) for the difference 3. These solutions are regarded initially as isolated examples. The correction of the preliminary solution $5-3=2$ demonstrates how relationships between numbers can be used to create or change differences: If the difference should be increased by one, either the minuend has to be increased by one or the subtrahend has to be decreased by one, $(a+1)-b = a-(b-1)=(a-b)+1$. At the same time, the children only mention the difference and the necessary changes; the single numbers to construct the difference are not relevant. In the second section, a more general way of interpreting the difference of 3 can be reconstructed in the sense of three overhanging crosses. In this view, the relation "3 more" is independent of two concrete and certain numbers but applies for all number pairs which meet this relation. In this scene, it seems that more than determining, comparing, composing, decomposing, increasing, and decreasing *numbers*, it is about determining, comparing, composing, decomposing, increasing, and decreasing *differences*.

1. Determine numbers: (a) Numbers are determined to specify both the number on the imaginary die and the numeral that has to be written. (b) The difference is determined.
2. Compare numbers: Numbers are compared to determine if the correct difference of 3 is constructed. By comparing with the desired difference of 3, differing differences can be identified.
3. Decompose numbers: The difference is seen as a part of the minuend. In order to create number pairs with the difference 3, the subtrahend has to be smaller by 3, so three fields have to be left empty, i.e., three crosses have to overhang.
4. Compose numbers: After increasing the minuend, the new difference is the sum of the increased number and the old difference.
5. Increase and decrease numbers: (a) Increasing the minuend in order to increase the difference by the same amount, (b) decreasing the subtrahend in order to increase the difference, and (c) to every number x ($\geq a$), which should be greater than a, a corresponding number can be found to have the property to be smaller by a, i.e., $x - a$.

Comparison of the Learning Situation in Kindergarten and Primary School

Both selected and interpreted scenes of kindergarten and primary school represent how dealing with complementary learning environments (CLEs) offers children *space for mathematical experience* as well as *space for mathematical play*. Both learning situations, the game in kindergarten, and the following more systematic formal learning in school create room for exploring and discussing elementary relationships of differences. In the following, the observed learning situations are compared with regard to the process of playing and the process of exploring and using number relations.

Playing in kindergarten and primary school In kindergarten, children get acquainted with "who has more?" as a game which can be played repeatedly with other children and with a guiding adult. From the process and the aim of playing, occasions arise for interpreting numbers in relation to other numbers and for discussing them with others. Thus, playing the game offers children rich *space for mathematical experiences* in order to construct number relations especially to interpret differences; this indeed is "mathematics embedded in play" (Ginsburg 2006). In primary school, however, the task "find three more results" opens *space for mathematical play*. This is particularly evident in the identified context "increase and decrease numbers": children take advantage of the opportunity to vary the number of crosses, choose bigger or smaller numbers, or change numbers and explore the effects. Consequently, "play centering on mathematics" (Ginsburg 2006) is enacted, since the difference of two numbers becomes the object of play.

Exploring and using number relations Children in kindergarten learn a variety of ways and strategies to determine, compare, compose, and decompose numbers, i.e., they explore rich number relations by dealing with the material. In primary school, one special mathematical aspect is in focus: the relation of differences. But here, too, rich number relations can be explored and used. For example, the conception of the difference as a kind of "entity" is developed, which is included in the minuend (as a part), but is independent of concrete numbers. Further systematic exploration and experience of number pairs with the difference 3 can lead children to deeper insights in compensation: $a - b = (a + x) - (b + x)$.

Complementary learning environment in transition The analysis shows that each learning situation represents specific institutional characteristics in terms of play and number relation. But at the same time, they are mutually linked: So already in kindergarten arises the potential for mathematical play about differences, which becomes more nuanced by the focused mathematical activity in primary school.

In view of the activity in primary school, children *construe differences* in the (game) material in kindergarten. The confident dealing with the representation and the language of the children ("roll," "I win," etc.) in primary school indicate the use of the gaming experience. The game gains in relevance as a meaningful and motivating context for mathematical activities (van Oers 2010). The experience of construing differences in the material can be used in order to create and record number pairs of a given difference up to 10, i.e., to *construct differences.* For that purpose, the materials have to be reinterpreted. The recording sheets are no longer primarily used to construe differences, but rather to construct differences. This can also be seen in the term "rolling" used by the children: "Roll the die" no longer means to generate number pairs randomly, but rather to find compatible number pairs by themselves. Both the "rolled" and the recorded numbers can be increased and decreased accordingly. Thus, the *space for mathematical play* is enhanced. By means of complementary learning environments based on mathematically rich games with structured game materials, playful mathematical handling can be stimulated by reinterpretation of the materials as mathematical objects. So the particular interactive learning situation provides incidental mathematical experience on the one hand and deeper insights in relations between numbers on the other hand. In this regard, they are complementary.

Conclusion

One focus of this chapter is to investigate and describe the potential of complementary learning environments (CLEs) to link more informal with more formal learning situations in the period of transition from kindergarten to primary school. Furthermore, the children's process of understanding differences and their playing activities in this context help to describe discontinuities and continuities of mathematical learning[4]:

[4] The highlighted disparities of playful mathematics in games and mathematical play can be seen in other playing and learning environments of the project.

1. *CLEs link individual learning processes.*
   In the context of a CLE, children can develop their individual mathematical insights from kindergarten by deepening, enhancing, and generalizing them in primary school. Invented strategies can be used further on but in a reflected way. The example of the CLE "who has more?" provides an insight in children's possible learning pathways from the challenge of construing differences (interpreting the difference of two numbers) to constructing differences (find number pairs with a given difference) to generalize relations of differences.
2. *CLEs link experiences with materials with more symbolic representations.*
   For CLEs, materials are chosen which can be used further on in school. Therefore, children gain first experience in dealing with the materials and its possible usages. The added recording sheets in school link the concrete materials with symbolic representations: the materials are represented in a "drawn" version and numerals are added. Later, number relations can only be represented symbolically by numerals or algebraic terms.
3. *CLEs link concrete playing experience with more abstract insights in number relations.*
   Basic requirements for CLEs are mathematically rich games which request children to construe number relations by playing the game. In doing so, children informally gain insights in number relations. A guiding adult, who involves children in mathematical discussions, can enhance their mathematical potential. Specific tasks in primary school and reflections over recordings lead children to more abstract, more general, and more flexible ideas in number relations.
4. *CLEs link different institutional learning conditions.*
   The approach of play-based and action-oriented learning provides occasions for a wide range of mathematical experiences that are consistent with the more informal learning conditions in kindergarten. The more formal learning conditions in primary school require specific learning goals and planned steps to reach it. In CLEs, each playing environment for kindergarten is linked to a learning environment in primary school through the materials and mathematical ideas as mentioned in aspects (1), (2), and (3).
5. *CLEs link shared learning activities and discussions with individual differentiated learning activities.*
   On the one hand, CLEs are intended to enable *all* children to reason about mathematical relations in the same meaningful context. At the same time, they are open for individual insights and strategies. As players of a mathematically rich game, children meet on the same level. Each player wants to win the game. Regarding their mathematical insights and strategies, these can differ widely. Nevertheless, it is possible to play together and benefit from each other (Stebler et al. 2013). The same can be said for school. There are always some children who did not have the chance to gain the desired experience due to several reasons. The mathematical richness of the learning environment enables differentiated learning processes on different levels. While some children continue to or newly learn how to construe differences, for example, other children have opportunities to reason more deeply about relationships between differences.

6. *CLEs link educators from kindergarten and teachers from primary school.*
   The idea of CLEs, the idea of linking informal and formal learning situation, provides occasions for cooperative activities between the educators from kindergarten and teachers from primary school. Communication about how to arrange learning situations with a special focus on continuity and discontinuity in each learning location is possible. Children's learning processes can be discussed to better coordinate the learning arrangements.

CLEs ensure continuity over the period of transition by using the discontinuities caused by the different learning conditions in a productive way. However, it is only one possibility to accompany the period of transition besides other mathematical activities like traditional games, picture books, and everyday experiences.

Furthermore, account should be taken of the fact that CLEs only work if there are enough children in primary school class who already gained the earlier experiences in kindergarten. If there are enough experienced children, children without experience can benefit from them and vice versa. Otherwise, teachers have to provide more time for gaining experience as a basis before they can continue with the lessons.

Nevertheless, the approach to enable children in kindergarten to gain rich experiences via mathematical games, which will be taken up and continued in primary school in the form of "mathematical play," poses a promising way to arrange linked learning processes from informal to formal learning situations found in kindergarten and primary school.

References

Ahtola, A., Silinskas, G., Poikonen, P.-L., Kontoniemi, M., Niemi, P., & Normi, J.-E. (2011). Transition to formal schooling. *Early Childhood Research Quarterly, 26*, 295–302. doi:10.1016/j.ecresq.2010.12.002.

Clements, D. H. (2004). Major themes and recommendations. In D. H. Clements, J. Sarama, & A.-M. DiBase (Eds.), *Engaging young children in mathematics: Standards for early childhood mathematics education* (pp. 7–72). Mahwah, NJ: Lawrence Erlbaum.

Clements, D. H., Sarama, J., & DiBase, A.-M. (2004). *Engaging young children in mathematics: Standards for early childhood mathematics education*. Mahwah, NJ: Lawrence Erlbaum.

Cobb, P., Confrey, J., diSessa, A., Lehrer, R., & Schauble, L. (2003). Design experiments in educational research. *Educational Researcher, 32*, 9–13. doi:10.3102/0013189X032001009.

Dockett, S., & Perry, B. (2010). What makes mathematics play? In L. Sparrow, B. Kissane & C. Hurst (Eds.), *Shaping the future of mathematics education: Proceedings of the 33rd annual conference of the Mathematics Education Research Group of Australasia* (pp. 715–718). Fremantle, WA: MERGA.

Flexer, R. J. (1986). The power of five: The step before the power of ten. *The Arithmetic Teacher, 34*(3), 5–9. doi:10.2307/41193003.

Gasteiger, H. (2012). Fostering early mathematical competencies in natural learning situations—Foundation and challenges of a competence-oriented concept of mathematics education in kindergarten. *Journal für Mathematik-Didaktik, 33*, 181–201. doi:10.1007/s13138-012-0042-x.

Ginsburg, H. P. (2006). Mathematical play and playful mathematics: A guide for early education. In D. G. Singer, R. M. Golinkoff, & K. Hirsh-Pasek (Eds.), *Play = learning. How play moti-*

vates and enhances children's cognitive and social-emotional growth (pp. 145–165). New York, NY: Oxford University Press.

Hasemann, K. (2005). Word problems and mathematical understanding. *Zentralblatt für Didaktik der Mathematik, 37*, 208–211. doi:10.1007/s11858-005-0010-8.

Hirsh-Pasek, K., Golinkoff, R. M., Berk, L. E., & Singer, D. G. (2009). *A mandate for playful learning in preschool. Presenting the evidence*. New York, NY: Oxford University Press.

Krajewski, K., & Schneider, W. (2009). Early development of quantity to number-word linkage as a precursor of mathematical school achievement and mathematical difficulties: Findings from a four-year longitudinal study. *Learning and Instruction, 19*, 513–526. doi:10.1016/j.learninstruc.2008.10.002.

Krummheuer, G. (2000). Interpretative classroom research in primary mathematics education. *Zentralblatt für Didaktik der Mathematik, 32*, 124–125. doi:10.1007/BF02655650.

Langhorst, P., Ehlert, A., & Fritz, A. (2012). Non-numerical and numerical understanding of the part-whole concept of children aged 4 to 8 in word problems. *Journal für Mathematik-Didaktik, 33*, 233–262. doi:10.1007/s13138-012-0039-5.

Nührenbörger, M., & Steinbring, H. (2009). Forms of mathematical interaction in different social settings: Examples from students', teachers' and teacher-students' communication about mathematics. *Journal of Mathematics Teacher Education, 12*, 111–132. doi:10.1007/s10857-009-9100-9.

Nührenbörger, M., & Tubach, D. (2012). Mathematische Lernumgebungen. Komplementäre Lerngelegenheiten in Kita und Grundschule. *Die Grundschulzeitschrift, 26*(255/256), 87–89.

Pramling Samuelsson, I., & Asplund Carlsson, M. (2008). The playing learning child: Towards a pedagogy of early childhood. *Scandinavian Journal of Educational Research, 52*, 623–641. doi:10.1080/00313830802497265.

Ramani, G. B., & Siegler, R. S. (2008). Promoting broad and stable improvements in low-income children's numerical knowledge through playing number board games. *In Child Development, 79*, 357–394. Retrieved from http://www.jstor.org/stable/27563489.

Resnick, L. B. (1983). A developmental theory of number understanding. In H. P. Ginsburg (Ed.), *The development of mathematical thinking* (pp. 109–151). New York: Academic Press.

Roßbach, H.-G. (2010). Bildungs- und Lernverläufe im Übergang. In A. Diller, H. R. Leu, & T. Rauschenbach (Eds.), *Wie viel Schule verträgt der Kindergarten?* (pp. 75–90). München: DJI.

Schuler, S. (2011). *Playing and learning in early mathematics education—Modelling a complex relationship*. Paper presented at the CERME 7, Rzeeszów.

Stebler, R., Vogt, F., Wolf, I., Hauser, B., & Rechsteiner, K. (2013). Play-based mathematics in kindergarten. A video analysis of children's mathematical behaviour while playing board game in small groups. *Journal für Mathematik-Didaktik, 34*, 149–175. doi:10.1007/s13138-013-0051-4.

Steinbring, H. (2005). *The construction of new mathematical knowledge in classroom interaction: An epistemological perspective*. Berlin, New York: Springer.

Steinweg, A. S. (2001). *Zur Entwicklung des Zahlenmusterverständnisses bei Kindern [Children's understanding of number pattern]*. Münster: LIT.

van Oers, B. (2010). Emergent mathematical thinking in the context of play. *Educational Studies in Mathematics, 74*, 23–37. doi:10.1007/s10649-009-9225-x.

van Oers, B. (2014). Cultural-historical perspectives on play: Central ideas. In L. Brooker, M. Blaise, & S. Edwards (Eds.), *The Sage handbook of play and learning in early childhood* (pp. 56–66). Los Angeles, CA: Sage.

Wittmann, E. C. (1995). Mathematics education as a 'design science'. *Educational Studies in Mathematics, 29*, 355–734. doi:10.1007/BF01273911.

Wittmann, E. C. (2001). Developing mathematics education in a systemic process. *Educational Studies in Mathematics, 48*, 1–20. doi:10.1023/A:1015538317850.

Wittmann, E. C., & Müller, G. N. (2009). *Das Zahlenbuch. Handbuch zum Frühförderprogramm [Zahlenbuch. Handbook of the early learning programme]*. Leipzig: Klett.

Investigating the Potential of the Home Learning Environment for Early Mathematics Learning: First Results of an Intervention Study with Kindergarten Children

Julia Streit-Lehmann and Andrea Peter-Koop

Abstract The context of the study reported in this chapter is a combined family literacy and family numeracy project for preschoolers and their parents, which aims to foster early mathematical competencies and its relevant language for children in the year prior to school enrollment, i.e., Grade 1. Special attention is given to children from families with a low socioeconomic and educational background that in Germany frequently correlates with a migration background. In preparation for a large-scale intervention study, a pilot study has been conducted with 57 preschoolers and their families from 3 kindergartens. The study followed a pre-/post-test design with a follow-up test. First results suggest that the majority of the sample demonstrated benefits from the intervention, irrespective of migration background or nationality. However, at the end of Grade 1, these positive results only had a lasting effect on the performance of children from families without migration background.

Introduction

It is widely acknowledged that parents play an important role in the early learning of their children (e.g., Bronfenbrenner 2000; Cross et al. 2009). With respect to mathematics, over the last 30 years internationally, a number of projects focusing on parental involvement in the mathematics education of their children have been implemented. An Internet search revealed that projects such as "Family Math" (Stenmark et al. 1986) and "Families Count" (Robinson and Fowler 1990) have been conducted with over 100 000 participants across the United States, Canada, Sweden, South Africa, Australia, New Zealand, and other countries. Participants were often families with low socioeconomic and educational background frequently with a migrant background or first nation families with preschool and/or primary

J. Streit-Lehmann • A. Peter-Koop (✉)
Faculty of Mathematics, University of Bielefeld, Bielefeld, Germany
e-mail: streit-lehmann@uni-bielefeld.de

T. Meaney et al. (eds.), *Mathematics Education in the Early Years*,
DOI 10.1007/978-3-319-23935-4_6

children. The aims of these and other similar projects on a smaller scale are to communicate core ideas of the curriculum and to introduce strategies and materials to help families support their children and foster their mathematics learning. In contrast to the large effort that has been put into the development and implementation of projects for parents, to date there has been little large-scale research on the impact of parental support on their children's learning in the early years. However, there is empirical evidence in the context of the PISA studies that suggests that educational success is strongly related to families' financial resources (see Schwarz and Weishaupt 2014). In addition, Ehmke and Siegle (2008) found that the mathematical competencies of parents are a predictor for their children's (15-year-olds) mathematical achievements as they support processes helpful for learning mathematics.

Research (e.g., Epstein 1995) further suggests that programs for parents (of school children) which have positive effects on children's mathematical learning and achievement:

- Provide suggestions for setting up a home environment supporting their children's school learning
- Allow regular exchange between parents and teachers about the content treated in class and students' learning and achievements
- Recruit parents to assist with school-based intervention
- Acknowledge parents' heterogeneous educational backgrounds and inform them about how they can assist their children with their homework and other curricular activities and tasks

In this context, activities that stimulate and facilitate parents to discuss mathematical problems with their children and collaboratively find a solution were the most effective (Sheldon and Epstein 2005). However, with increasing year levels and correspondingly increasingly complex mathematics content, parents find it harder to support their children as they themselves may not have the mathematical skills and understanding of school mathematics (Gal and Stout 1995). Hence, it seems easier to actively involve the parents of preschool and primary school children in the home as well as school-based mathematics learning of their children, especially when considering that the first mathematics learning will take place prior to formal schooling.

Theoretical Background: Number Concept Development

Children start developing mathematical knowledge and abilities a long time before they enter formal education (e.g., Anderson et al. 2008; Ginsburg et al. 1999). In their play as well as in their everyday life experiences at home and in kindergarten, they develop a foundation of skills, concepts, and understandings related to early numeracy (Baroody and Wilkins 1999). Anderson et al. (2008) reviewing international studies on preschool children's development and knowledge conclude that research "points to young children's strong capacity to deal with number prior to school, thus diminishing the value of the conventional practice that pre-number activities are more appropriate for this age group upon school entry" (p. 102). However, the range of mathematical competencies children develop prior to school varies quite substantially. While most preschoolers manage to develop a wide range of informal mathematical knowledge and skills prior to school, there is a small number of children who, for various reasons, struggle with the acquisition of number skills (e.g., Clarke et al. 2008). Furthermore, clinical psychological studies suggest that children potentially at risk in learning mathematics can already be identified 1 year prior to school entry by assessing their number concept development (e.g., Aunola et al. 2004; Krajewski and Schneider 2009; Peter-Koop and Kollhoff 2015). Findings from intervention studies also indicate that these children benefit from an intervention prior to school focused on helping them to develop a foundation of number and quantity-related knowledge and skills for successful school-based mathematics learning (Peter-Koop and Grüßing 2014). This seems to be of crucial importance as findings from the SCHOLASTIK project (Helmke and Weinert 1999) indicate that students who are low achieving in mathematics at the beginning of primary school generally tend to stay in this position. In most cases, sadly, school does not have a compensational effect for these children. In addition, Stern (1997) emphasizes that with respect to success at school, subject-specific knowledge prior to school is more important than general cognitive factors such as intelligence.

While educational practice in (German) kindergartens in many cases still draws on Piaget's (1952) theoretical model that emphasizes the importance of pre-number activities for number concept development (Anderson et al. 2008), post-Piagetian research provides evidence suggesting that the development of number skills and concepts results from the integration of number skills, such as counting, subitizing, and comparison (e.g., Fuson et al. 1983; Clements 1984; Sophian 1995; Peter-Koop and Grüßing 2014).

Krajewski and Schneider (2009) provided a theoretical model (see Fig. 1) that is based on the assumption that the linkage of imprecise nonverbal quantity concepts with the ability to count forms the foundation for understanding several major principles of the number system. The model describes how early mathematical competencies are acquired via three developmental levels.

At the *first level* (*basic numerical skills*), number words and number-word sequences are isolated from quantities. In the sense of Resnick's (1989) "proto-quantitative comparison schema" (p. 163), children compare quantities without counting by using words like "less," "more," or "the same amount." At the age of 3–4 years, most children start to link number words to quantities; they develop awareness of numerical quantity (Dehaene 1992) and hence enter the *second level* (*quantity–number concept*). According to Krajewski and Schneider (2009), the understanding of the linkage between quantities and number words (level II) is acquired in two phases:

Level IIa: Imprecise quantity to number-word linkage, i.e., the development of a rather imprecise conception of the attribution of number words to quantities and the connection of number words to rough quantity categories (e.g., three (two) is "a bit," eight (twenty) is "much," and hundred (thousand) is "very much")

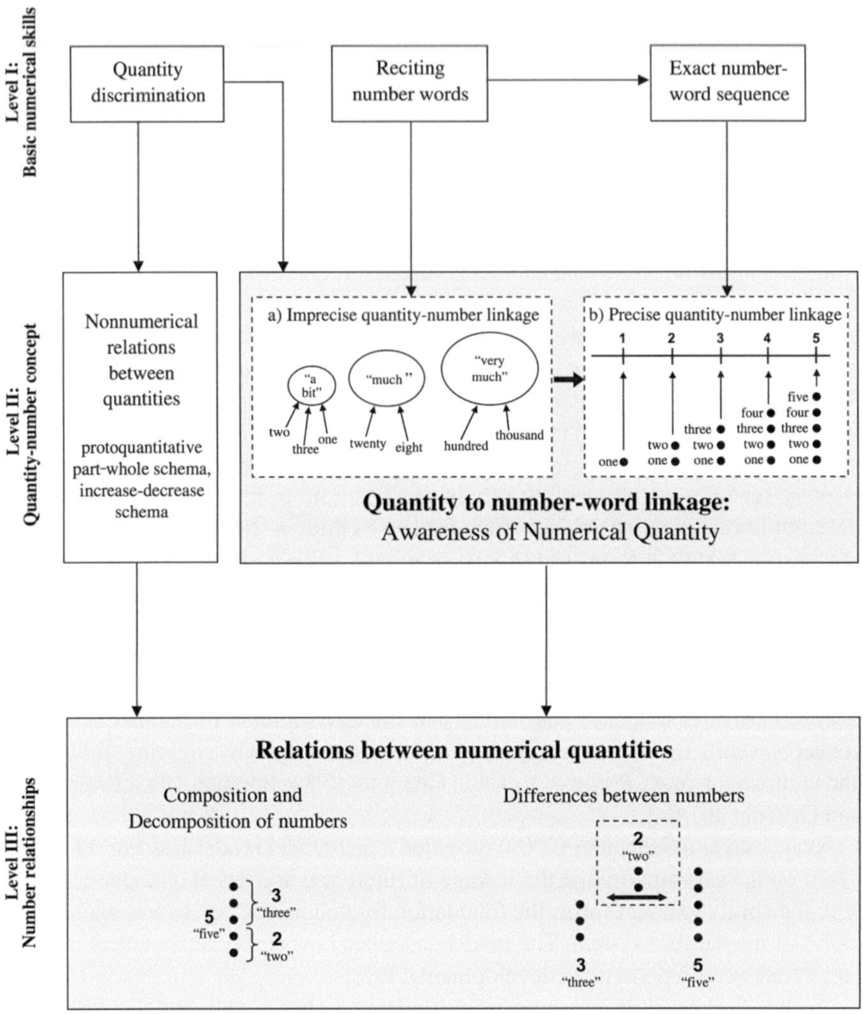

Fig. 1 Model of early mathematical development (Krajewski and Schneider 2009, p. 515)

Level IIb: Precise quantity to number-word linkage, i.e., the ability to distinguish close number words such as "two" and "three" and the linkage of number words to exact quantities "where counting is linked with quantity discrimination" (p. 514)

At the *third level* (*linking quantity relations with number words*) children then understand "that the relationship between quantities also takes on a number-word reference. They realize that numerically indeterminate quantities, e.g., "all" lollies, can be divided into smaller amounts, e.g., "a few" lollies, and "also understand that this can also be represented with precise numbers" (p. 516).

However, it is important to acknowledge that children are not necessarily at the same developmental stage with respect to number words and number symbols. Some children might have already reached the third level when dealing with smaller numbers, while they still operate with larger numbers on the second level. Furthermore, the use of manipulatives also affects children's performances. Hence, with respect to numerical development, it is very difficult to classify a child precisely at one level. While competencies on the third level reflect first computation skills and initial arithmetic understanding, Krajewski and Schneider (2009) describe the first two levels (basic numerical skills/quantity–number concept) as "mathematical *precursor* skills" (p. 516).

In addition, Krajewski and Schneider (2009) investigated the predictive validity of the quantity–number competencies of these developmental levels for mathematical school achievement. The results of their 4-year study indicate that quantity–number competencies measured in kindergarten predict about 25 % of the variance in mathematical school achievement in Grades 3 and 4. Moreover, a subgroup analysis indicated that low-performing fourth graders had already shown substantially lower early mathematical competencies with respect to number and counting skills in the year prior to them starting school (i.e., Grade 1 in Germany) than their better performing peers, both prior to school and in Grade 4. It can be concluded that these quantity–number competencies constitute an important prerequisite for the understanding of school mathematics. These results conform to findings of other longitudinal studies (e.g., Aunola et al. 2004; Peter-Koop and Kollhoff 2015).

Context of the Study: The KERZ Project

KERZ is a combined family literacy and family numeracy project addressing preschoolers in their final year of kindergarten (5-year-olds) and their families. Hence, KERZ is an abbreviation for "**K**inder (**er**)**z**ählen," which means "children count" and "children tell." In German, these are very similar sounding verbs. KERZ is a joint development/research project conducted by mathematics education researchers from three German universities (i.e., Bielefeld University, Bremen University, and University of Education Karlsruhe), involving data from different German states and regions. Special attention in this project is given to children from families with migration backgrounds and/or a low socioeconomic and educational background, because both groups have been identified as educationally disadvantaged in Germany (Baumert and Schümer 2002). On the one hand, research suggests that a migration background is not necessarily problematic with respect to school mathematics learning and that achievement in mathematics is rather influenced by the socioeconomic and educational family background. On the other hand, research by Prediger et al. (2013) on factors for underachievement in high stakes test in mathematics suggests that academic language proficiency in the language of assessment is more relevant than other background factors. With respect to our study, migrant kindergarten children's language acquisition in (at least) their first and second language has to be taken into consideration. When assessing language proficiencies, Cummins (1979) distinguishes between "cognitive academic language proficiency" (CALP) and "basic interpersonal communication skills" (BICS). BICS is related to language competencies used in everyday situations; in contrast, CALP is required in educational contexts in order to communicate about rather abstract contents and concepts. Furthermore, while CALP builds on BICS, it is less specific for a certain language than BICS. Cummins found that well-developed BICS in at least one language (either the first or second language that a child acquires) is essential for the development of CALP in this language, as educational success is related to academic language proficiency (Prediger et al. 2013). Naturally, kindergarten children are at the very beginning with respect to their development of CALP, and children who have restricted basic communication skills in either their first or their second language may have extra challenges. Development of these basic communication skills is frequently related to the educational background of the parents/families in regard to their socioeconomic status (Schmitman gen. Pothmann 2008).

The primary aim of the associated research study is to investigate how family-based activities related to (informal) early childhood mathematics can support early mathematical learning and to monitor possible long-term effects of the intervention in the first years of school. Parental involvement in this context includes dialogical family reading of mathematics-related picture books and playing board and dice games that require knowledge and abilities with respect to counting and comparing sets, enumerating, number words and symbols, as well as spatial visualization.

A second research interest of the main study is to investigate the potential of such a home learning environment in contrast to kindergarten-based mathematical activities that are supposed to foster number-concept development. Hence, the study will follow a control group design with group 1 being the treatment group in which children experience early mathematics training at kindergarten by specially trained kindergarten teachers without parental involvement and group 2 being the control group with a focus on the home learning environment and no additional mathematical instruction at kindergarten.

Since the KERZ project is addressing families with a low socioeconomic and educational background, one key obstacle that needed to be overcome was the lack of resources, i.e., children's books and games suitable to foster early mathematics learning, in the families. In solving this logistical problem, a strategy developed in the ENTER project[1] by Dagmar Bönig and Jochen Hering at Bremen University was adopted. A treasure chest (see Fig. 2) was provided for the kindergartens involved in the project. This treasure chest contains a number of selected books, games, and activities that are made available for the children to borrow and take home for a week. In order to assist non-German-speaking families, translations of rules and text-reduced picture books were provided to encourage the families to talk in their native language(s) as well as in German.

[1] For details about the ENTER project, see http://opus4.kobv.de/opus4-bamberg/frontdoor/index/index/docId/5697.

Fig. 2 Treasure chest provided in the KERZ project

While the kindergarten is the place where the materials can be borrowed (and returned), it is made very clear to children, parents, as well as the kindergarten teachers that these materials are for home use only in order to foster the home learning environment. With respect to the tension between mathematics instruction and the construction of mathematics, most of the books and games that were chosen for the treasure chest encourage a constructive approach on different levels of development. While instructions for games are by nature instructive, in most cases on an individual level, they allow for differentiation. Instructive elements are further related to the teaching and learning of concepts and facts, i.e., the number-word sequence, when parents support their child by repeating and correcting the sequence in either the first and/or second language. However, the materials chosen clearly support and challenge constructive approaches and foster the collaboration of children, their parents, and siblings with respect to developing their mathematical understanding and skills.

The idea is that the children in the final year of kindergarten keep the materials for a few days (usually 1 week), use them with their families, return them, then select something new, and so on. The borrowing process is monitored by the kindergarten staff, who also assist the children and families with explaining the contents and rules, if needed. In order to document what individual children have borrowed, each week the kindergarten teachers complete a chart, which was provided and hung up in the home group (see Fig. 3).

In addition, once a week the kindergarten teachers get the participating preschoolers together to talk about and share their experiences and to create interest in borrowing materials that peers have enjoyed reading and/or playing at home. Respective questions are, for example:

What was the book/game about?
How did you like it?
Who did you read it/play it with?
How often did you read/play it?
Which numbers do you have to know to play the game?

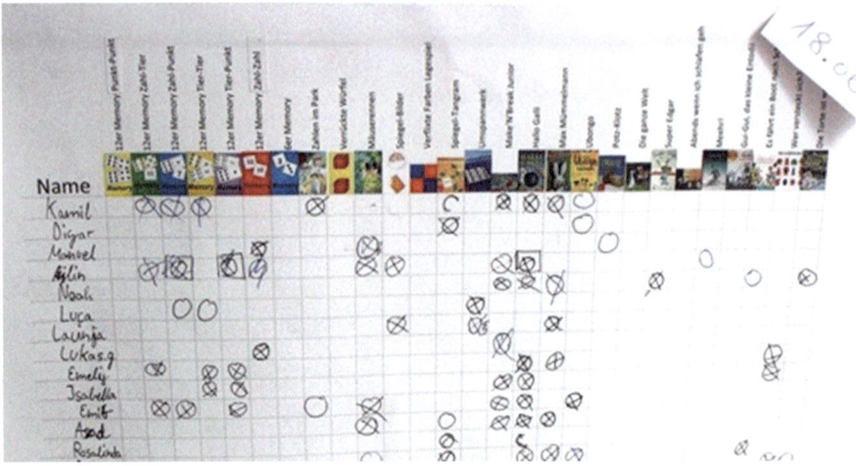

Fig. 3 Chart that documents the borrowing of materials during the project

We anticipated that this would help to clarify if the materials have been used at home (1), to monitor the frequency of the use (2), and to develop children's mathematical language by introducing and using relevant terminology such as number words, prepositions, and adjectives (3). For example, prepositions help to describe the structure of the number line in terms of the location of numbers (i.e., "number *before* 9/number *after* 9," or "39 is the number *above* 49 on the arithmetic rack"). The use of adjectives helps to communicate comparisons such as "bigger/smaller," "the same," or "different."

In order to trial the design and research instruments of a future large-scale study—involving around 1000 children from 50 kindergartens and their families across Germany—a pilot study was conducted in 2012.

Design of the Pilot Study and Methodology

The pilot study was conducted with 57 preschoolers from 3 kindergartens and their families in north west Germany. While one kindergarten is mainly attended by children from German middle-class families (Kiga 1), the other two are predominantly attended by children from families with a lower socioeconomic status (Kiga 2 and Kiga 3). Kiga 2 and Kiga 3 were deliberately chosen, as the majority of the children are from migrant families with rather low German language competencies and low socioeconomic background.[2]

[2] Information about the level of education, the German language skills (based on a self-assessment and assessment by the kindergarten teachers), and the duration of the parents' living in Germany had been obtained by a parent questionnaire at the start of the project. Kiga 2 and Kiga 3 are located in a suburb with predominantly social housing, available only for families with a low income.

Between February and June 2012, the children in their final year in kindergarten had access to the treasure chest. The borrowing process and its documentation followed the procedure described above. Parental information about the study and parents' questionnaires were provided in German, Arabic, Polish, Russian, and Turkish—acknowledging the main migration groups in this area of Germany.

Similar to what was anticipated for the main study in preparation, the pilot study followed a pre-/post test design. All participating children ($n=57$) completed two tests: the EMBI-KiGa (Peter-Koop and Grüßing 2011) and the TEDI-MATH (Kaufmann et al. 2009). In addition, all participating children performed on the CPM (colored progressive matrices intelligence test) (Raven et al. 2010) in order to control this variable with respect to the impact of the intervention.

The EMBI-KiGa is a semi-standardized one-on-one early numeracy interview based on the "First Year at School Mathematics Interview" developed in the context of the Australian "Early Numeracy Research Project" (Clarke et al. 2006) which had been published as a German adaptation (Peter-Koop and Grüßing 2011). It documents early mathematical competencies of children ages 3–6 years old. The EMBI-KiGa addresses mathematical *precursor* skills as identified by Krajewski and Schneider (2009). On an operational level, the EMBI-KiGa provides information on two subtests. The first subtest involves 11 items related to the first two levels of the model of early mathematical development (see Fig. 1), such as comparison, part-whole schema, and number–word sequence, while the second subtest explicitly focuses on developing counting skills. The EMBI-KiGa is task based and supported by manipulatives in order to allow children, who for various reasons might struggle with their language, to articulate their developing mathematical understanding through the use of specific materials provided for each task.

The TEDI-MATH, originally developed by French psychologists, is a one-on-one clinical interview which compares the mathematical performance of 4- to 8-year-olds with their age group, standardized in half-year sequences (Kaufmann et al. 2009). The TEDI-MATH covers counting skills, one-to-one correspondence, number words, part-whole relations, and initial addition and subtraction skills. Both instruments, EMBI-KiGa and TEDI-MATH, were conducted with the complete sample at both measuring points, *before* the borrowing from the treasure chest starts and *after* the 4 months intervention period.

A follow-up test was conducted at the end of Grade 1 with the DEMAT 1+ (Krajewski et al. 2002)—a standardized paper and pencil test, based on the curriculum for first graders in German primary schools to be conducted either at the end of Grade 1 or at the beginning of Grade 2. Like the TEDI-MATH, the DEMAT 1+ uses percentiles to rank children's mathematical competencies. The percentiles cover the whole range of abilities. For a child to reach percentile 90, for example, means that only 10 % of his/her peers perform better.

In addition, data was also collected from the parents before and after the intervention with respect to personal data about family background, education, language and migration background, as well as their individual attitudes and beliefs with respect to mathematics and mathematics learning. The questionnaire developed for

Table 1 Design of the pilot study

	Preschoolers	Parents	Kindergarten teachers
February 2012 *Measuring Point 1 (MP1)*	**EMBI-KiGa TEDI-MATH CPM**	Questionnaire: Family and educational background, reflection of their mathematics learning	Questionnaire: Information about the kindergarten and its pedagogical approach
March 2012		Parent information night	Monthly visits by the researcher
April 2012			
May 2012			
June 2012 *Measuring Point 2 (MP2)*	**EMBI-KiGa TEDI-MATH**	Questionnaire: Feedback on the use of the materials at home	Questionnaire: Feedback on the borrowing process
June 2013 *Measuring Point 3 (MP3)*	DEMAT 1+		

this purpose addressed parents' knowledge about content and curricula in school mathematics, their understanding of what mathematical competencies (if at all) a child should acquire before school entry, and a self-assessment of their own mathematical competencies. All questionnaires were disseminated in the parents' first language if necessary.

The participating kindergarten teachers answered questionnaires concerning their institutions, the handling of the treasure chest, and language skills of the participating children. Table 1 provides an overview of the stages and instruments of the data collection. The gray cells indicate the timing of the intervention period in which children could borrow books and games and explore them with their families.

Results

The first results of the pilot study suggest a relationship between the effectiveness of the KERZ project and the migration background of the children. However, it has to be taken into account that in our sample, a family migration background correlated with low socioeconomic status and low educational background. In order to refer to the development of the mathematical competencies of the children from the first to the second measuring point (MP1 to MP2), a distinction between "strong enhancement," "slight enhancement," and "no enhancement" was made. However, "enhancement" does not only mean an absolute increase in competencies, because an increase could be completely explained by the increasing age and corresponding intellectual development of the children. The term "enhancement" in this context rather refers to an enhancement relative to the particular peer group.

The three labels "strong, slight, and no enhancement" correspond to the percentile ranks of the TEDI-MATH and were confirmed by the data of the EMBI-KiGa. The distinctions between these labels are quite severe to avoid false-positive interpretations: For example, "strong enhancement" means an increase of at least 50 percentiles from MP1 to MP2, or alternatively an increase between 25 and 49 percentiles, while moving out of the lowest fifth into the midrange or moving out of the midrange into the highest fifth. Consequently, all EMBI values[3] had to improve to get that label. "Midrange" means in this context the middle three-fifths. When children improved between 25 and 49 percentiles, while staying in the midrange or having stagnating EMBI values, they were considered to have a "slight enhancement," and it is the same when children improved between 5 and 24 percentiles with all EMBI values improving.

As shown in large blue and green areas in the bar diagram on the very left of Fig. 4, in all three kindergartens more than half of the sample clearly improved their mathematical competencies slightly or strongly.

Eight of the 15 (53 %) children in Kiga 1 demonstrated an improvement from MP1 to MP2; of these eight children, three showed a "strong enhancement." Fifteen of the 20 children (75 %) in Kiga 2 improved from MP1 to MP2; again three of them showed a "strong enhancement." Thirteen of the 22 children in Kiga 3 (59 %) demonstrated an improvement, while 7 of these 13 children showed "strong enhancement." The orange sections of the diagrams in Fig. 4 represent those groups of children that showed "no enhancement." It is important to notice that the group of children who showed no enhancement includes those who already performed highly prior to the intervention and therefore could not demonstrate further substantial gains. In each of the three kindergartens, there were a few very highly performing children, before the start of the intervention. For example, a child reaching percentile 92 at MP 1 and percentile 95 at MP 2 is labeled "no enhancement."

The immediate effects are irrespective of the existence of a migration background, and no significant differences concerning the mean of intelligence between the three kindergarten groups have been found. In addition, the graph on the right side (see Fig. 4) shows that around one-third of *all* children in the sample are not affected by KERZ, independent of a migration background. Thirty-nine of the 57 children in the sample have a migration background. Thirteen of the remaining 18 children (72 %) without a migration background demonstrated improvement (see blue and green sections in the bar diagrams). Of the 39 children with a migration background 23 children (77 %) showed improvement. Roughly two-thirds demonstrated "slight" or "strong enhancement," which we interpret as a positive result.

However, this picture changes when considering the results of the follow-up test at the end of Grade 1. The analysis of the performances on the DEMAT 1+ clearly shows that mainly children from families without migration background could

[3]The EMBI provides two kinds of information for each child. Firstly, a (numerical) point score between 0 and 11 that shows how many of the 11 items of first subtest (*mathematical precursor skills*) have been solved correctly and secondly with respect to the second subtest (*counting*), a (ordinal) growth point that identifies his/her level of counting skills on a range between 0 and 6. These two measures are called "EMBI values."

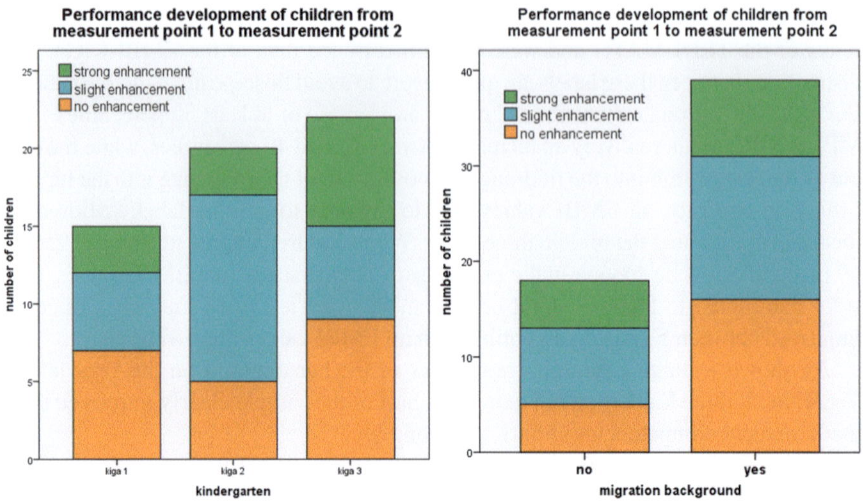

Fig. 4 Immediate impact on mathematical competencies ($n=57$)

reach percentile ranks higher than those of MP 1. Although this finding suggests a sustainable success of the intervention, children with migration background, however, mainly reached percentile ranks equal to or even lower than at MP 1.

Figure 5 shows the results of the follow-up test at the end of Grade 1, one year after the intervention. Only 42 out of the 57 children participated in the follow-up test at the end of Grade 1 as some children had moved away, did not attend school during the test, or did not start school in the first place. Twenty-nine of the 42 children who participated had a migration background, while 13 children did not. In order to characterize their development, four categories were used: "performance like MP 2 or better" (green), "performance better than at MP 1" (light blue), "performance like MP 1 or lower" (red), and "negligible change" (gray). Hence, the green and light blue sections represent positive results. The gray sections symbolize the absence of changes, irrespective of the different performance levels (however, this only applied to very few children). The red sections represent all of those children who after initial improvement observed from MP1 to MP2, 1 year later at MP3 showed results equal or even lower than the percentile reached at MP1.

With respect to the transition to school, the data indicates that a sustainable benefit of the intervention was related to family background. Predominantly children without a migration background maintained their progress, while this does not hold true for the children with migration background (see Fig. 5). The two bar diagrams on the right-hand side of Fig. 5 show that only 2 of the 13 children (15 %) without a migration background are represented in the red section (i.e., performance at $MP3 \leq MP1$) but 20 out of 29 children (69 %) with a migration background.

While children from all three kindergartens showed similar engagement and cooperation, the migration status explains the variance between Kiga 1 (children mainly from middle-class families) and the other two institutions. Children with

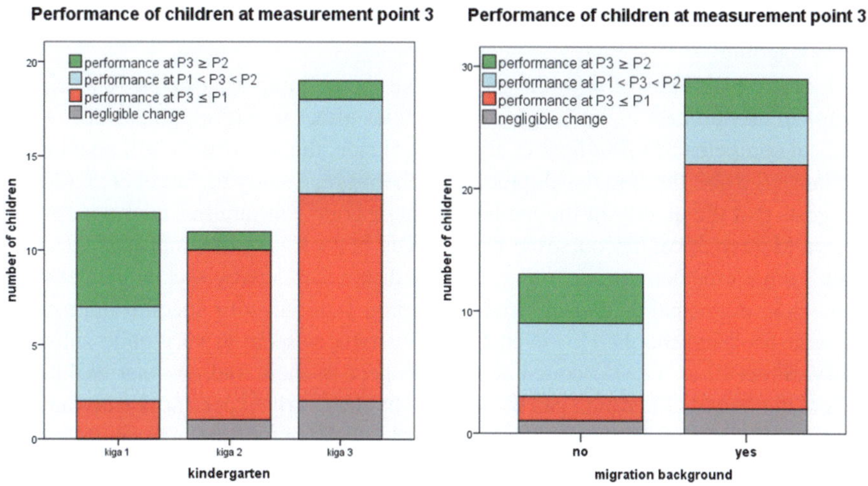

Fig. 5 Long-term impact on mathematical competencies ($n = 42$)

migration backgrounds (in our sample mainly from families with low socioeconomic and educational background) in this study clearly demonstrated less mathematical achievement at the end of Grade 1 than their peers without a migration background (from predominantly middle-class families). The main study will help to clarify whether this only holds true for the sample used in the pilot study or whether this will prove to be a broader effect.

Discussion and Implications

Since about two-thirds of the participating children benefitted slightly or strongly from the intervention and this effect was irrespective of a migration background, the KERZ project can be considered a success. However, the fact that the children with a migration background in the sample did not seem to experience long-lasting benefits from the intervention, as their results at MP3 suggest, is of concern. Reasons for this could be that classroom instruction in school mathematics is differently effective for children with and without migration background (who also come from disadvantaged families). In this context, our findings confirm the results of a national study focusing on children's achievements in the subjects German and Mathematics at the end of Grade 4 (Stanat et al. 2012) that constitutes migration background-related disparities in the areas reading, text understanding, and mathematics in all German states (Haag et al. 2012). When controlling the variable socioeconomic status however, the disadvantages in competence growth of children with migration background were clearly reduced. Another reason that might have had a negative influence on the performance of children from this group is the fact that the DEMAT 1+ as a paper and pencil test has high demands with respect to text understanding, which serves as a disadvantage when it comes to testing mathematics skills and understanding.

Language abilities of both the parents and the children are obviously a critical factor in this process as language competencies affect the development of mathematical competencies (Prediger et al. 2013). Hence, the main study will also monitor language abilities and development. Furthermore, a study by Street et al. (2005) suggests that the quality of the mathematical discourse in families is also a critical factor. Mathematical discourses as experienced in the family are important for children's school mathematics learning, as these discourses convey special images with respect to mathematics and can make children familiar with specific interaction patterns used in schools. However, these experiences seem to vary quite substantially. Street et al. (2005) conclude with respect to their findings that children's school mathematics learning can be impeded by their early experiences with family-based mathematical discourse, because they do not have the language requirements needed to master the change from family-based to school-based discourses, which are based on different values, rules, and behavioral patterns. Projects such as KERZ may help both children and parents to overcome this barrier, if they provide special opportunities for parents that help them to better prepare their children for school mathematics learning in Germany. In this context, it is important to acknowledge that the authors are aware of the fact that we take a special, i.e., deficit-oriented, perspective on certain families when discussing our results (and even before, when planning our design). We acknowledge that one could take the opposite perspective and perceive the school system and school mathematics in particular deficit as it obviously does not cater well enough for individual needs. Hence, in our roles as mathematics teacher educators, we certainly portray the perspective that schools and teachers should be adapting better to the actual children and their parents. However, in order to help to change the system and improve access to school mathematics for all children, it is necessary to identify the problem and to develop research-based solutions.

The rather disappointing results of the pilot study in terms of a lasting effect in school, i.e., that only children without migration background tended to sustain their earlier achievements, may be related to the specific standardized paper and pencil test used for the follow-up. Hence, for the main study, alternative instruments will be considered that might be more suitable for children with a migration background who might have underperformed based on their language competencies rather than their mathematical competencies. Another aspect that the main study will incorporate is considering whether early intervention is more effective in a home learning environment or at kindergarten.

References

Anderson, A., Anderson, J., & Thauberger, C. (2008). Mathematics learning and teaching in the early years. In O. N. Saracho & B. Spodek (Eds.), *Contemporary perspectives on mathematics in early childhood education* (pp. 95–132). Charlotte, NC: Information Age.

Aunola, K., Leskinen, E., Lerkkanen, M.-K., & Nurmi, J.-E. (2004). Developmental dynamics of mathematical performance from preschool to grade 2. *Journal of Educational Psychology, 96*, 762–770.

Baroody, A. J., & Wilkins, J. (1999). The development of informal counting, number, and arithmetic skills and concepts. In J. Copley (Ed.), *Mathematics in the early years* (pp. 48–65). Reston, VA: NCTM.

Baumert, J., & Schümer, G. (2002). Familiäre Lebensverhältnisse, Bildungsbeteiligung und Kompetenzerwerb im nationalen Vergleich. In Deutsches PISA-Konsortium (Ed.), *PISA 2000—Die Länder der Bundesrepublik Deutschland im Vergleich* (S. 159–202). Opladen: Leske + Budrich.

Bronfenbrenner, U. (2000). Ecological system theory. In A. E. Kazdin (Ed.), *Encyclopedia of psychology* (Vol. 3, pp. 129–133). Washington, DC: American Psychological Association.

Clarke, B., Clarke, D., & Cheeseman, J. (2006). The mathematical knowledge and understanding young children bring to school. *Mathematics Education Research Journal, 18*(1), 78–103.

Clarke, B., Clarke, D., Grüßing, M., & Peter Koop, A. (2008). Mathematische Kompetenzen von Vorschulkindern: Ergebnisse eines Ländervergleichs zwischen Australien und Deutschland. *Journal für Mathematik-Didaktik, 29*(3/4), 259–286.

Clements, D. (1984). Training effects on the development and generalization of Piagetian logical operations and knowledge of number. *Journal of Educational Psychology, 76*, 766–776.

Cross, C. T., Woods, T., & Schweingruber, H. (Eds.). (2009). *Mathematics learning in early childhood. Paths towards excellence and equity*. Washington, DC: National Academies Press.

Cummins, J. (1979). Linguistic interdependence and the educational development of bilingual children. *Review of Educational Research, 49*(2), 222–251.

Dehaene, S. (1992). Varieties of numerical abilities. *Cognition, 44*, 1–42.

Epstein, J. L. (1995). School/family/community partnerships: Caring for the children we share. *Phi Delta Kappan, 76*, 701–712.

Ehmke, T., & Siegle, T. (2008). Einfluss elterlicher Mathematikkompetenz und familialer Prozesse auf den Kompetenzerwerb von Kindern in Mathematik. *Psychologie in Erziehung und Unterricht, 55*, 253–264.

Fuson, K. C., Secada, W. G., & Hall, J. W. (1983). Matching, counting, and the conservation of number equivalence. *Child Development, 54*, 91–97.

Gal, I., & Stout, A. (1995). *Family achievement in mathematics*. Philadelphia, PA: National Center on Adult Literacy, University of Pennsylvania.

Ginsburg, H., Inoue, N., & Seo, K. (1999). Young children doing mathematics: Observations of everyday activities. In J. Copley (Ed.), *Mathematics in the early years* (pp. 88–99). Reston, VA: NCTM.

Haag, N., Böhme, K., & Stanat, P. (2012). Zuwanderungsbezogene Disparitäten. In P. Stanat, H. A. Pant, K. Böhme, & D. Richter (Eds.), *Kompetenzen von Schülerinnen und Schülern am Ende der vierten Jahrgangsstufe in den Fächern Deutsch und Mathematik. Ergebnisse des IQB Ländervergleichs 2011* (pp. 209–235). Berlin: Waxmann.

Helmke, A., & Weinert, F. E. (1999). Schooling and the development of achievement differences. In F. E. Weinert & W. Schneider (Eds.), *Individual development from 3 to 12: Findings from the Munich Longitudinal Study* (pp. 176–192). Cambridge, UK: Cambridge University Press.

Kaufmann, L., Nuerk, H.-C., Graf, M., Krinzinger, H., Delazer, M., & Willmes, K. (2009). *TEDI-MATH. Test zur Erfassung numerisch-rechnerischer Fertigkeiten vom Kindergarten bis zur 3. Klasse*. Göttingen: Hogrefe.

Krajewski, K., Küspert, P., & Schneider, W. (2002). *DEMAT 1+. Deutscher Mathematiktest für erste Klassen*. Göttingen: Hogrefe.

Krajewski, K., & Schneider, W. (2009). Early development of quantity to number-word linkage as a precursor of mathematical school achievement and mathematical difficulties: Findings from a four-year longitudinal study. *Learning and Instruction, 19*(6), 513–526.

Peter-Koop, A., & Grüßing, M. (2011). *ElementarMathematisches BasisInterview für den Einsatz im Kindergarten*. Offenburg: Mildenberger.

Peter-Koop, A., & Grüßing, M. (2014). Early enhancement of kindergarten children potentially at risk in learning school mathematics—Design and findings of an intervention study. In U. Kortenkamp, B. Brandt, C. Benz, G. Krummheuer, S. Ladel, & R. Vogel (Eds.), *Early mathematics learning. Selected papers of the POEM 2012 conference* (pp. 307–321). New York: Springer.

Peter-Koop, A., & Kollhoff, S. (2015, June 28–July 3). *Exploring the influence of early numeracy understanding prior to school on mathematics achievement at the end of Grade 2.* Paper presented at the mathematics education research conference of Australasia, University of the Sunshine Coast, Queensland, Australia.

Piaget, J. (1952). *The child's conception of number.* London: Routledge.

Prediger, S., Renk, N., Büchter, A., Gürsoy, E., & Benholz, C. (2013). Family background or language disadvantages? Factors for underachievement in high stakes tests. In A. Lindmeier, & A. Heinze (Eds.), *Proceedings of the 37th conference of the International Group for the Psychology of Mathematics Education* (Vol. 4, pp. 49–56). Kiel: PME.

Raven, J. C., Raven, J., & Court, J. H. (2010). *Coloured progressive matrices.* Frankfurt/Main: Pearson.

Resnick, L. (1989). Developing mathematical knowledge. *American Psychologist, 44*(2), 162–169.

Robinson, I., & Fowler, H. (1990). Families counting on. In K. Clements (Ed.), *Whither mathematics?* (pp. 285–286). Melbourne: Mathematical Association of Victoria.

Schmitman gen. Pothmann, A. (2008). *Mathematiklernen und Migrationshintergrund. Quantitative Analysen zu frühen mathematischen und (mehr)sprachlichen Kompetenzen.* Doctoral thesis, University of Oldenburg, Faculty of Education.

Schwarz, A., & Weishaupt, H. (2014). Veränderungen in der sozialen und ethnischen Zusammensetzung der Schülerschaft aus demografischer Perspektive. *Zeitschrift für Erziehungswissenschaft, 17*(2), 9–35.

Sheldon, S. B., & Epstein, J. (2005). Involvement counts: Family and community partnership and mathematics achievement. *The Journal of Educational Research, 98*(4), 196–206.

Sophian, C. (1995). Representation and reasoning in early numerical development. *Child Development, 66*, 559–577.

Stanat, P., Pant, H. A., Böhme, K., & Richter, D. (Eds.). (2012). *Kompetenzen von Schülerinnen und Schülern am Ende der vierten Jahrgangsstufe in den Fächern Deutsch und Mathematik. Ergebnisse des IQB Ländervergleichs 2011.* Berlin: Waxmann.

Stenmark, J., Thompson, V., & Cassey, R. (1986). *Family math.* Berkeley, CA: University of California.

Stern, E. (1997). Ergebnisse aus dem SCHOLASTIK-Projekt. In F. E. Weinert & A. Helmke (Eds.), *Entwicklung im Grundschulalter* (pp. 157–170). Weinheim: Beltz.

Street, B., Baker, D., & Tomlin, A. (2005). *Navigating numeracies: Home/school numeracy practices.* Dordrecht: Springer.

The Impact on Learning When Families and Educators Act Together to Assist Young Children to Notice, Explore and Discuss Mathematics

Ann Gervasoni and Bob Perry

Abstract *Let's Count* is a new Australian early mathematics initiative that aims to promote positive mathematical experiences for young children (3–5 years) as part of families' everyday activities. The 154 children who experienced Let's Count in 2013 demonstrated noteworthy growth in their mathematical knowledge from the beginning of their preschool year to its end. On almost every measure, the Let's Count cohort bettered the performance of the comparison groups, with some measures showing statistically significant differences. This suggests that educators and families working in partnership to assist young children to notice, explore and discuss the mathematics they encounter during everyday experiences had a positive effect on children's construction of mathematics.

Introduction

There is serious debate within the international community about whether or not it is appropriate to engage children in formal mathematics education prior to formal schooling, including during preschool. This debate acknowledges that children's formal mathematics knowledge and dispositions vary considerably when they begin school and that, among other factors, this is likely due to their differing experiences and opportunities to engage with mathematical ideas prior to school. Significantly, many children starting school are more mathematically capable than many mathematics curricula and textbook writers assume (Bobis 2002; Clarke et al. 2006; Ginsburg and Seo 2000; Hunting et al. 2012). This means that these children's experiences

A. Gervasoni (✉)
Monash University, Clayton, VIC, Australia
e-mail: ann.gervasoni@monash.edu

B. Perry
Charles Sturt University, Albury, NSW, Australia

T. Meaney et al. (eds.), *Mathematics Education in the Early Years*,
DOI 10.1007/978-3-319-23935-4_7

throughout early childhood have enabled them to learn formal mathematical ideas. It also suggests that some children may be inadequately challenged by the mathematics tasks and instruction they experience in their first year of school. This may have a negative impact on their opportunity to thrive mathematically (Perry and Dockett 2008). Variation in children's mathematics knowledge also means that some may be less favourably positioned than others to profit from mathematics instruction when they begin school. This raises issues about how families, educators and communities can best approach mathematics learning in the early years so that all children benefit and about how to assist children who are less favourably positioned than others when they begin school. These issues are explored in this chapter by drawing on the experience of the Australian *Let's Count* initiative (Perry and Gervasoni 2012). In particular the chapter will consider how the findings of the *Let's Count* Longitudinal Evaluation contribute to the debate about whether or not it is appropriate to engage children in formal mathematics education prior to formal schooling.

The *Let's Count* Longitudinal Evaluation

Let's Count is a new early mathematics initiative commissioned by The Smith Family, an Australian children's charity, to assist early childhood educators to work in partnership with families living in financially disadvantaged communities to promote positive mathematical experiences for young children (3–5 years). The initiative aims to foster opportunities for children to engage with the mathematics encountered as part of their everyday lives and talk about it, document it and explore it in ways that are fun and relevant to them and that enable them to learn powerful mathematical ideas in ways that develop positive dispositions to learning and mathematical knowledge and skills. The Smith Family aim is 'helping disadvantaged Australian children to get the most out of their education, so they can create better futures for themselves' (The Smith Family 2013). It was piloted in 2011 in five sites across Australia whose communities were identified as experiencing social and economic disadvantage. In 2013–2014, The Smith Family delivered a revised *Let's Count* programme in four additional sites that were also participating in a longitudinal evaluation of the programme (Gervasoni and Perry 2013).

The *Let's Count* approach initially involves two professional learning modules for early childhood educators:

Module 1: Noticing and exploring everyday opportunities for mathematics
Module 2: Celebrating mathematics

Between modules, the educators connect with families to consider ways that they can encourage children to notice, explore and discuss the mathematics that they encounter in everyday situations, including through games, stories and songs. The educators used a range of strategies for connecting with parents, depending on what was most effective for their community. Most typically, educators organised group meetings or met one on one with parents to discuss the *Let's Count* ideas. Throughout the year the educators and parents then continued to discuss the children's mathematics investigations and learning. The following example illustrates how educators supported parents and children to explore mathematics:

> One little boy came in today and said, 'I really want to measure my bed'. So we made a measuring tape for him. I said, 'You could use your hands' and he said 'No, I want a measuring tape'. So we made a measuring tape. … the information came from his mother first and then we discussed it with the child. The mother came in and said, 'Oh he really wants to measure his bed' and I went 'Ok, we can do that. We can work out a way to do that for you'. (Excerpt from a 2013 *Let's Count* educator interview)

A key aim of *Let's Count* was for parents to explore and discuss mathematics with their children in everyday situations. The following example from a participating parent in 2013 demonstrates how parents would build on children's 'noticing' and support them to explore:

> She comes out with things every day, basically. Something that really surprised me … Oh, [?] were talking about my birthday and that I'm turning 22 and she said, 'Oh mummy, you're turning 22, isn't that two two?', as in … 2-2' and I was like 'Yes, that's a number' and then she's just like, 'So how do we add …?' like 'What do we do to get to that number?'…. She was just trying to work out how to get to 22, like all different scenarios on how to get to the number 22!

One aspect of evaluating the effectiveness of *Let's Count* was to measure participating children's mathematical growth across their preschool year and also to compare their knowledge just prior to beginning school with a comparison group of 125 children whose families had not participated in *Let's Count*. This comparison group was from the same economically disadvantaged communities, and the children were assessed in December 2012 in order to provide baseline data about the range of children's mathematical knowledge in these communities prior to the introduction of *Let's Count* in 2013–2014 and prior to children beginning school. It was decided also to compare the mathematical knowledge of the *Let's Count* group with that of a more representative cohort from the *Early Numeracy Research Project* (Clarke et al. 2002). The 1438 children participating in the *Early Numeracy Research Project* in 2001 were assessed in March 2001 just after they started school and were from 34 schools in the State of Victoria, Australia. They were selected to provide a representative sample of the Victorian population.

Let's Count evaluation data was also gathered at multiple points from early childhood educators (surveys and interviews), parents and other adult members of families (interviews). However, the focus of this chapter is the development of children's formal mathematics knowledge in the year prior to their beginning school.

Assessing Children's Knowledge of School Mathematics

The tool selected to assess children's mathematical knowledge for the *Let's Count Longitudinal Evaluation* was the *Mathematics Assessment Interview* (Gervasoni et al. 2010, 2011). This assessment was designed for young children, is task based and interactive, is derived from extensive research and enables mathematical learning to be measured in nine domains. All tasks are presented orally. One section of the assessment focuses on early mathematics concepts for children who are beginning school (the Foundation Detour). This assessment was originally developed as part of the *Early Numeracy Research Project* (ENRP) (Clarke et al. 2002; Department of Education, Employment and Training 2001). Following refinement, it was renamed the *Mathematics Assessment Interview* (MAI) (Gervasoni et al. 2010, 2011).

The principles underlying the construction of the tasks in the MAI and the associated mathematics growth point framework were to:

- Describe the development of mathematical knowledge and understanding in the first three school years in a form and language that was useful for teachers.
- Reflect the findings of relevant international and local research in mathematics (e.g. Fuson 1992; Gould 2000; Mulligan 1998; Steffe et al. 1983; Wright et al. 2000).
- Reflect, where possible, the structure of mathematics.
- Allow the mathematical knowledge of individuals and groups to be described.
- Enable a consideration of children who may be mathematically vulnerable (Gervasoni and Lindenskov 2011).

The interview includes the whole number domains of counting, place value, addition and subtraction strategies and multiplication and division strategies; the measurement domains of time, length and mass; and the geometry domains of properties of shape and visualisation. The assessment tasks in the interview take between 30 and 45 min for each child and were administered in this study by independent, trained assessors who followed a detailed script and recorded responses on a standardised record sheet. Each child completed about 30 tasks in total, and given success with one task, the assessor continued with the next tasks in a domain for as long as a child was successful, according to the script. At the end of each interview, the record sheet was analysed to determine the growth point reached by a child in each domain. Each growth point represents substantial expansion in knowledge along paths to mathematical understanding (Clarke 2001). An example of the addition and subtraction growth points is provided in Table 1.

The processes for validating the growth points, the interview items and the comparative achievement of students are described in full in Clarke et al. (2002). A critical role for the assessor throughout the interviews was to listen and observe the children, noting their responses, strategies and explanations while completing each task. These responses were independently coded to determine whether or not a response was correct, to identify the strategy used to complete a task and to identify the growth point reached by a child overall in each whole domain. This information was entered into an SPSS database for analysis. Of particular interest for this study were the children's responses to tasks in the early mathematics concept section and the initial tasks in the other domains.

Table 1 Addition and subtraction strategies growth points

Growth points	Title	Descriptions
1	Count all (two collections)	Counts all items from one to find the total of two collections
2	Count on	Counts on from one number to find the total of two collections
3	Count back/count down to/count up from	Given a subtraction situation, chooses appropriately from strategies including count back, count down to and count up from
4	Basic strategies (doubles, commutativity, adding 10, tens facts, other known facts)	Given an addition or subtraction problem, strategies such as doubles, commutativity, adding 10, tens facts and other known facts are evident
5	Derived strategies (near doubles, adding 9, build to next ten, fact families, intuitive strategies)	Given an addition or subtraction problem, strategies such as near doubles, adding 9, build to next ten, fact families and intuitive strategies are evident
6	Extending and applying addition and subtraction using basic, derived and intuitive strategies	Given a range of tasks (including multi-digit numbers), solves them mentally using the appropriate strategies and a clear understanding of key concepts

Preschool Children's Mathematics Knowledge

The preschool children participating in the *Let's Count* Longitudinal Evaluation in 2013 were eligible to begin school in January 2014 and were aged between 4.5 and 5.5 years by the end of preschool. They were presented with all the tasks in the early mathematics concepts or 'Foundation Detour' Section of the *Mathematics Assessment Interview* and also with tasks from each of the whole number, measurement and geometry domains for as long as each child was successful. All children in the 2012 comparison group were assessed using the Foundation Detour, the four whole number domains, two measurement domains (time and length) and two geometry domains to establish comparison data. In 2013, children were assessed using the Foundation Detour, the four whole number domains and then randomly assigned for assessment in one measurement and one geometry domain. This reduced the length of the interview for each child but maintained the opportunity for the research to compare children's growth in measurement and geometry during their preschool year. Summaries of the children's responses are presented in the tables below. Task results are grouped in tables according to the associated *Australian Curriculum: Mathematics* school foundation year standard (Australian Curriculum, Assessment and Reporting Authority (ACARA) 2013). This enables some consideration of how well the curriculum guidelines match children's mathematics capabilities. Each table shows the percentage of children who were successful with each task for the *Let's Count* group in both March and December 2013, for the 2012 comparison group, and for 1438 children in the February/March 2001 ENRP First

Year at School cohort (Clarke et al. 2006). The December percentage scores for each task for the 2012 and 2013 groups were compared for statistical significance. These results are presented in the first column of each table. Results for some tasks were not available for the ENRP comparison group as indicated with 'na' in the tables.

Table 2 describes children's success with tasks involving small sets of objects, usually small plastic teddies. The tasks were all associated with the Australian Curriculum Foundation Standard: students make connections between number names, numerals and quantities up to 10. LC refers to the children involved in *Let's Count*.

The results for the *Let's Count* preschoolers suggest that about three-quarters of the children in December could demonstrate the associated Australian Curriculum Standard even before they started school. The ENRP comparison group results are similar. Large numbers of the *Let's Count* children in March were also successful with these tasks, but less so than in December. The *Let's Count* group was significantly different to the comparison group with respect to tasks involving making sets of objects and also reducing a set by one.

Table 3 shows the percentage of children able to recognise immediately the number of dots on a card without counting them and match a numeral to the number of dots. These results highlight that the majority of students can recognise quantities up to about four items without counting and about one-sixth of preschoolers can conceptually do the same for nine dots. There is some variation in success rates between the *Let's Count* children in March and December. The majority of children can match numerals to the number of dots, although nine was much harder to match than the other numbers. This ability to recognise quantities without counting is important for teachers to build upon when planning instruction and is important for exploring pattern and structure. There was little difference between the comparison and intervention groups for these tasks, except for recognising 0 and 2 dots.

The importance of pattern and structure in young children's mathematical learning is gaining increased attention. The Australian Curriculum Foundation proficiencies of fluency and reasoning focus on continuing and creating patterns. The data presented in Table 4 suggest that for the comparison and ENRP groups, about three-quarters of children could match patterns when they began school and about one-third of children could continue and explain a pattern. The success rate is significantly greater for the 2013 *Let's Count* group compared to the comparison groups.

While continuing, creating and describing patterns are likely to be a profitable aspect of instruction for most children in their first year at school, many children need further challenges. The Foundation Standard also focuses on students counting to and from 20 and ordering small collections. Several MAI tasks focused on sequence counting, counting a larger collection of at least 20 items and ordering numerals. The percentage of children able to complete these tasks is presented in Table 5.

The data suggest that the majority of preschoolers can count to 10 and many can forward count to 20, but not back from 20. However, few children could count 20 teddies successfully and also identify how many teddies were left when one teddy

Table 2 Percentage success on tasks with small sets

Tasks	Significance: Comp (Dec 2012) to LC (Dec 2013) (χ^2, p)	LC Mar 2013 ($n=141$)	LC Dec 2013 ($n=116$)	Comp Dec 2012 ($n=125$)	ENRP Feb 2001 ($n=1438$)	Australian Curriculum Foundation Standard
Tasks with small sets						
Count 4 teddies	NS	88	96	95	93	Students make connections between number names, numerals and quantities up to 10
Identify one of two groups as 'more'	NS	68	92	90	84	
Make a set of 5 teddies when asked	7.043, $p<0.01$	63	90	77	85	
Conserve 5 when rearranged by child	NS	67	88	79	58	
Combine 5 + 3 blue teddies and total	NS	57	71	75	na	
Make collection of 7 (when shown number 7)	11.016, $p<0.01$	27	84 $n=92$	63	na	
Know one less than 7 when 1 teddy removed	12.018, $p<0.01$	23	82 $n=85$	61	na	
Know one less than 7 without recounting	12.018, $p<0.01$	10	40 $n=85$	25	na	
Part/whole tasks						
Show 6 fingers—often 5 and 1	NS	51	78	79	78	
6 fingers second way	4.566, $p<0.05$	5	40	27	20	
6 fingers third way	NS	1	15	10	8	
One-to-one correspondence						
Know 5 straws needed to put a straw in 5 cups	NS	74	87	88	92	

Table 3 Percentage success in subitising tasks and matching numerals to dots

Tasks	Significance: Comp (Dec 2012) to LC (Dec 2013) (χ^2, p)	LC Mar 2013 ($n=141$)	LC Dec 2013 ($n=116$)	Comp Dec 2012 ($n=125$)	ENRP Feb 2001 ($n=1438$)	Australian Curriculum Foundation Standard
Subitising tasks						
Recognise 0 without counting	$\chi^2=4.690$, $p<0.05$	65	91	81	82	Students make connections between number names, numerals and quantities up to 10
Recognise 2 without counting	$\chi^2=5.133$, $p<0.05$	91	99	94	95	
Recognise 3 without counting	NS	63	91	83	84	
Recognise random 3 without counting	NS	66	87	86	na	
Recognise 4 without counting	NS	52	79	70	71	
Recognise random 4 without counting	NS	39	53	50	na	
Recognise 5 without counting	NS	33	41	44	43	
Recognise 9 without counting	NS	14	9	16	9	
Tasks matching numerals to dots						
Match numeral to 0 dots	NS	38	81	73	63	
Match numeral to 2 dots	NS	69	89	90	86	
Match numeral to 3 dots	NS	53	83	73	79	
Match numeral to 3 random dots	NS	57	83	82	na	
Match numeral to 4 dots	NS	49	75	73	77	
Match numeral to 4 random dots	NS	49	76	69	na	
Match numeral to 5 dots	NS	44	68	65	67	
Match numeral to 9 dots	NS	28	47	38	41	

Table 4 Percentage success in pattern tasks

Tasks	Significance: Comp (Dec 2012) to LC (Dec 2013) (χ^2, p)	LC Mar 2013 (n = 141)	LC Dec 2013 (n = 116)	Comp Dec 2012 (n = 125)	ENRP Feb 2001 (n = 1438)	Australian Curriculum Foundation Standard
Pattern tasks						Fluency proficiency includes continuing patterns Reasoning proficiency includes creating patterns
Name colours in pattern	NS	90	99	98	94	
Match pattern	$\chi^2=5.623$, $p<0.05$	49	85	72	76	
Continue pattern	$\chi^2=5.102$, $p<0.05$	16	48	34	31	
Explain pattern	NS	16	42	34	31	

was removed. This focus on the cardinal value of 20 is a profitable area for instruction in the first year at school, but is not highlighted in the Foundation Standard. There was a noted difference between the *Let's Count* children's ability in March and December to order numeral cards. The *Let's Count* 2013 group in December was statistically significantly more able to count to 20 and order one-digit numbers than was the 2012 group.

Several tasks in the interview focused on measuring length and time (see Table 6). The data highlights that many children beginning school are able to compare and order lengths, thus meeting the Foundation Standard. Most children in all cohorts were aware of the purpose of a clock, but few children who were about to begin school were able to both name some days of the week and identify 2 o'clock on an analogue clock.

Spatial reasoning is a key aspect of learning mathematics. The data presented in Table 7 shows children's success with tasks involving describing and interpreting locations, recognising the properties of shapes and using mental imagery to manipulate shapes.

The data suggest that the *Let's Count* preschoolers were proficient in these aspects of mathematics and almost all children assessed in December could meet the Foundation Standard prior to beginning school. The most difficult task involved dynamic imagery, with at least 16 % of the children successful in December. Some of the *Let's Count* 2013 group were able to trace hidden shapes by the end of the year. The difference to the comparison group was statistically significant.

The interview also includes a range of tasks involving calculations, although few students progressed far in these domains. Results for four calculation tasks are presented in Table 8. All tasks were presented orally and involved the use of materials.

Most children who were successful with the first three tasks worked out the answers by counting all the items one by one. A small number of students used the count-on strategy. Most children solved the division task through grouping rather

Table 5 Percentage success with counting and ordering numerals

Tasks	Significance: Comp (Dec 2012) to LC (Dec 2013) (χ^2, p)	LC Mar 2013 ($n=141$)	LC Dec 2013 ($n=116$)	Comp Dec 2012 ($n=125$)	ENRP Feb 2001 ($n=1438$)	Australian Curriculum Foundation Standard
Counting tasks						Students count to and from 20 and order small collections
Rote count to 10	NS	66	93	87	na	
Rote count to 20	6.117, $p<0.05$	17	55	45	na	
Count a collection of at least 20 and when one item is removed, know total without recounting	8.079, $p<0.05$	2	16	8	na	
Ordering numbers tasks						
Order numeral cards 1–5	N/A	30	68	na	na	
Order numeral cards 1–9	NS	16	60	48	46	
Order numeral cards 0–9	10.354, $p<0.01$	10	52	31	38	
Order 3 one-digit numbers	NS	4	52	47	na	
Order 3 two-digit numbers	NS	1	15	14	na	

Table 6 Percentage success with length and time measurement tasks

Tasks	Significance: Comp (Dec 2012) to LC (Dec 2013) (χ^2, p)	Mar 2013 (n = 14[illegible])	LC Dec 2013 (n = 116)	Comp Dec 2012 (n = 125)	ENRP Feb 2001	Australian Curriculum Foundation Standard
Ordering length tasks		(n = 109)	(n = 116)	(n = 125)	(n = 1438	
Order 3 candles from smallest to largest	NS	43	78	73	61	Students compare objects using mass, length and capacity
Order 4 candles from smallest to largest	NS	23	63	54	50	
Length measurement tasks		(n = 77)	(n = 60)	(n = 125)		
Accurately compare two lengths of string and stick	NS	43	73	65	na	
Measure length—informal units	NS	1	7	8	na	
Time measurement tasks		(n = 69)	(n = 60)	(n = 125)		Students connect events and the days of the week
Aware of the purpose of a clock	NS	74	83	83	na	
Know some days/months and 2 o'clock	NS	0	5	17	na	

Table 7 Percentage success on spatial tasks

Tasks	Significance: Comp (Dec 2012) to LC (Dec 2013) (χ^2, p)	LC Mar 2013 ($n=141$)	LC Dec 2013 ($n=116$)	Comp Dec 2012 ($n=125$)	ENRP Feb 2001 ($n=1438$)	Australian Curriculum Foundation Standard
Language of location tasks						
Beside	NS	77	96	94	88	Students use appropriate language to describe location
Behind	4.304, $p<0.05$	76	96	87	87	
In front of	NS	63	88	91	83	
Properties of shapes tasks		($n=81$)	($n=59$)	($n=125$)		Students group objects based on common characteristics and sort shapes and objects
Knows square	NS	75	92	85	na	
Knows circle	NS	90	95	92	na	
Knows rectangle	NS	54	76	74	na	
Knows some triangles	9.548, $p<0.01$	81	95	83	na	
Knows all triangles	9.548, $p<0.01$	59	58	63	na	
Visualisation tasks		($n=83$)	($n=58$)	($n=125$)		
Recognises static images (rectangle) in embedded situations	7.894, $p<0.05$	68	95	90	na	
Identifies a reoriented rectangle in room and traces possible shapes when a shape is partially hidden	7.894, $p<0.05$	5	17	16	na	

Table 8 Percentage success on calculation tasks involving materials (teddies)

Tasks	Significance: Comp (Dec 2012) to LC (Dec 2013) (χ^2, p)	LC Mar 2013 ($n = 141$)	LC Dec 2013 ($n = 116$)	Comp Dec 2012 ($n = 125$)	Australian Curriculum Foundation Standard
Calculation tasks					Problem-solving proficiency: using materials to model authentic problems, sorting objects, using familiar counting sequences to solve unfamiliar problems and discussing the reasonableness of the answer
Adds 5 + 3 when screen over 5 is removed	NS	27	63	49	
Adds 9 + 4 when screen over 9 is removed	9.664, $p < 0.01$	11	42	25	
Calculates total for 2 teddies in 4 cars	NS	39	64	58	
Divides 12 teddies between 4 mats	15.497, $p < 0.01$	1	61	31	

than sharing by ones. The results and the children's strategies indicate that a large group of children are able to meet the Foundation's *problem-solving proficiency* before beginning school. It was statistically significant that the 2013 *Let's Count* group was more likely than the 2012 comparison group to successfully add 9+4 teddies and successfully divide 12 teddies between 4 mats.

Whole Number Growth Point Distributions

The data presented in the previous tables showed the percentage of children who were successful with each assessment task. Analysis of children's performance on the MAI also enables the associated growth points that children have reached in the whole number, measurement and spatial reasoning domains to be determined. The growth points represent key milestones in children's learning, and there are typically six growth points in each domain. The following figures show the growth point distributions in each domain for the 2013 *Let's Count* group, as '4-year-olds' in March and '5-year-olds' in December, and the December 2012 comparison group.

As may be expected of preschool children, Figs. 1, 2, 3 and 4 show that the majority of children are on the emerging growth points in the four whole number domains. About 10 % of the 4-year-olds and about one-quarter of the 5-year-olds have reached Growth Point 1 or Growth Point 2 in most domains. It is clear that a larger percentage of the 5-year-old *Let's Count* group have reached higher growth points than the comparison group in both the counting and addition and subtraction domains. This corresponds with statistically significant differences in task performance described in Tables 5 and 8.

Measurement and Spatial Reasoning Growth Point Distributions

Figures 5, 6, 7 and 8 show the growth point distributions for the measurement and spatial reasoning domains. These indicate quite a range in knowledge in each domain for both the '4-year-old' *Let's Count* group in March and '5-year-old' groups in December. The growth point distributions highlight that the majority of children, including the '4-year-olds' in March, have progressed to at least Growth Point 1 in most domains. This is in contrast to the growth points reached by children in the whole number domains. There are some apparent differences between the groups assessed in December. With time, fewer children in the *Let's Count* group knew some times, days and months than for the comparison group. With length, 10 % more children in the *Let's Count* group could compare, order and match lengths than for the comparison group. In properties of shape, almost 20 % more children in the *Let's Count* group could sort and compare shapes than could children

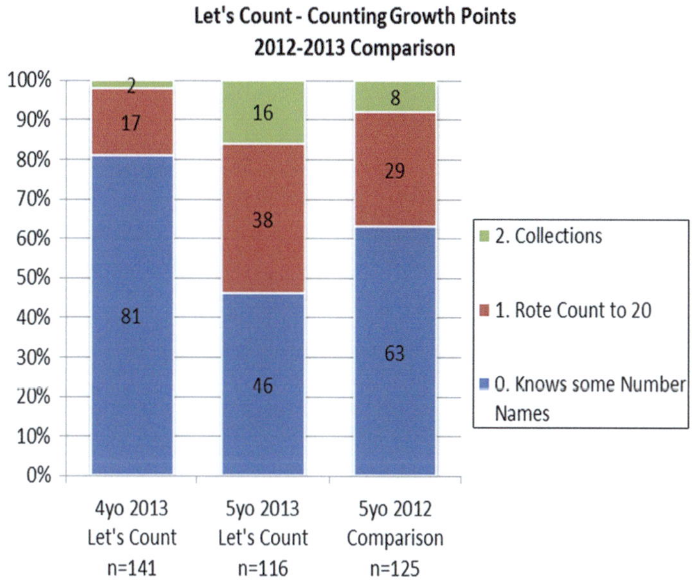

Fig. 1 Counting growth point distributions

Fig. 2 Place value growth point distributions

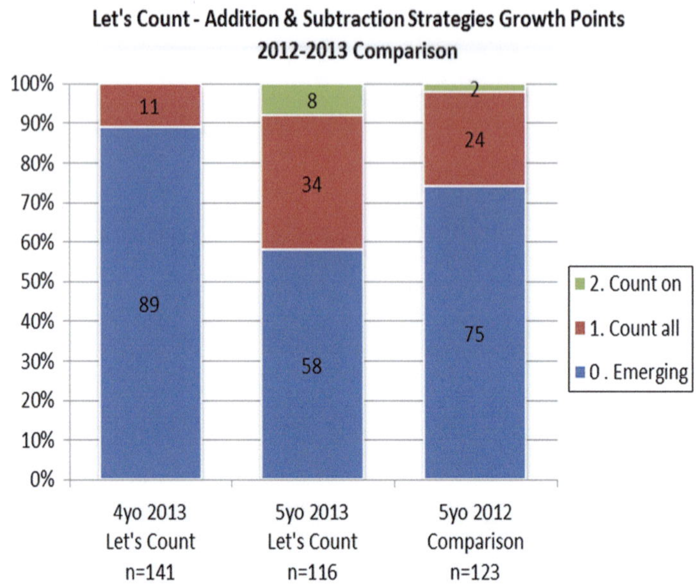

Fig. 3 Addition and subtraction strategies growth point distributions

Fig. 4 Multiplication and division strategies growth point distributions

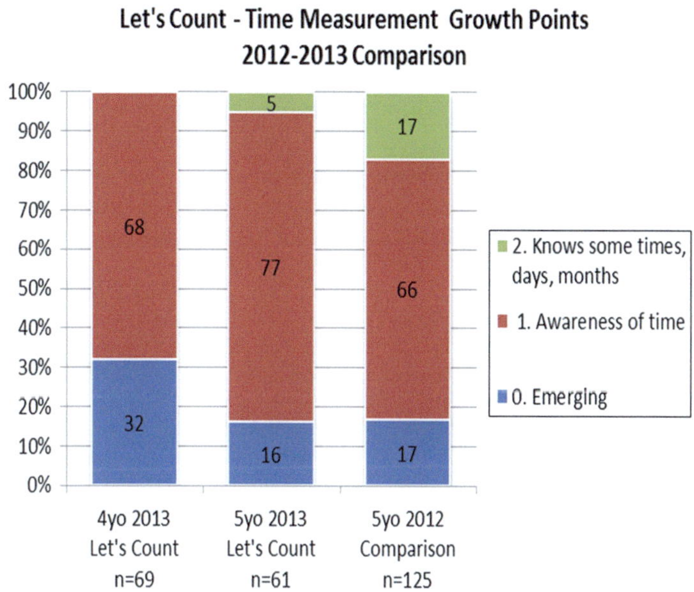

Fig. 5 Time growth point distributions

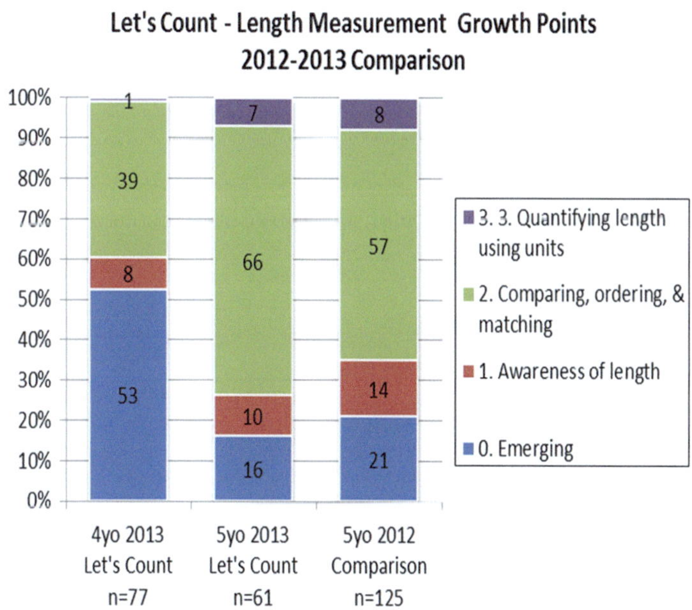

Fig. 6 Length growth point distributions

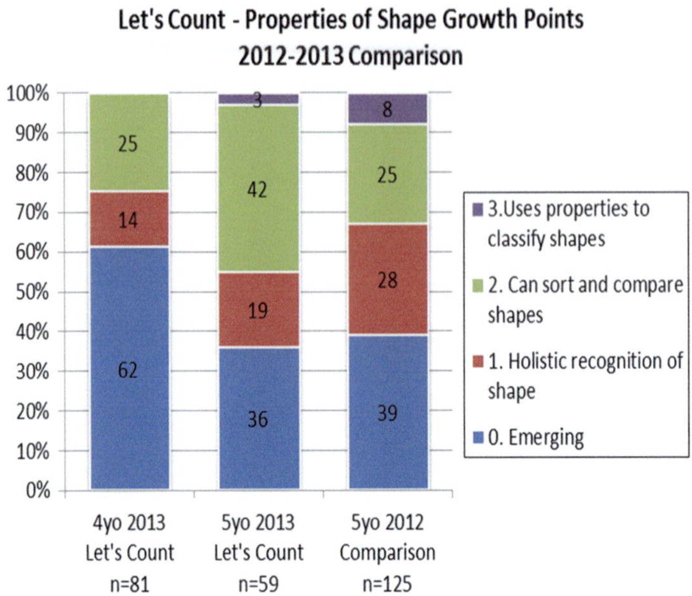

Fig. 7 Properties of shape growth point distributions

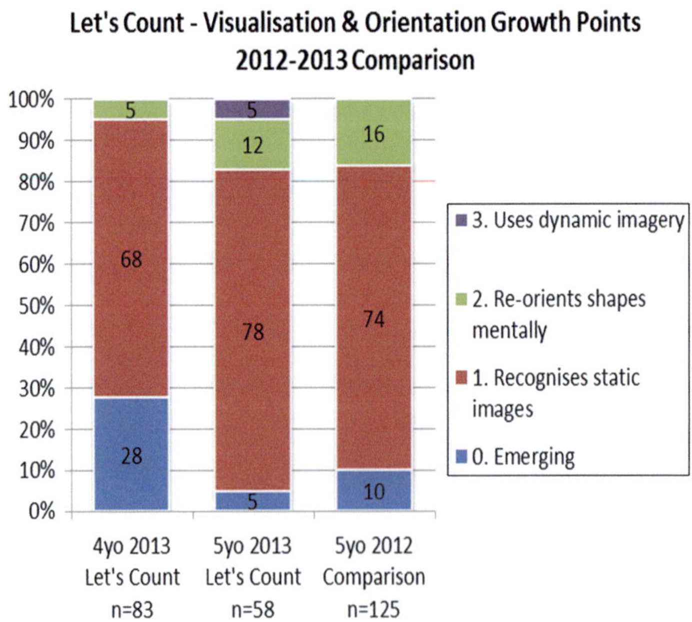

Fig. 8 Visualisation and orientation growth point distributions

in the comparison group. In visualisation and orientation, the percentage of children on the emerging Growth Point 0 was half that of the comparison group, and 5 % of *the Let's Count* group could use dynamic imagery (Growth Point 3), while no students in the comparison group could do this.

Discussion

The findings highlight the broad range of formal mathematics knowledge that many children construct prior to beginning school, either through their own play or through the more intentional teaching prompted by their families and early childhood educators. This supports the findings of earlier research (e.g. Bobis 2002; Clarke et al. 2006; Gervasoni and Perry 2013, 2015; Ginsburg and Seo 2000; Gould 2012; Hunting et al. 2012).

As part of the *Let's Count* approach, at least one family member and the early childhood teacher of each child who experienced *Let's Count* in 2013 acted together to consider ways in which they could support each child to notice, explore and discuss the mathematics they encountered during everyday activities.

The MAI data show that all the children in the 2013 cohort could demonstrate some formal mathematics during their 4-year-old assessment at the beginning of their preschool year. These children also showed noteworthy growth in their mathematical knowledge from the beginning of their preschool year to its end. The extent of this growth is further emphasised by comparing the *Let's Count* group's end-of-year performance with the 2012 comparison group and the cohort assessed at the beginning of primary school during the *Early Numeracy Research Project* in 2001 (Clarke et al. 2002). On almost every measure, the *Let's Count* cohort bettered the performance of the two comparison groups, with some comparisons showing statistically significant differences. This suggests that educators and families acting in partnership to assist young children to notice, explore and discuss the mathematics they encounter during everyday experiences had a positive effect on their construction of mathematics.

Overall, the findings from the *Let's Count* Longitudinal Evaluation highlight that young children construct many formal mathematical concepts during their everyday family and preschool experiences. Comparisons between the 2013 *Let's Count* group and the 2012 comparison cohort highlight some statistically significant differences. This suggests that the informal but intentional engagement of adults with children as they notice, explore and discuss mathematics as part of everyday experiences is associated with children's construction of mathematical ideas. In the case of the *Let's Count* children, most were better prepared to take advantage of the more formal mathematics education and instruction that they encounter when they begin school.

We conclude that it is both appropriate and important to engage children with mathematical explorations prior to school, but suggest that this does not need to take the shape of formal mathematics education. The results from the *Let's Count* Longitudinal Evaluation show that informally exploring and discussing the mathematics encountered as part of everyday life is effective in facilitating mathematics learning and more in keeping with preschool children's development. Our findings also highlight the variability in children's mathematics knowledge at the beginning and end of preschool. It could be that if preschool teachers more intentionally discussed and explored mathematics with the children who less often spontaneously notice, explore and discuss the mathematics in everyday experiences, then their mathematics learning may be enhanced. This could position them to benefit more favourably from instruction at school. This is a profitable area for further research.

Acknowledgement *Let's Count* was commissioned by The Smith Family and is supported by the Origin Foundation and developed in partnership with BlackRock Investment Management.

References

Australian Curriculum, Assessment and Reporting Authority (ACARA). (2013). *The Australian curriculum: Mathematics v2.4*. Retrieved from: http://www.australiancurriculum.edu.au/Mathematics/Curriculum/F-10.

Bobis, J. (2002). Is school ready for my child? *Australian Primary Mathematics Classroom, 7*(4), 4–8.

Clarke, D. M. (2001). Understanding, assessing and developing young children's mathematical thinking: Research as powerful tool for professional growth. In J. Bobis, B. Perry, & M. Mitchelmore (Eds.), *Numeracy and beyond: Proceedings of the 24th annual conference of the Mathematics Education Research Group of Australasia* (pp. 9–26). Sydney: MERGA.

Clarke, D., Cheeseman, J., Gervasoni, A., Gronn, D., Horne, M., McDonough, A., et al. (2002). *ENRP final report*. Melbourne: ACU.

Clarke, B., Clarke, D. M., & Cheeseman, J. (2006). The mathematical knowledge and understanding young children bring to school. *Mathematics Education Research Journal, 18*(1), 78–102.

Department of Education, Employment and Training. (2001). *Early numeracy interview booklet*. Melbourne: Department of Education, Employment and Training.

Fuson, K. (1992). Research on whole number addition and subtraction. In D. A. Grouws (Ed.), *Handbook of research on mathematics teaching and learning* (pp. 243–275). New York: Macmillan.

Gervasoni, A., & Lindenskov, L. (2011). Students with 'special rights' for mathematics education. In B. Atweh, M. Graven, W. Secada, & P. Valero (Eds.), *Mapping equity and quality in mathematics education* (pp. 307–323). Dordrecht: Springer.

Gervasoni, A., Parish, L., Hadden, T., Turkenburg, K., Bevan, K., Livesey, C., et al. (2011). Insights about children's understanding of 2-digit and 3-digit numbers. In J. Clark, B. Kissane, J. Mousley, T. Spencer & S. Thornton (Eds.), *Mathematics: Traditions and [new] practices*. Proceedings of the 23rd biennial conference of The AAMT and the 34th annual conference of the Mathematics Education Research Group of Australasia (Vol. 1, pp. 315–323). Alice Springs: MERGA/AAMT.

Gervasoni, A., Parish, L., Upton, C., Hadden, T., Turkenburg, K., Bevan, K., et al. (2010). Bridging the numeracy gap for students in low SES communities: The power of a whole school approach. In L. Sparrow, B. Kissane, & C. Hurst (Eds.), *Shaping the future of mathematics education*. Proceedings of the 33rd annual conference of the Mathematics Education Research Group of Australasia (pp. 202–209). Fremantle: MERGA.

Gervasoni, A., & Perry, B. (2013). Children's mathematical knowledge prior to starting school. In V. Steile (Ed.), *Mathematics education: Yesterday, today and tomorrow*. Proceedings of the 36th annual conference of the Mathematics Education Research Group of Australasia (pp. 338–345). Melbourne: MERGA.

Gervasoni, A., & Perry, B. (2015). Children's mathematical knowledge prior to starting school and implications for transition. In B. Perry, A. MacDonald, & A. Gervasoni (Eds.), *Mathematics and transition to school—International perspectives*. Dordrecht: Springer.

Ginsburg, H. P., & Seo, K.-H. (2000). Preschoolers' mathematical reading. *Teaching Children Mathematics, 7*(4), 226–229.

Gould, P. (2000). Count me in too: Creating a choir in the swamp. In *Improving numeracy learning: What does the research tell us?* Proceedings of the ACER research conference (pp. 23–26). Melbourne: Australian Council for Educational Research.

Gould, P. (2012). What number knowledge do children have when starting kindergarten in NSW? *Australasian Journal of Early Childhood, 37*(3), 105–110.

Hunting, R., Bobis, J., Doig, B., English, L., Mousley, J., Mulligan, J., et al. (2012). *Mathematical thinking of preschool children in rural and regional Australia: Research and practice*. Melbourne: Australian Council for Educational Research.

Mulligan, J. (1998). A research-based framework for assessing early multiplication and division. In C. Kanes, M. Goos & E. Warren (Eds.), *Teaching mathematics in new times*. Proceedings of the 21st annual conference of the Mathematics Education Research Group of Australasia (Vol. 2, pp. 404–411). Brisbane: MERGA.

Perry, B., & Dockett, S. (2008). Young children's access to powerful mathematical ideas. In L. D. English (Ed.), *Handbook of international research in mathematics education* (2nd ed., pp. 75–108). New York: Routledge.

Perry, B., & Gervasoni, A. (2012). *Let's Count educators' handbook*. Sydney: Smith Family.

Steffe, L., von Glasersfeld, E., Richards, J., & Cobb, P. (1983). *Children's counting types: Philosophy, theory, and application*. New York: Praeger.

The Smith Family (2013). *Who we are*. Retrieved from: http://www.thesmithfamily.com.au/.

Wright, R., Martland, J., & Stafford, A. (2000). *Early numeracy: Assessment for teaching and intervention*. London: Paul Chapman.

Part III
Mathematical Processes

When Is Young Children's Play Mathematical?

Ola Helenius, Maria L. Johansson, Troels Lange, Tamsin Meaney, Eva Riesbeck, and Anna Wernberg

Abstract One of Bishop's six mathematical activities is playing which includes modelling, hypothetical thinking and abstraction. These can be in young children's play, but do they by their presence make this play mathematical? In this chapter, we explore this question by first defining play and then comparing its features with what is known about mathematicians' academic play and how mathematics education researchers have described young children's play. From this theoretical discussion, we discuss the features of play, which can enable it to be described as mathematical. We use these features to analyse a small episode of children playing to discuss if and how their play could be considered to be mathematical.

Introduction

Mathematics and play are often combined, especially in discussing young children's engagement in mathematical tasks in preschools (see, e.g. Ginsburg 2006; Sarama and Clements 2009; Lange et al. 2014). In these discussions, mathematics and play are connected in three different ways. In regard to young children, often play is considered as a vehicle for learning, while for mathematicians play is described as a necessary component of their creativity in problem-solving. The third relationship is that which considers playing as a mathematical activity. In this chapter, we compare these perspectives in order to identify the features of play that can be

O. Helenius (✉)
National Centre for Mathematics, Gothenburg University, Gothenburg, Sweden
e-mail: ola.helenius@ncm.gu.se

M.L. Johansson
Luleå Technical University, Luleå, Sweden
e-mail: maria.l.johansson@ltu.se

T. Lange • T. Meaney
Bergen University College, Bergen, Norway
e-mail: troels.lange@hib.no; tamsin.meaney@hib.no

E. Riesbeck • A. Wernberg
Malmö University, Malmö, Sweden
e-mail: eva.riesbeck@mah.se; Anna.wernberg@mah.se

T. Meaney et al. (eds.), *Mathematics Education in the Early Years*,
DOI 10.1007/978-3-319-23935-4_8

considered mathematical. This is important because often what young children are engaged in is recognised as play but dismissed as not being mathematical unless it includes obvious mathematical content, such as numbers. For example, in regard to older children, Carraher and Schliemann (2002) suggested that "there seems to be relatively little mathematical activity in children's out-of-school activities, and when it does come into play, it does not seem to call for a deep understanding of mathematical relations" (p. 150).

Kamii et al. (2004) suggested that it is more appropriate for young children to learn to "make many mental relationships about objects, people, and events" (p. 56) which could be achieved through modelling, abstraction and hypothetical thinking than it is to learn "specific topics in mathematics" (p. 46), such as shape names. Therefore, it is surprising that although mathematical processes such as problem-solving are deemed important by mathematicians and mathematics educators alike, there is scant support for categorising young children's actions as mathematical processes.

Many people have identified features belonging to play (see, e.g. Huizinga 1976; Bruner 1975; Ugurel and Morali 2010). Incorporating features of other researchers, Fromberg (1999) defined young children's play as:

> *Symbolic*, in that it represents reality with an "as if" or "what if" attitude
> *Meaningful*, in that it connects or relates experiences
> *Active*, in that children are doing things
> *Pleasurable*, even when children are engaged seriously in activity
> *Voluntary and intrinsically motivated*, whether the motive is curiosity, mastery, affiliation, or something else
> *Rule-governed*, whether implicitly or explicitly expressed
> *Episodic*, characterized by emerging and shifting goals that children develop spontaneously and flexibly. (p. 28)

Features such as these can be seen in the Swedish preschool curriculum, in which play is considered the foundation for children's learning, including the learning of mathematics:

> Play is important for the child's development and learning. Conscious use of play to promote the development and learning of each individual child should always be present in preschool activities. Play and enjoyment in learning in all its various forms stimulate the imagination, insight, communication and the ability to think symbolically, as well as the ability to co-operate and solve problems. (Skolverket 2011, p. 6)

Having play as the vehicle for learning affects many aspects of the interactions between children and between children and the teacher. For example, from examining an activity where preschool children explored glass jars, we found that although the teacher could offer suggestions about activities, the children did not have to adopt them and could suggest alternatives (Lange et al. 2014). The importance of children's ability to control their environment in a play situation has been acknowledged by others—"I suggest that the success of the physical manipulations, and ultimate mathematical conceptualisations, is very much dependent upon the successful self-regulation of the social context" (Macmillan 1995, p. 123).

Although the features of play that are connected to mathematics learning are often undefined, the mathematics of young children is generally equated with school mathematics topics. For example, Vogel (2014) stated:

> The conception of the mathematical situations of play and exploration provides that the arrangement has its root in one of the following five mathematical domains: number and operations, geometry and spatial thinking, measurements, patterns and algebraic thinking or data and probability (including combinatorics). (p. 224)

In discussing everyday mathematics that occurs in young children's play, Ginsburg (2006) drew on the work of John Dewey, to provide a less extensive list of counting, measuring and rhythmic sequencing. Even when a mathematical process, such as argumentation (Perry and Dockett 1998), is discussed in regard to children's play, generally it is seen only from the perspective of what it contributes to children's learning of mathematical content, particularly school mathematical content. In van Oers's (2014) discussion of a play-based curriculum, play is seen as providing possibilities for children to become aware of "quantitative and spatial dimensions of reality" (p. 115) within problem-solving situations—"Mathematics emerges in children's development, not as an elaboration of implicit mathematics in play, but as an attribution from outside of mathematical meanings to children's actions or utterances" (p. 114).

The focus on mathematical content is somewhat surprising given that the attributes of play seem, at least at first glance, to be more closely connected to mathematical processes than to content. However, the lack of definitions of play in many of the articles that promote play as an approach to mathematical learning may go some way to explaining this anomaly.

In contrast to mathematics educators' perceptions that play is useful as a viaduct for learning mathematical content, mathematicians' views on play focus more on mathematical processes. In discussing the childhood memories of adults working in the mathematics, science and technology industries, Bergen (2009) found that many of them had spent time involved in construction play. She suggested that:

> The "worlds" children construct, either with concrete materials such as blocks or interlocking pieces or with virtual-reality simulation games, give them the imaginative experiences and the interest in "seeing what might happen" to prepare them to create new worlds of design in later work experiences. (p. 419)

Creativity and imagination, rather than content, seem to have been the impetus for mathematical understanding. This is supported by other studies, which looked at the long-term, mathematical achievement implications of making constructions with Lego (Wolfgang et al. 2003). In their longitudinal study of the impact of block play on school mathematics achievement, Wolfgang et al. (2001) found that the complexity and adaptiveness of children's block play in preschool correlated with their mathematics achievement in high school. The more complexity in their building play in preschool, the more likely the children were to have higher mathematical achievement in high school. Similarly, Morsanyi et al. (2013) found that 10-year-old children who displayed higher mathematics performance also had better ability to reason logically about belief-inconsistent fantasy content. For example, they were better able to deduce that the mouse was bigger than an elephant from the two statements, *the elephant is smaller than the dog* and *the dog is smaller than the mouse*.

From this viewpoint, play is a factor for developing mathematical creativity and imagination. However, although mathematicians might view it in this way, they also viewed school mathematics as not valuing the necessity of play. From interviews with research mathematicians, Benjamin Bloom suggested that the way that most children were introduced to mathematics, through precision and accuracy, actually stifled their development (Brandt 1985). He suggested that a playful approach, as recommended by Alfred North Whitehead (1959), would be a better way to encourage children to respond to mathematics like mathematicians.

Taking the mathematician's point of view, Holton et al. (2001) defined mathematical play as "that part of the process used to solve mathematical problems, which involves both experimentation and creativity to generate ideas, and using the formal rules of mathematics to follow any ideas to some sort of a conclusion" (p. 403). Thus, Holton et al. also situated play as a necessary component for doing mathematics. They identified six criteria that they saw as essential components of mathematical play:

(1) it is a solver-centred activity with the solver in charge of the process;
(2) it uses the solver's current knowledge;
(3) it develops links between the solver's current schemata while the play is occurring;
(4) it will, via 3, reinforce current knowledge;
(5) it will, via 3, assist future problem solving/mathematical activity as it enhances future access to knowledge;
(6) it is irrespective of age (Holton et al. 2001, p. 404).

These criteria have some resemblance to Fromberg's (1999) attributes of play and were included in some way or the other in her literature review. For example, Fromberg's identification of play as needing to be meaningful is similar to Holton et al.'s (2001) discussion of using the solver's current knowledge. Fromberg also acknowledged that it was valuable that children could control the intensity of their play, such as in play fighting situations, as well as using current knowledge for exploration of past or future experiences.

Nevertheless, there are also differences. In Fromberg's (1999) review, she emphasised the social aspects of play, "for young children, play is a way to strengthen worthwhile, meaningful learning and co-operation with others rather than merely acquiring facts alone" (p. 45). In Holton et al.'s (2001) mathematical play, the focus is on the individual solver of problems, and the important role of social interactions in research mathematicians' problem-solving is not emphasised. In Meaney (2005), an exchange between two mathematicians showed how the ways that they interacted allowed them to put forward and discuss the merits of different ideas in what Holton et al. (2001) would label as mathematical play. It therefore seems somewhat problematic to have a definition of mathematical play that focuses only on the role of the individual problem-solver.

Belonging to a shared social situation, mathematicians continuously negotiate what can and cannot be mathematics. For example, theoretical computer science was initially an area within mathematics but was not considered "mathematical enough". In recent times, there have been indications that other areas of applied mathematics are also being pushed out of mathematical departments (Osgood 1998; Garfunkel and Young 1990). However, budget cuts in higher education may be forcing the reversal of some of these trends.

The similarities between features of play and features of playful mathematics raise questions about whether all play can be considered mathematical or only specific features of some play situations. Considering Bishop's (1988) description of the mathematical activity, playing, provides a response to this question.

Bishop (1988) considered playing to be one of the six mathematical activities that all cultures engaged in, with both adults and children as possible participants. He considered participation in these activities to be the equivalent of participation in mathematics. Academic Mathematics, which he distinguished by writing with a capital M, was one version of these six activities. For him, playing provided an answer to how mathematics is done and thus is strongly connected to mathematical processes. Playing is the social procedures and rules of performance, "the 'as if' of imagined and hypothetical behaviour" (p. 24). Consequently, he described the features of play as:

- To imagine something—which is the basis for thinking hypothetically and beginning to think abstractly
- To model—which means abstracting certain features from reality
- To formalise and ritualise rules, procedures and criteria
- To predict, guess, estimate and assume what could happen
- To explore numbers, shapes, dimensions, positions and arguments (i.e. engage in the other five mathematical activities in playful ways)

Although Bishop's six activities have been used in regard to research in to preschool mathematics (Macmillan 1995, 1998; Flottorp 2011; Johansson et al. 2012; Helenius et al. 2014), there has been little discussion about playing as a mathematical activity. In Macmillan's (1998) research, play is seen as the situation in which children participate. The closest to a discussion of playing as a mathematical activity came in descriptions of episodes which involved a "play on words" (p. 60) and where the children negotiated and regulated the play situation. In her (1995) article, Macmillan summarised Bishop's (1988) description but did not operationalise it in regard to her data. Although Johansson et al. (2012) and Helenius et al. (2014) identify and describe examples of playing as a mathematical activity, the discussion of what counted as playing relied on Bishop's own definition. Although there seems to be some overlap with aspects of the two previous kinds of relationships (play as a vehicle for learning mathematics and play as a necessary component of creativity for mathematicians), playing as a mathematical activity has rarely been used in empirical research.

In the next section, we synthesise the features connected to the different approaches that link mathematics to play, before using those features to analyse an interaction in a group of 6-year-old children.

The Features of Play Which Are Mathematical

In order to determine the features that could contribute to play being considered mathematical, we group the common features across the three approaches discussed in the previous section as creative, participatory and rule negotiation. In identifying

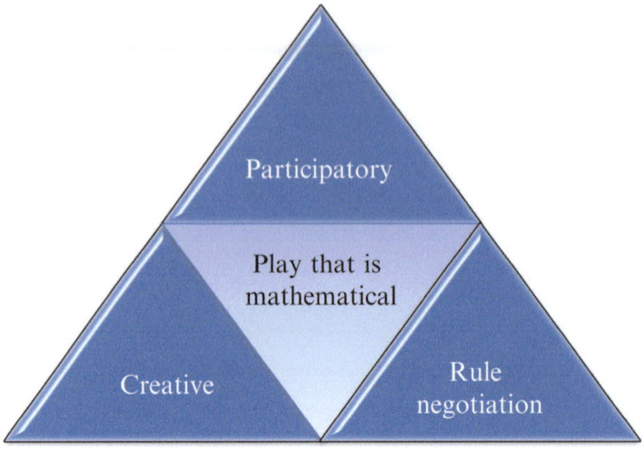

Fig. 1 The features of play that is mathematical for young children

the groups, we extend Bishop's (1988) definition of the mathematical activity, playing, to reflect both what it could be for young children and how it relates to what playing as a mathematical activity might be for other groups. Given that Bishop (1988) did not originally explicitly discuss the needs of young children (see Bishop 2016), it seems important to clarify when young children's play is mathematical.

Our expectation is that in play, both that of children and adult, these groups of features are interrelated and that it is the interrelationship that contributes to mathematical ideas being developed, perhaps for the first time by mathematicians or as a reproduction of culturally valued mathematics by children (see Fig. 1). Our argument is that unless aspects of all three features are in evidence, then the activity may not be playing in our extended version of Bishop's mathematical activity.

The interrelationship of the features is evident in the historical development of mathematical ideas. Although mathematics can be understood philosophically in many ways (Ernest 1991), if considered as a humanistic enterprise, then it is intelligible only in a social context (Hersh 1997). Creating new mathematical ideas, such as the introduction of negative numbers, requires old mathematical truths to be abandoned or reinterpreted through rule negotiation. It took several hundred years of social negotiations among mathematicians before negative numbers were accepted as a creative solution to certain kinds of mathematical problems. By participating in these negotiations, mathematicians eventually agreed on which rules needed changing and which rules could stay the same. Imre Lakatos' book *Proof and Refutations* (1976) describes how mathematics as a discipline can keep its internal coherence despite regularly and repeatedly dealing with new "rule-breaking" entities. Part of this comes about because mathematicians accept what it means to participate in both local debates but also in the wider community of practising mathematicians, as determined in the historical moment. Our point in highlighting the interrelationship of the features is that many elements of doing mathematics could be described as play and that in deliberating whether young children's play should be considered mathematical, it is relevant to consider how it is related to what mathematicians do.

Creative

In the earlier discussion, it seems that both playing and doing mathematics include being creative, and thus a creative element is necessary for play to be considered mathematical. Fromberg (1999) saw play as having a symbolic nature in that it is a representation of reality. Posing problems and finding solutions, which were recognised by Bergen (2009) as linking children's play to the work of scientists, mathematicians and engineers, can be considered expressions of the "as if" and "what if" aspects identified by Fromberg (1999). Similarly, Bishop (1988) suggested that hypothetical thinking grows out a requirement for play to be distanced from reality. Play models reality by referencing to it, but not including all aspects of it. This allows certain issues to be explored imaginatively, without the constraints that reality might require. Holton et al. (2001) acknowledged that it is the solver of the problem who is in charge of the solution process and thus can determine the components that should be considered as integral for solving it. Fromberg's (1999) definition of play included that it was voluntary and intrinsically motivated, often because children were curious about something, and this encouraged them to continue their play. When play occurs over a long period of time, participants can change the focus or problems that are being explored as other ideas become more interesting.

Therefore, an operational description of what creative aspects make play mathematical would include:

- Modelling a situation, by incorporating aspects of hypothetical thinking and abstraction, that includes some, but not all aspects of reality
- Participants determining or accepting the altered reality as a consequence of playing
- Posing and solving problems that participants set themselves

Participatory

Mathematics has long been recognised as a cultural activity and thus as constructed by groups of people (Bishop 1988). Participating in mathematics involves posing or solving a problem that others would recognise as mathematical. When solving an individual mathematical problem, participants are not free to do whatever they like; instead, they must agree to abide by the rules of mathematics (Holton et al. 2001). However, young children are unlikely to know the rules of mathematics. Therefore, for play to count as mathematical, they must abide by group-negotiated rules, but perhaps not mathematical rules.

In making participation an essential feature of play, we exclude individual play, such as block play, as being mathematical. Interactions with materials provided by adults could be considered as implicitly participating with others and therefore part of the social construction of mathematics. However, we consider that it is the explicit contributions of others, children and adults that enrich play through extending the possibilities for modelling, augmenting the altered reality and posing and solving problems. For example, learning how to produce more complex block arrangements is likely to come from observing the block building of others and reflecting on how this could be used in solving a problem such as making a tower bigger when only having triangular blocks.

In discussing playing as a mathematical activity, Bishop indicated that participation involves agreeing to suspend normal reality in order to take on the specific reality of a particular play situation. However, because the play situation models some aspect of reality, even if done in an imaginative way, there is also a recognition that participation is both at the local level of the immediate situation and also at the societal level which determines the rules and values that affect immersion in reality. As participants move backwards and forwards between the two levels, play situation is adapted when new problems become of interest. The acceptance of play situations as allowing for the modelling of interactions between people that occurs in real situations contributes to participants predicting, guessing, estimating and making assumptions about what could happen within its altered reality. Thus, participation through interactions with others ties the imaginative, creative aspects of play to reality. For example, playing being Martians living on Mars can only be done by basing that play on what is known about being humans on Earth.

For play to be considered a mathematical activity, participation occurs within the rules of the play situation but draws on what participants know of the wider societal reality. Therefore, it includes the following features:

- Participants show an awareness that their participation depends on others recognising that they are acting according to negotiate rules.
- Participation is both within the play situation and as part of the wider societal reality.

Rule Negotiation

Fromberg (1999) suggested that play was rule governed, although the rules can be implicit rather than explicit. When a group of children build a city with blocks, agreement about what to use to represent different types of buildings may be agreed upon, not as a result of verbal negotiation but by the labelling of the same kind of block as a house by different children during the play situation.

Bishop (1988) also acknowledged that during play participants have to agree to the suspending of some aspects of reality. If participants do not agree to the rule that some aspects of reality are suspended, then play cannot occur. However, rules can be changed as the play situation develops but only if all participants agree to the changes, although this may be possibly under threat that they will be excluded from the play if they do not agree. Rule negotiation is an essential component of play, but how inclusive the negotiations are will depend on power structures within the group of participants.

Although school students often regard mathematics as just a set of rules (Wong et al. 2002), the rules have been agreed to as a result of consensus amongst mathematicians. Indeed as van Oers (2001) stated:

> It is not the link with meaningful problem situations as such that defines the nature of 'real' mathematics, but the observance of particular rules, the use of particular concepts and tools, the engagement with certain values that define whether one is doing mathematics or not. (p. 71–72)

Like Fromberg (1999), Bishop (1988) identified rules as being a component of play. He considered that when play became incorporated into games, there is a formalising and ritualising of rules, procedures and criteria. This also contributes to the use of strategies when the focus shifts to winning the game. Often strategies require logical thinking, similar to that used in mathematics (Holton et al. 2001).

For play to be considered mathematical, it needs to include all or most of the following:

- Participants must abide by the implicit and/or explicit rules of the play.
- For rules to change, negotiation needs to occur between participants.
- Negotiating the rules contributes to forming the boundaries of the play situation and thus what aspects of reality can be suspended and what aspects are modelled in what ways.

Methodology

To ascertain if the criteria in the previous sections contribute in a meaningful way to determining if a play situation could be labelled mathematical, it is important to use them with empirical material. Consequently, we chose a short video, just over three and a half minutes long, to analyse a group of 6 year olds, attending a Swedish preschool class, engaged in free play. Children do not begin school in Sweden until they are around 7 years old, but in the year before, they attend preschool class which is usually situated in the school that they will later attend. The preschool class is considered a bridge between preschool (which is for 1–5-year-olds) and school. The video was collected as part of wider project investigating what mathematics is in preschool, with preschool class providing a contrast to this.

Videos were collected from the same preschool class over several months, but the video used here was the only one that included free play. The extract was chosen because it seemed to be typical of the free play situations we saw in preschool, and while it included the children discussing numbers as part of the buying and selling, this was circumstantial rather than the focus of their play. Although still young children, we considered that by 6 years, these children would have had many experiences of free play and so would be proficient participants in it. Analysing such a situation would indicate whether the features of mathematical play could be identified in the complex environments in which play generally occurs with young children.

The video was split into four short sections which seemed to have a natural beginning and end within themselves. Each of these sections was analysed by looking at what features were apparent. The features connected to participation focused our attention on whether all children were engaged equally in regard to the other features essential for this play to be considered mathematical. This allows the analysis to consider how different combinations of features appeared in the play.

To Be or Not to Be Mathematical

Four children, three boys and a girl, play with different Lego constructions. The situation is somewhat chaotic as the children leave and come back and move in and out of different storylines. Stills from the video are provided with the transcripts in order to clarify what was occurring. The transcripts are provided in the original Swedish with an English translation.

Episode 1. Buying a Popsicle

In this episode, Klara is not present as she has just left the group. It begins with Teo wanting to buy a pretend popsicle (piggelin), something which would be quite cheap with some pretend money (kroner bills). Tom takes on the role of seller and by doing this controls what is allowable in this version of reality. Figure 2 shows Teo (left) showing Tom (right) the pretend three and four kroner notes that he has.

Teo	Får man köpa nåt här?	Can you buy something here?
Patrik	Nej.	No.
Teo	Men får jag ändå köpa nåt?	But can I still buy something?
Tom	Men var är hundralapparna?	But where are the hundred kroner bills?
Teo	A men jag har bara såna här pengar, men kan jag, kan jag få köpa något?	Ah, but I only have these money, but can I, can I buy something?
Tom	Ja	Yes.
Teo	Jag vill köpa piggelinen.	I want to buy the popsicle.
Tom	Den kostar alla dom.	It costs all of those.
Teo	Nä inte alla mina pengar.	No, not all my money.
Tom	Jo, den kostar allt det.	Yes, it costs all that.
Teo	Nä!	No!

In this episode, the children are negotiating the rules in regard to modelling the reality that they want to explore. The problem of what does a popsicle cost is one that the children set up and try to resolve themselves. One of the Lego pieces, a blue plastic pole, comes to take on the role of a popsicle, a creative innovation brought into an existing play interaction. The pretend money takes the role of real money, although Tom queries why there are only 3 and 4 kroner bills and not hundred kro-

Fig. 2 Teo and Tom negotiating the price of a popsicle

ner bills. This suggests that Tom does not see these smaller bills as sufficient for the kind of buying and selling activity that he wants to model. Nonetheless, this situation fulfils the criteria of being mathematical, in that it models a real situation but does not try to include all aspects of buying and selling. Although Patrik initially does not accept that Teo can buy anything, Tom's continual engagement with Teo indicates that these two boys have accepted the conditions of this being a buying and selling exchange with Tom as the seller.

In regard to the criteria for participation, it seems that Patrik withdrew because he did not want to sell any of the Lego constructions. Teo, however, was able to continue in this play because of Tom's willingness to interact under the conditions invoked by Teo's desire to buy something with his money. Teo's querying of the need to give Tom all of his kroner bills draws on his understanding of the real world, in which popsicles rarely cost all of the money that his parents might have. Thus, he moved between the local play situation and his understanding of the real world.

The querying by Tom of the kind of money that Teo has as well as Teo's querying of Tom's claim that the popsicle costs all of Teo's money shows some negotiation about what rules should apply in this situation. However, the basic premises that a blue plastic Lego pole could stand for a popsicle and that it could be bought with pretend kroner bills was not queried. Therefore, the boundaries of the play situation remained in place.

Episode 2. Klara's Chocolate

Klara returned to the group with a small brown plastic block. She then showed it to the boys (see Fig. 3) and said "Have you had the chocolate?" (Har ni haft chokladen?). Tom said "No" before returning to his conversation with Teo. A few turns

Fig. 3 Klara showing her chocolate to Tom

later, Klara tried again to attract the boys' attention by saying "Check this out, a small chocolate" (Kolla in detta då, en liten choklad) but is not successful in having the boys take up this alternative play situation. She made one last attempt to gain the boy's interest before dropping this discussion.

Although Klara seemed to have been just as creative as Tom and Teo by turning a plastic block into a chocolate, she did not get an opportunity to present a problem because this situation was not accepted by the others. Therefore, there was no joint participation and no rule negotiation, so the situation did not fulfil the criteria of being mathematical, nor even play as it does not meet Fromberg's (1999) criteria of connecting meaningful experiences.

Episode 3. The Car

Having been unsuccessful in changing the direction of the play by introducing the chocolate, Klara tried to enter the buying and selling play situation by offering first a trade and then by supporting Patrik's offer to buy a Lego car. Although there was some interest in her offers, they were eventually rejected as not being appropriate. Figure 4 shows Klara and Patrik's interest in the car.

The following is the children's discussion:

Patrik	Bilen.	The car.
Klara	Byter ni den här bilen mot alla de här och mitt bygge?	Do you [plural] want to trade this car against all of these and my construction?
Patrik	Det här är inte vårt.	This is not ours.

(continued)

(continued)

Klara	Jo den här delen är min.	Yes, this part is mine.
Tom	Oh, kolla vad häftigt! Om man snurrar på denna, så snurrar de här däcken.	Oh, check this out, so cool! If you spin this one, the wheels spin.
Klara	Kolla!	Check this!
Patrik	Den här vill jag köpa.	I want to buy this.
Tom	Men var är pengarna då?	But where is the money then?
Klara	Här i …	Here in ….
Tom	Nä där är dom inte!	No, they are not there.

In this episode, the children do not seem to be able to enter the virtual reality of the buying and selling situation that Tom and Teo had been in. Ownership of the constructions, determined in the real world by who had built them, thwarts Klara's efforts to make a trade. She returned to the facts about what had been her contribution to the construction to support her claim that she could do the trade. Patrik also had no success in convincing Tom that he had a legitimate right to enter into the buying and selling play situation. Although the money was only play money, Teo held on to it, and Patrik had nothing else which was accepted as an appropriate model for real money. Thus, in spite of the fact that there was some joint participation, there was no seamless merging between the local and societal levels of the modelling. Tom positioned himself as the arbitrator of what was acceptable for the play situation and judged that Patrik and Klara's suggestions could be ignored. Thus, this situation could also not be seen as mathematical or playing.

Fig. 4 Klara and Patrik showing strong interest in the car

Episode 4. Popsicle Buying Successfully Negotiated

Towards the end of the video, Teo re-entered the discussion with another attempt to buy the popsicle. In the meantime, the teacher has given him some more kroner bills, although still only worth 3 and 4 kroners each. At one point, he copied Klara's trading attempt by picking up a small blue Lego block, labelling it as a popsicle and then trying to use it to exchange for the pretend popsicle that he wanted from the start. Eventually, Tom accepted the deal but took Teo's money to count out the 40,000 kroner that he said it would cost (Fig. 5).

Fig. 5 Teo with his popsicle trade and Tom counting out the 40,000 kroner

Teo	Men, ok, jag vill köpa piggelinen för de här.	But, ok, I want to buy the popsicle for these.
Tom	Nej, nej, nej.	No, no, no.
Teo	Här. Så får jag piggelinen	Here. Then I will get the popsicle.
Tom	Varför måste du ha piggelinen? Då dör du ju	Why do you have to have the popsicle? You will die then.
	(Ohörbart)	(Inaudible)
Tom	Vad köper du?	What do you buy?
Teo	Piggelinen.	The popsicle.
Patrik	Men då blir ni också sjuka.	But then you [plural] will get sick as well.
Teo	Men kan jag betala med den här piggelinen?	But can I pay with this popsicle?
Tom	Det där är fyrtiotusen. I så fall får du betala, Vänta, de här också.	That is forty thousand. In that case you will have to pay. Wait those too.
Teo	Men	But ....
Tom	Jo, väldigt mycket kostar det alltihop.	Yes, it cost very much all of that.
Teo	Ni är så elaka varför måste det kosta så mycket?	You [plural] are so mean, why does it have to cost that much?
Tom	Den kostar inte alls mycket. Den kostar bara en sån där.	It doesn't cost that much. It costs only one of these.

As with the earlier episode, Tom and Teo built up the play situation in which Teo wanted to buy a popsicle, first with the money he had and, when this initially did not work, to do a trade. Unlike the situation with Klara and Patrik, Tom seemed to accept that as possible. Although initially he rejected the sale and trade offers, a reference to the real world, by Teo labelling him as "mean", meant that he did accept Teo's money as being equivalent of the 40,000 kroner he had demanded as payment. Thus, the situation was creative and accepted by the two main participants. Although Tom attempted to change the play situation by suggesting that Teo would get sick if he ate the popsicle, something that Patrik also supported, Teo ignored this suggestion. It can be said that the offer to renegotiate the play situation was rejected by one of the main characters, Teo. Unlike the previous episode, this rejection did not affect the possibilities to continue playing; instead, it strengthened the boundaries of what was and what was not acceptable for this play situation. Consequently, it can be said that the criteria for mathematical play were all met within this episode, at least for Teo and Tom.

Conclusion

In this chapter, we present the case that play sometimes has features which could deem it to be mathematical. This is not to say that mathematics is play or that play is always mathematics, rather that some of the features of playful situations allow them to be classified as mathematical, in the sense of Bishop's playing as a mathematical activity. One point that can be made from the analysis of the video is that at times the three features, participatory, creative and rule negotiation, occurred together and were interrelated. As one of these features, or aspects of a particular feature, became more influential, then the influence of the other features was affected. This suggests that Bishop's mathematical activity of playing can be elaborated to clarify what it might be for young children, by searching for the presence of the three features and how they are related in a particular situation.

Our analysis suggests that by using the features to determine mathematical play, it is possible to see how playing can support children to develop their understandings of mathematical processes, such as modelling, hypothetical thinking and abstraction identified by Bishop (1988) as included in playing as a mathematical activity. Our approach of having a set of three features—creative, participatory and rule negotiation—is based on research on playful aspects of the work of mathematicians (Holton et al. 2001) and Fromberg's (1999) definition of play.

The analysed play situation indicated that interactions by themselves did not produce mathematical play situations. A problem, such as how to negotiate the buying and selling of a popsicle, and its solution, needed to be the common objective for the children in order for them to be engaged in mathematical play. When Teo and Tom positioned themselves as having the only valid contributions to the problem posing and problem-solving, then what became the creative components of the problem-solving was restricted to what they contributed. Moreover, the rule negotiation was based on what they considered to be genuine considerations, for example, whether the money that Teo had was sufficient. Teo was able to affect the rule negotiation component by appealing to his friendship with Tom. On the other hand, Klara and Patrik were not successful in participating in the mathematical play. They tried to enter it through offering creative alternatives to the current discussions—Klara's chocolate—and trying to affect the rule negotiation—Patrik's interest in buying the car. This suggests that apart from the components identified in the model, there is also an issue of power to do with who can control what becomes mutually accepted within the play situation. (The issue of power is discussed more fully in Helenius et al. (2015a, b)).

The fact that some play does include the features which make it mathematical is important because, generally, play is not recognised as having anything in itself which could add to children's mathematical understandings. This is often because mathematics is reduced to mathematical content, such as counting or naming shapes. Nevertheless, even when the focus does seem to be on mathematical processes, or as suggested below by Lee and Ginsburg (2009), "mathematising", these are not seen as possible to develop in children's free play:

> Children do indeed learn some mathematics on their own from free play. However, it does not afford the extensive and explicit examination of mathematical ideas that can be provided only with adult guidance. … Early mathematics is broad in scope and there is no guarantee that much of it will emerge in free play. In addition, free play does not usually help children to mathematise; to interpret their experiences in explicitly mathematical forms and understand the relations between the two. (Lee and Ginsburg 2009, p. 6)

Although we disagree with Lee and Ginsburg's (2009) point that play does not encourage children to mathematise, we acknowledge the importance of the role of the teacher in children's play. Given the issues of power that arose in the video, it may be that the teacher could facilitate all children to be creative, participatory and contribute to the negotiation of rules within the play situation. Notwithstanding that more research is needed both to test out the features for what makes play mathematical and to see how they affect children's later performance in mathematics, we suggest that a teacher's active participation in the play could contribute to children learning more about mathematics as suggested by Lee and Ginsburg.

As already indicated, we acknowledge that identifying these features is only the start of a research programme to better understand the contribution that mathematical play makes to children's later understandings of mathematics. The criteria for what makes play mathematical need to be tested with other kinds of play as well as considering how the teacher's role in the play could contribute to more equitable learning opportunities. It is also important to consider how to conduct longitudinal studies to ascertain whether there are benefits for children from engaging in this kind of play in their later lives. With the schoolification of preschools (Alcock and Haggerty 2013; Garnier 2012; Sofou and Tsafos 2010), there is a need to know whether the reduction of play in preschools will be detrimental to children's mathematical learning, rather than valuable as politicians and policymakers suggest. We do not want to be "throwing the baby out with the bathwater", so that increasing the amount of formal mathematics education that children receive in preschool actually decreases their interest and actual learning of mathematics.

References

Alcock, S., & Haggerty, M. (2013). Recent policy developments and the "schoolification" of early childhood care and education in Aotearoa New Zealand. *Early Childhood Folio, 17*(2), 21–26.

Bergen, D. (2009). Play as the learning medium for future scientists, mathematicians, and engineers. *American Journal of Play, 1*(4), 413–428.

Bishop, A. J. (1988). *Mathematical enculturation: A cultural perspective on mathematics education*. Dordrecht: Kluwer.

Bishop, A. J. (2016). Can values awareness help teachers and parents transition preschool learners into mathematics learning? In T. Meaney, T. Lange, A. Wernberg, O. Helenius, & M. L. Johansson (Eds.), *Mathematics education in the early years—Results from the POEM conference 2014*. Cham: Springer.

Brandt, R. S. (1985). On talent development: A conversation with Benjamin Bloom. *Educational Leadership, 43*(1), 33–35.

Bruner, J. S. (1975). Play is serious business. *Psychology Today, 8*(8), 80–83.

Carraher, D. W., & Schliemann, A. D. (2002). Is everyday mathematics truly relevant to mathematics education. In M. E. Brenner & J. N. Moschkovich (Eds.), *Journal for Research in Mathematics Education Monograph: Everyday and academic mathematics in the classroom (monograph)* (pp. 131–153). Reston, VI: National Council of Teachers of Mathematics.

Ernest, P. (1991). *The philosophy of mathematics education*. London: Falmer Press.

Flottorp, V. (2011, February 9–13). How and why do children classify objects in free play? A case study. In M. Pytlak, T. Rowland, & E. Swoboda (Eds.), *Proceedings from seventh congress of the European Society for Research in Mathematics Education, Rzeszów, Poland* (pp. 1852–1862). European Society for Research in Mathematics. Available from http://www.mathematik.uni-dortmund.de/~erme/index.php?slab=proceedings.

Fromberg, D. P. (1999). A review of research on play. In C. Seefeldt (Ed.), *The early childhood curriculum: Current findings in theory and practice* (3rd ed., pp. 27–53). New York: Teachers College Press.

Garfunkel, S., & Young, G. (1990). Mathematics outside of mathematics departments. *Notices of the American Mathematical Society, 37*, 408–411.

Garnier, P. (2012). Preschool education in France Scholarisation of the École maternelle and Schoolification of Family Life. *Pedagogy—Theory & Praxis, 5*, 43–53.

Ginsburg, H. P. (2006). Mathematical play and playful mathematics: A guide for early education. In D. G. Singer, R. M. Golinkoff, & K. Hirsh-Pased (Eds.), *Play = learning: How play motivates and enhances children's cognitive and social-emotional growth* (pp. 145–165). New York: Oxford University Press.

Helenius, O., Johansson, M. L., Lange, T., Meaney, T., Riesbeck, E., & Wernberg, A. (2014, February 4–5). Preschool teachers' awareness of mathematics. In O. Helenius, A. Engström, T. Meaney, P. Nilsson, E. Norén, J. Sayers, & M. Österholm (Eds.), *Evaluation and comparison of mathematical achievement: Dimensions and perspectives: Proceedings from Madif9: Nionde forskningsseminariet med Svensk Förening för Matematikdidaktisk Forskning, Umeå*. Forthcoming.

Helenius, O., Johansson, M. L., Lange, T., Meaney, T., & Wernberg, A. (2015a, June 21–26). Beginning early: Mathematical exclusion. In S. Mukhopadhyay & B. Greer (Eds.), *Proceedings of the eighth international mathematics education and society conference, Portland State University, Oregon, USA* (pp. 596–609). Portland, OR: MES8.

Helenius, O., Johansson, M. L., Lange, T., Meaney, T., & Wernberg, A. (2015b, February 4–8). Mathematical exclusion with the every day. In *Proceedings of the 9th Congress of European Research in Mathematics Education, Prague*. Prague: CERME9 and Charles University in Prague. Forthcoming.

Hersh, R. (1997). *What is mathematics, really?* Oxford: Oxford University Press.

Holton, D. D., Ahmed, A., Williams, H., & Hill, C. (2001). On the importance of mathematical play. *International Journal of Mathematical Education in Science and Technology, 32*(3), 401–415.

Huizinga, J. (1976). Nature and significance of play as a cultural phenomenon. In R. Schechner & M. Schuman (Eds.), *Ritual, play, and performance: Readings in the social sciences/theatre* (pp. 46–66). New York: Seabury Press.

Johansson, M. L., Lange, T., Meaney, T., Riesbeck, E., & Wernberg, A. (2012, July 8–15). What maths do children engage with in Swedish preschools? In *Proceedings from TSG1: Mathematics education at preschool level at ICME-12 the 12th International Congress on Mathematics Education, Seoul, Korea*. Available from http://www.icme12.org/sub/tsg/tsgload.asp?tsgNo=01.
Kamii, C., Miyakawa, Y., & Kato, Y. (2004). The development of logico-mathematical knowledge in a block-building activity at ages 1-4. *Journal for Research in Childhood Education, 19*(1), 44–57.
Lakatos, I. (1976). *Proof and refutations: The logic of mathematical discovery*. Cambridge, UK: Cambridge University Press.
Lange, T., Meaney, T., Riesbeck, E., & Wernberg, A. (2014). Mathematical teaching moments: Between instruction and construction. In U. Kortenkamp, B. Brandt, C. Benz, G. Krummheuer, S. Ladel, & R. Vogel (Eds.), *Early mathematics learning: Selected papers of the POEM 2012 conference* (pp. 37–54). New York: Springer.
Lee, J. S., & Ginsburg, H. P. (2009). Early childhood teachers' misconceptions about mathematics education for young children in the United States. *Australasian Journal of Early Childhood, 34*(4), 37–45.
Macmillan, A. (1995). Children thinking mathematically beyond authoritative identities. *Mathematics Education Research Journal, 7*(2), 111–131.
Macmillan, A. (1998). Pre-school children's informal mathematical discourses. *Early Child Development and Care, 140*(1), 53–71.
Meaney, T. (2005). Mathematics as text. In A. Cronaki & I. M. Christiansen (Eds.), *Challenging perspectives in mathematics classroom communication* (pp. 109–141). Westport, CT: Information Age.
Morsanyi, K., Devine, A., Nobes, A., & Szücs, D. (2013). The link between logic, mathematics and imagination: Evidence from children with developmental dyscalculia and mathematically gifted children. *Developmental science, 16*(4), 542–553.
Osgood, B. (1998). Mathematics as engineering: Notes from a foreign correspondent. *Focus on Calculus. A Newsletter for the Calculus Consortium Based at Harvard University,* (14).
Perry, B., & Dockett, S. (1998). Play, argumentation and social constructivism. *Early Child Development and Care, 140*(1), 5–15.
Sarama, J., & Clements, D. H. (2009). Building blocks and cognitive building blocks. *American Journal of Play, 1*(3), 313–337.
Skolverket. (2011). *Curriculum for the preschool Lpfö 98: Revised 2010*. Stockholm: Skolverket.
Sofou, E., & Tsafos, V. (2010). Preschool teachers' understandings of the national preschool curriculum in Greece. *Early Childhood Education Journal, 37*(5), 411–420.
Ugurel, I., & Morali, S. (2010). A short view on the relationship of mathematics and game from literature context and concept of the (educational) mathematics game. *World Applied Sciences Journal, 9*(3), 314–321.
van Oers, B. (2001). Educational forms of initiation in mathematical culture. *Educational Studies in Mathematics, 46*(1-3), 59–85.
van Oers, B. (2014). The roots of mathematising in young children's play. In U. Kortenkamp, B. Brandt, C. Benz, G. Krummheuer, S. Ladel, & R. Vogel (Eds.), *Early mathematics learning: Selected papers of the POEM 2012 conference* (pp. 111–123). New York: Springer.
Vogel, R. (2014). Mathematical situations of play and exploration as an empirical research instrument. In U. Kortenkamp, B. Brandt, C. Benz, G. Krummheuer, S. Ladel, & R. Vogel (Eds.), *Early mathematics learning: Selected papers of the POEM 2012 conference* (pp. 223–236). New York: Springer.
Whitehead, A. N. (1959). *The aims of education and other essays*. London: Ernest Benn.
Wolfgang, C. H., Stannard, L. L., & Jones, I. (2001). Block play performance among preschoolers as a predictor of later school achievement in mathematics. *Journal of Research in Childhood Education, 15*(2), 173–180.
Wolfgang, C. H., Stannard, L. L., & Jones, I. (2003). Advanced constructional play with LEGOs among preschoolers as a predictor of later school achievement in mathematics. *Early Child Development and Care, 173*(5), 467–475.
Wong, N.-Y., Marton, F., Wong, K.-M., & Lam, C.-C. (2002). The lived space of mathematics learning. *Journal of Mathematical Behaviour, 21*(1), 25–47.

Two Frameworks for Mathematical Reasoning at Preschool Level

Lovisa Sumpter

Abstract In this chapter, young children's mathematical reasoning is explored using two different frameworks. Two cases of reasoning are analysed and discussed in order to illustrate how the mathematical foundation is used in young children's arguments and choices that they make when solving mathematical problems. The first framework focuses on arguments and warrants and is used to analyse individual reasoning. The second identifies strategy choices and categorises different types of reasoning that are developed in groups. In both frameworks, the mathematical foundation is central.

Introduction

I'm sitting on the train. My seat is one of four sharing a table. The other three seats are occupied by a mum and two small children, a boy and a girl. The boy, who is the oldest of the two, turns to me and says: 'I'm four!'. I smile and ask the little girl how old she is. 'I'm four!', she replies. The boy laughs and says: 'No, she is two!' and shows me two fingers to illustrate. 'Ok. So you are four and your sister is two. How much older are you than your sister?', I ask the boy. He looks at me a bit puzzled. Then he holds up four fingers on his left hand and two fingers on his right and places the hands opposite each other, so he can compare the number of fingers. I can see him nodding when he is counting the fingers on his left hand which do not match a finger on the right hand. One nod. One more nod. 'Two!', he says with a smile. The mother looks at me and says, 'I have never seen him doing that before'. The boy turns to me again: 'You are a big girl, aren't you?'

Research has shown that young children are more capable of developing mathematical concepts and processes than previously thought (Clements and Sarama 2007; Mulligan and Vergnaud 2006). This is further emphasised by studies focusing on general mathematical processes such as problem solving, argumentation and justification (Perry and Dockett 2007), early algebraic reasoning (Papic et al. 2011), and

L. Sumpter (✉)
Dalarna University, Falun, Sweden
e-mail: lsm@du.se

T. Meaney et al. (eds.), *Mathematics Education in the Early Years*,
DOI 10.1007/978-3-319-23935-4_9

modelling and statistical reasoning (English 2012). Recent Swedish research shows how young children can use different mathematical competencies, alternatively labelled processes (NCTM 2000), in their mathematical reasoning (Säfström 2013). A mathematical competence is defined as:

> the ability to understand, judge, do, and use mathematics in a variety of intra- and extra-mathematical contexts and situations in which mathematics plays or could play a role. (Niss 2003, p. 7)

In Säfström's (2013) research the children questioned other children's arguments and justified their own. One of the conclusions from this research is that mathematical reasoning is something that children can use from an early age; other skills do not have to be developed before this competence can be used. We also know that mathematical reasoning predicts mathematical achievement later in school (Nunes et al. 2012).

Mathematical reasoning is also a social activity. The negotiation carried out by children described by Säfström (2013) was a central part of the interaction structure when creating collective mathematical reasoning (Voigt 1994). As most preschools in Sweden focus on social skills such as learning to cooperate and share, mathematics and mathematical problem solving could provide opportunities for learning mathematical reasoning as well as learning constructive social learning.

Although the development of process and sense making of mathematical concepts can take place without explicit guidance (McMullen et al. 2013), there seems also to be evidence that children do not develop these competencies without support (Bobis et al. 2005, 2008). Children need to be part of situations which provide opportunities to learn (Hiebert 2003). Guidance from an adult is more likely to support children to gain more extensive and explicitly investigated mathematical ideas (Björklund 2008; Lee and Ginsburg 2009; van Oers 1996). For example, the boy on the train faced a mathematical problem but was supported in solving it by the questions asked by a guide (a 'big girl'). Nevertheless, the solution strategy was a product of his own creativity.

Although there is a growing body of research about preschool children's mathematical reasoning, few studies incorporate theories about mathematical reasoning and theoretical concepts are seldom discussed explicitly. In this chapter, I explore young children's mathematical reasoning with two theoretical frameworks. The two frameworks highlight different aspects of reasoning in relationship to individual and collective reasoning.

Mathematical Reasoning

Mueller (2009) suggested 'mathematical understanding and thus mathematical knowledge depend upon reasoning' (p. 138). Therefore, it is not surprising to find mathematical reasoning included in several frameworks that describe teaching/learning pathways, such as curricula (e.g. NCTM 2000; Niss 2003) including the

Swedish curriculum from preschool up to upper secondary school level (National Agency for Education 2011a–c). One of the goals that Swedish preschools should aim for is that children 'develop their mathematical skill in putting forward and following reasoning' (National Agency for Education 2011a, p. 10). This appears to be a challenging goal, given that Swedish children struggle with mathematical reasoning and problem solving later, as documented in international tests such as TIMSS (National Agency for Education 2012).

Despite this central role, few theoretical frameworks characterise reasoning in detail (Lithner 2008; Yackel and Hanna 2003). For instance, Skemp (1978) described two different kinds of understandings, instrumental and relational understanding, that is, the base for student's reasoning, but gave no further specification. When Wyndhamn and Säljö (1997) studied children's mathematical reasoning when solving mathematical problems, they focused on the content and rules in students' reasoning. However, no definition of reasoning was provided. At the very least, in a framework about mathematical reasoning, it could be expected that the mathematical content that is the basis for the reasoning should be explicit.

One of the few frameworks providing a definition of reasoning was that of Ball and Bass (2003). They described mathematical understanding as founded on mathematical reasoning. Reasoning is comprised of a 'set of practises and norms that are collective not merely individual or idiosyncratic, and rooted in the discipline' (Ball and Bass 2003, p. 29). Such a framework is helpful when distinguishing between the body of public knowledge and language. However, Ball and Bass (2003) also seem to imply that mathematical reasoning is rooted in logic.

In order to study reasoning based on subjective, rather than mathematical, knowledge including arguments such as 'I do this because my teacher says so', a different approach would be needed. Therefore, the question arises, if mathematical reasoning is thought of as logical thinking, should preschool children be expected to produce such thinking? Yet Ball and Bass (2003) concluded that 'mathematical reasoning is no more than a basic skill' (p. 28). This implies that mathematical reasoning could be found at all levels of mathematical understanding including preschool. To study mathematical reasoning, there is a need for appropriate tools and theories.

For young children, mathematical reasoning is often related to oral language skills (Charlesworth 2005), and therefore one way of studying reasoning is to use understandings of argumentation. By studying individuals' argumentation and the choices that they make when solving tasks (e.g. Lithner 2008; Sumpter 2013), it is possible to identify different types of reasoning.

Säfström (2013) analysed different definitions of mathematical reasoning in order to define this competence as:

> Explicitly justifying choices and conclusions by mathematical arguments. Select, use and create informal and formal arguments. Interpreting and evaluating one's own and others' reasoning. Reflecting on the role of reasoning. Knowing what a proof is. (p. 36)

This definition as with much of the research mentioned earlier was about the reasoning of individuals. However, it is plausible to expect other forms of reasoning than just individual reasoning when preschool children are trying to solve mathematical tasks and exercises, mainly because of social contexts where activities are taking place. Reasoning as a social activity occurs when 'learners participate as they interact with one another' (Yackel and Hanna 2003, p. 228). Therefore, in this chapter I discuss examples of individual and collective reasoning from the perspective of two different frameworks. Two frameworks are used because differences appear when one person makes all the central decisions, in contrast to the situation where several participants contribute to the development of the reasoning.

Individual Reasoning

In Lithner's (2008) framework, an individual's reasoning is defined as the line of thought adopted to produce assertions and reach conclusions in tasks. This line of thought does not have to be based on formal logic; it could even be incorrect. It is produced from starting with a task and ending with some sort of answer. To structure the data, task solving is seen as occurring in four steps:

1. A problematic situation (PS) is met where it is not obvious for the individual as to how to proceed.
2. A strategy choice (SC) is made, a choice that can be supported by a predictive argument.
3. The strategy is implemented (SI) and the implementation can be supported by verifying arguments.
4. A conclusion (C) is obtained.

This is not necessarily a linear structure; an individual can jump between the steps in his or her reasoning. Argumentation is the part of reasoning which aims to convince the individual or others that the reasoning is appropriate. A strategic choice may not result from a conscious decision, but can include actions that are more subconscious.

An important part of this framework is the content of the argument. Lithner (2008) suggested that the argument is anchored in the reasoning by the individual referring to relevant mathematical properties of the components. These components are objects, transformations, and concepts. Objects are fundamental entities, "the 'thing' that one is doing something with" (Lithner 2008, p. 261), e.g. numbers, variables, and functions. A transformation is the process done to an object, with a sequence of these transformations being a procedure, e.g. finding a polynomial maxima or a division algorithm. Concepts are central mathematical ideas built on a set of objects, transformations, and their properties, for example, the infinity concept. However, in the same way objects can be transformed, a transformation can be transformed into an object. As Lithner (2008) discussed, the 'status of a component depends on the situation' (p. 261). Also, some properties are more relevant than others, and the division of surface and intrinsic properties indicates what is relevant, depending on the context:

> In deciding if 9/15 or 2/3 is largest, the size of the numbers (9, 15, 2, 3) is a *surface* property that is insufficient to resolve the problem, while the quotient captures the *intrinsic* property. (Lithner 2008, p. 261)

When the boy on the train solved the problem of the difference between 4 and 2, he did a comparison which is a transformation of the objects 'cardinal number 4' and 'cardinal number 2'. He could also have used subtraction in the meaning of 'take away', for instance, counting down from 4 to 2. This would have been another transformation to the same objects.

In Lithner's (2008) framework, creative and imitative mathematical reasoning are separated. Reasoning is defined as Creative Mathematically Founded Reasoning (CMR) if it fulfils the following conditions (Lithner 2008):

1. Novelty
2. Plausibility
3. Mathematical foundation

Novelty means that a new reasoning sequence is created or re-created. To do this the arguments supporting the strategy choice and/or the implementation of the strategy need to be true or plausible. The mathematical foundation is created when the arguments are anchored in intrinsic mathematical properties. For example, the strategy choice could be about constructing or reconstructing an algorithm where the construction, or more specifically the arguments for the construction, is based on mathematical properties. Global decisions about the strategy choice could be based on CMR, but in the process of solving the problem, a specific local step could involve imitative rather than CMR. Alternatively, the global reasoning about strategy choice could be imitative reasoning with local steps that are CMR (Bergqvist 2006).

It is important to stress that creative mathematical thinking is not restricted to people with an exceptional ability in mathematics, but it can be hard to perform without appropriate interconnected competencies. The competencies are knowledge (the mathematical foundation), heuristics, beliefs, and control (Schoenfeld 1985). They are both cognitive (e.g. the mathematical knowledge) and affective (beliefs). Therefore, students might not even try to produce a CMR (Lithner 2008) even in situations when they easily could have made progress (Bergqvist et al. 2007; Sumpter 2013).

Imitative reasoning is a family of different types of reasoning: *memorised reasoning* (MR) where the strategy choice is founded on recalling an answer and the strategy implementation consists of writing this answer down with no other consideration and *algorithmic reasoning* (AR) where the strategy choice is recalling a certain algorithm (set of rules) that will probably solve the problematic situation. Algorithmic reasoning has three subcategories: familiar AR, delimiting AR, and guided AR. In this chapter, the focus is on the two main categories, CMR and imitative reasoning. (For a longer discussion and further explanations, see Lithner 2008.)

Collective Reasoning

Reasoning, especially when it is a result of social interaction, can also be seen as a collective process. The decisions and arguments are created between a group of people, not just by one person. Mueller (2009) and Mueller et al. (2012) highlighted the importance for collaboration when constructing mathematical arguments. Cobb et al. (1992) argued that it is through participation in the practice of collective argumentation that students learn mathematics. Therefore, this process is social and 'comprises a set of practices and norms that are collective' (Ball and Bass 2003, p. 29). Through collective reasoning, it is possible to study the dynamics of mathematical reasoning such as negotiations of mathematical meaning (Voigt 1994).

Krummheuer (2007) studied students learning mathematics through participation in processes of collective argumentation. Based on Toulmin's (2003) description of argumentation, Krummheuer (2007) described mathematical arguments as consisting of four main components: data, conclusion, warrant, and backing. Depending upon which components are present, it is possible to see how arguments are used, for instance, how they are directed. Previous research has shown that during free outdoor play, Swedish preschool children use a variety of products and procedures in their argumentation when they challenge, support, and take the reasoning forward (Sumpter and Hedefalk 2015). When needed, they use concrete materials to strengthen their arguments and also as an aid for reaching a conclusion but also included abstract social constructs such as jokes as part of their reasoning.

However, argumentation is not the same as reasoning, and so there is a need to establish the relationship between them. As stated previously, arguments are considered to have four components: conclusion, data, warrant, and backing. Data are the facts or the things that are being reasoned about. Warrants can be defined as the statements that legitimise the reasoning. Backings are about what are permitted, representing 'unquestionable basic convictions' (Krummheuer 2007, p. 65). Together, arguments can be linked to each other creating a chain or reasoning sequence which leads to an accepted conclusion that can act as data for a new argument. A chain of arguments is in alignment with Lithner's (2008) definition of mathematical reasoning as the line of thought. A chain of arguments does not necessarily need to be based on logic and may even be incorrect. Based on this, mathematical reasoning is not restricted to deductive logic. Toulmin's diagram shows the relation between the four objects (Krummheuer 2007) (see Fig. 1).

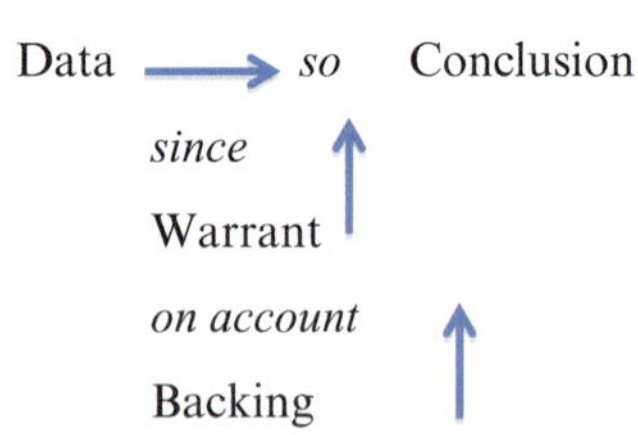

Fig. 1 Toulmin's diagram of argumentation; the implications of *arrows* are given in italics

Table 1 Structure of data

	Data	Argument	Conclusion
Person A			

In this chapter, I analyse children's arguments from a mathematical point of view. Therefore, I use Krummheuer's (2007) structure and notions as a starting point for studying the conclusions, warrants, and backing developed by a collective but add Lithner's (2008) concept of anchoring arguments in mathematical properties. When analysing collective reasoning, a revised version of Krummheuer's (2007) analysis of argumentation (AA) will be used (see Table 1).

In Table 1, the implication arrows 'since' and 'on account' are presented in the column 'Argument'. These arguments are analysed using the notions of objects, transformations, and concepts. Conclusions are analysed in the same way, with references back to arguments when appropriate, e.g. 'Since X, therefore Y'. In this way, data, analysis, and conclusions are presented in the same table. This alternative structure and coding scheme was first tested and used in Sumpter and Hedefalk (2015).

Tom and Jim

This interaction comes from a set of observations made when the author spent 5 days (different times over a period of 2 years; three outdoors sessions and two indoors sessions) in a preschool. As the data were collected by note taking, some details would be missed. However, in this case the interaction was quite short making it more likely that the important details were captured. There was no specific aim with the observations more than the focus on mathematical activities and child–child interactions. This episode comes from an occasion when the preschool took the 5-year-olds, the oldest children of the group, to play in the woods. Tom and Jim (both 5 years) are playing with sticks. The sticks are in their game laser swords and the question, brought up by the boys themselves, is: Which stick is the longest? (Table 2).

In this episode, the boys are allowed to explore, negotiate, and support their own and each other's argument without an adult. Tom and Jim use several mathematical properties concerning measurement when taking their reasoning forward. They compare magnitudes and order, they may have used conservation, and there is a possible use of transitivity when working with three objects. These function as a mathematical foundation; fulfilling the role of warrants, they are the grounds for the arguments. The accepted conclusion $B > A$ is later used as data for a new argument: when they establish $C > B$ and $C > A$. The arguments are directed and there are elements of challenge ('No, it is not!') and support ('It is the longest'.). In this way Tom and Jim arrive at a conclusion they both agree on.

Table 2 Tom and Jim solving 'which stick is the longest?'

	Data	Mathematical properties of the argument	Conclusion
Tom	Look! A laser sword! [picks up a stick from the ground]		
Jim	I got one, too. [picks up a stick from the ground]		
[Tom and Jim are playing with the sticks for a few minutes]			
Tom	My sword is long	The length of Tom's stick is large	
Jim	My sword is longer	The length of Jim's stick is longer than the length of Tom's stick	B>A
Tom	No, it is not!	Objection to B>A	
Jim	Look! [holds up his stick next to Tom's stick]	Comparing magnitudes, here lengths. The length of Jim's stick is longer than the length of Tom's stick	Since my stick is longer than yours, B>A
Tom	Ok. But mine is thicker. Boom boom! [Pretend shooting]	(Identification of another property of the stick, although not relevant to the question, 'which stick is the longest?')	B>A
Jim [looking around for other sticks]	Look at this one then! [Drags out a large branch]		
Tom	That one is longer than mine! That one is a laser cannon!	The branch is longer than Tom's stick. This is concluded without a direct measure. He may be using an understanding of conservation of length	C>A
Jim	It is the longest. Ha, ha! It is a cannon!	Possible use of transitivity: C>B>A	C>B, C>A

Heidi

This episode comes from a set of video recordings. The author asked parents of children from three preschools if their children could participate in a problem-solving session. The first ones to agree were Heidi's parents. This recording was made in their home to simplify the permissions for recording. Since the focus is on

Heidi's reasoning and not on the social context, the assumption is that this episode provides an illustration of a preschool child's mathematical reasoning when solving a mathematical task. It is not considered a play session, although Heidi might look at it as play. The task was provided by the interviewer as a stimuli to generate strategy choice, strategy implementation, and conclusion. Heidi is at the time 3 years and 5 months old. According to her parents, she has not been engaged in any specific mathematical activities at home nor at preschool.

As an introduction to the problem-solving session, she solved three tasks with the help of blocks: one addition task ($4+3$), one subtraction task ($5-2$), and one division task ($4\div 2$). In the division task, the interviewer tried to give all the blocks to one teddy. Heidi objected to this and then gave each teddy equal amounts.

Heidi then worked on how nine blocks should be divided by three toy animals (Teddy Bluebear, Rabbit-y, and George the Dog). How many blocks do they get each? There were nine blocks, consisting of three of each colour, green, blue, and yellow, but are mixed up.

Interviewer	Shall we see if George the Dog can count the blocks?
Heidi	Yes. [Counts when the interviewer points to the blocks one by one with George the Dog's paw] One, two, three, four, five, six, seven, eight, nine. Nine!
Interviewer	Nine blocks! Look, George the Dog is really happy!
Heidi	[laughs] Why is he shaking?
Interviewer	He is happy! That is what George the Dog is doing when he is happy. Shall we divide the blocks? Shall we divide so they get a few blocks each? Do you want to do that?
Heidi	Yes.
Interviewer	Shall we do it together? Who should have this one? [Points at a yellow brick that is closest to the interviewer.]
Heidi	Rabbit-y!
Interviewer	Then…?
Heidi	[Points at another yellow block]
Interviewer	Who should have this one?
Heidi	Bluebear. [points at the remaining yellow block]
Interviewer	Who should have this yellow block?
Heidi	The dog.
Interviewer	What should we do now?
Heidi	The green and the blue ones.

Heidi distributes first the green blocks and then the blue blocks to the toy animals.

Reasoning Structure

The data are organised using the reasoning structure suggested by Lithner (2008): problematic situation (PS), strategy choice (SC), strategy implementation (SI), and conclusion (C). Choice should here be interpreted in a wide sense (choose, guess, etc.) and could also include subconscious preferences.

PS: Nine blocks should be divided by three teddies.

SC: Identify property of the blocks: three colours. Group the blocks after colour: $9=3+3+3$. Then perform division: $9/3=(3+3+3)/3=3/3+3/3+3/3$.

SI: Straightforward. First yellow, second green, last blue.
C: Each teddy gets 1 blue, 1 green, and 1 yellow resulting in 3 blocks.

In this sequence, although the interviewer asked questions, Heidi was the one making all of the central decisions. She decided which blocks are going to be shared out (except for the first one), in which order they should be shared, and how many at a time. As a strategy choice, Heidi recognised the colour of the blocks and used this property when grouping the blocks into smaller subsets. Then she performed division for each of these subsets, one colour at the time, without any observed hesitation. The task was considered a new problem for Heidi in that sense that she did not have memorised knowledge (e.g. $9 \div 3 = 3$) or used a familiar algorithm based on surface arguments ('this is the algorithm we normally use'). In that way her reasoning was novel. She could have divided objects before in other activities and maybe even solved 9 divided by 3 previously. However, in solving this task, it seemed more likely that the algorithm was reconstructed than Heidi used imitative reasoning. Also, the grouping of the objects was an added transformation rather than 'just sharing'. Her choice to group the blocks, a local step in the reasoning, is plausible and has a mathematical foundation. If she, for instance, used quotient division instead of partitioning, it would have been a different strategy choice from a mathematical point of view. This reasoning is categorised as CMR.

Discussion

The starting point of this chapter was that although there is a growing body of research studying preschool children's mathematical reasoning, few studies use and anchor their analysis in theories and frameworks about mathematical reasoning. Mathematical reasoning is often considered to represent a high quality of thinking (Lithner 2008), and with such a definition, it is hard to see how Swedish preschools should work in order to help children 'develop their mathematical skill in putting forward and following reasoning' (National Agency for Education 2011a, p. 10). Therefore, there is a need to understand the different types of reasoning that children produce and/or how arguments are used and directed and what they are based on.

In this chapter, two cases of reasoning are described, individual and collective reasoning, and an example of each kind is analysed using two different types of frameworks. In the case of Tom and Jim, analysing their reasoning as a collective process highlighted the content and direction of their arguments. Tom and Jim used a mathematical foundation, the properties of measurement, to reach a shared conclusion. Similar behaviour has been observed in previous studies (Sumpter and Hedefalk 2015). Just as in Säfström (2013), the children showed that they could use different mathematical competencies and the ability to challenge and justify arguments.

In the case of Heidi, individual reasoning was the focus. Even though she interacted with the interviewer (mainly through a teddy), she made all the central decisions, making her reasoning individual. Her strategy choice and conclusion allow her reasoning to be categorised as CMR. It seemed that this type of reasoning was helpful for solving a mathematical problem when a specific solution method both at a global and local level was not known.

It would have been surprising if Heidi performed an imitative reasoning considering that she has not yet been involved in formal mathematical training. Imitative reasoning is more likely to be something produced when an individual has had access to and learnt a lot of mathematical procedures which were linked to specific tasks, such as occurs when working alone with a textbook (Lithner 2008). Heidi has not experienced this yet. Most likely, she does not have a belief that a certain task should be solved with a specific algorithm. Her reasoning, at the moment, is limited to her mathematical knowledge, her creativity, and the milieu in which she operated on a daily basis. Similar to collective reasoning, as part of a social process with practices and norms (Ball and Bass 2003), her 'mathematical world view' (Schoenfeld 1985) would be a factor determining what could be seen as possible and not. The same conclusion can be drawn about Tom and Jim's reasoning. They were not restrained by the idea of trying to find the right algorithm, a behaviour observed, for instance, in upper secondary school students' reasoning (Bergqvist et al. 2007; Sumpter 2013). The implication for preschool mathematics is that there is a need to support children to explore and produce their own reasoning, both individually and collectively, instead of telling what is 'the correct one' or just establishing 'the correct answer'.

The two frameworks highlighted different aspects of reasoning, each of them suitable to the different types of data, individual and collective. In the case of collective reasoning, the direction of arguments can be helpful in order to understand the social processes, such as participation and contributions. If mathematical objects, transformations, and concepts are present, the warrants and backings can be analysed from the point of view of mathematical content. However, it is not possible to state anything about different types of reasoning, CMR or imitative reasoning. Neither does this framework stress different types of strategy choices. In the case of individual problem solving, focusing on strategy choice and the conclusion using the structure of data made it possible to categorise different types of reasoning. It could highlight different strategy choices for different problematic situations. Although it seems suitable for doing an analysis of the solving of task, it does not seem suitable for reflecting on the reasoning used in interactions (Säfström 2013). However, the results are still interesting if it could help us explain and predict behaviour.

The use of one framework in doing an analysis does not exclude also using the other because their foci are on different things. The results from the two analyses will highlight different aspects of the reasoning.

In the curriculum, preschool teachers are responsible for ensuring that children in their preschools 'are stimulated and challenged in their mathematical development' (National Agency for Education 2011a, p. 11). Moreover, research has indicated that students learn when they have the opportunity to learn (Bobis et al. 2005, 2008; Hiebert 2003) and that young children that are guided can expand their mathematical thinking further than without a guide (Björklund 2008; Lee and Ginsburg 2009; van Oers 1996). Given this, it would be interesting to see how Tom's, Jim's, and Heidi's reasoning could be developed through being stimulated, so that they were supported to do what it says in the curriculum 'to develop their mathematical skill in putting forward and following reasoning' (National Agency for Education 2011a, p. 10). What reasoning could they perform, now and later, especially since mathematical reasoning predicts mathematical achievement later in school (Nunes et al. 2012)? Tom, Jim, and Heidi show creativity and skills for putting forward arguments based on mathematical properties. These are good qualities. It is possible to recognise their competencies in this area: the question is, what is being done with it?

References

Ball, D., & Bass, H. (2003). Making mathematics reasonable in school. In J. Kilpatrick, G. Martin, & D. Schifter (Eds.), *A research companion to principles and standards for school mathematics* (pp. 27–44). Reston, VA: National Council of Teachers of Mathematics.

Bergqvist, E. (2006). *Mathematics and mathematics education two sides of the same coin. Some results on positive currents related to polynomial convexity and creative reasoning in university exams in mathematics.* PhD thesis, Umeå University, Umeå.

Bergqvist, T., Lithner, J., & Sumpter, L. (2007). Upper secondary students' task reasoning. *International Journal of Mathematical Education in Science and Technology, 39*(1), 1–12.

Björklund, C. (2008). Toddlers' opportunities to learn mathematics. *International Journal of Early Childhood, 40*(1), 81–95.

Bobis, J., Clarke, B., Clarke, D., Thomas, G., Wright, R., Young-Loveridge, J., & Gould, P. (2005). Supporting teachers in the development of young children's mathematical thinking: Three large scale cases. *Mathematics Education Research Journal, 16*(3), 27–57.

Bobis, J., Mulligan, J., & Lowrie, T. (2008). *Mathematics for children: Challenging children to think mathematically* (3rd ed.). Sydney: Pearson Education.

Charlesworth, R. (2005). Prekindergarten mathematics: Connecting with national standards. *Early Childhood Education Journal, 32*(4), 229–236.

Clements, D. H., & Sarama, J. (2007). Effects of a preschool mathematics curriculum: Summative research on the *Building Blocks* project. *Journal for Research in Mathematics Education, 38*, 136–163.

Cobb, P., Yackel, E., & Wood, T. (1992). A constructivist alternative to the representational view of mind in mathematics education. *Journal for Research in Mathematics Education, 23*(1), 2–33. doi:10.2307/749161.

English, L. D. (2012). Data modelling with first-grade students. *Educational Studies in Mathematics, 81*(1), 15–30.

Hiebert, J. (2003). What research says about the NCTM standards. In J. Kilpatrick, G. Martin, & D. Schifter (Eds.), *A research companion to principles and standards for school mathematics* (pp. 5–26). Reston, VA: National Council of Teachers of Mathematics.

Krummheuer, G. (2007). Argumentation and participation in the primary mathematics classroom: Two episodes and related theoretical abductions. *The Journal of Mathematical Behavior, 26*, 60–82.

Lee, J. S., & Ginsburg, H. P. (2009). Early childhood teachers' misconceptions about mathematics education for young children in the United States. *Australasian Journal of Early Childhood, 34*(4), 37–45.

Lithner, J. (2008). A research framework for creative and imitative reasoning. *Educational Studies in Mathematics, 67*(3), 255–276.

McMullen, J. A., Hannula-Sormunen, M. M., & Lehtinen, E. (2013). Young children's recognition of quantitative relations in mathematically unspecified settings. *The Journal of Mathematical Behavior, 32*(3), 450–460.
Mueller, M. F. (2009). The co-construction of arguments by middle-school students. *The Journal of Mathematical Behavior, 28*, 138–149.
Mueller, M. F., Yankelewitz, D., & Maher, C. A. (2012). A framework for analyzing the collaborative construction of arguments and its interplay with agency. *Educational Studies in Mathematics, 80*, 369–387.
Mulligan, J. T., & Vergnaud, G. (2006). Research on children's early mathematical development: Towards integrated perspectives. In A. Gutiérrez & P. Boero (Eds.), *Handbook of research on the psychology of mathematics education: Past, present and future* (pp. 261–276). London: Sense.
National Agency for Education. (2011a). *Curriculum for the Preschool Lpfö 98: Revised 2010.* Stockholm: Skolverket.
National Agency for Education. (2011b). *Curriculum for compulsory school, preschool class and recreation centre 2011.* Västerås: Edita.
National Agency for Education. (2011c). *Curriculum for the upper secondary school.* Västerås: Edita.
National Agency for Education. (2012). *TIMSS 2011 Svenska grundskoleelevers kunskaper i matematik och naturvetenskap i ett internationellt perspektiv [TIMSS 2011 Swedish compulsory school students' knowledge in mathematics and science in an international perspective].* Stockholm: Skolverket.
NCTM [National Council of Teachers of Mathematics]. (2000). *Principles and standards for school mathematics.* Reston, VA: The Council.
Niss, M. (2003). Mathematical competencies and the learning of mathematics: The Danish KOM project. In *Third Mediterranean conference on mathematics education* (pp. 115–124).
Nunes, T., Bryant, P., Barros, R., & Sylva, K. (2012). The relative importance of two different mathematical abilities to mathematical achievement. *British Journal of Educational Psychology, 82*(1), 136–156.
Papic, M., Mulligan, J. T., & Mitchelmore, M. C. (2011). Assessing the development of preschoolers' mathematical patterning. *Journal for Research in Mathematics Education, 42*(3), 237–269.
Perry, B., & Dockett, S. (2007). *Play and mathematics.* Adelaide: Australian Association of Mathematics Teachers (Retrieved January 17, 2013, from http://www.aamt.edu.au/Documentation/Statements/Early-Childhood-Mathematics-support-paper-Play).
Säfström, A. I. (2013). *Exercising mathematical competence. Practising representation theory and representing mathematical practice.* PhD thesis, Göteborgs Universitet, Göteborg.
Schoenfeld, A. H. (1985). *Mathematical problem solving.* Orlando, FL: Academic Press.
Skemp, R. (1978). Relational understanding and instrumental understanding. *Arithmetic Teacher, 26*(3), 9–15.
Sumpter, L. (2013). Themes and interplay of beliefs in mathematical reasoning. *International Journal of Science and Mathematics Education, 11*(5), 1115–1135.
Sumpter, L., & Hedefalk, M. (2015). Preschool children's collective mathematical reasoning during free outdoor play. *The Journal of Mathematical Behavior, 39*, 1–10.
Toulmin, S. E. (2003). *The uses of argument. Updated edition.* Cambridge, UK: Cambridge University Press.
van Oers, B. (1996). Are you sure? Stimulating mathematical thinking during young children's play. *European Early Childhood Education Research Journal, 4*(1), 71–87.
Voigt, J. (1994). Negotiation of mathematical meaning and learning mathematics. *Educational Studies in Mathematics, 26*(2/3), 275–298.
Wyndhamn, J., & Säljö, R. (1997). Word problems and mathematical reasoning—a study of children's mastery of reference and meaning in textual realities. *Learning and Instruction, 7*(4), 361–382.
Yackel, E., & Hanna, G. (2003). Reasoning and proof. In J. Kilpatrick, W. G. Martin, & D. Schifter (Eds.), *A research companion to principles and standards for school mathematics* (pp. 227–236). Reston, VA: National Council of Teachers of Mathematics.

Adaptability as a Developmental Aspect of Mathematical Thinking in the Early Years

Götz Krummheuer and Marcus Schütte

Abstract For the purpose of analyzing the longitudinal development of mathematical thinking, statuses of participation of a child are reconstructed and theoretically described in terms of the framework of the "Interactional Niche in the Development of Mathematical Thinking." In this chapter, we deal with the concept of adaptability within this framework, which we consider as having a twofold characteristic. The interactive process of negotiation of meaning adapts more or less to the needs of the participating children, thus producing specific options of participation and conversely the children adapt more or less successfully to the then accomplished content-related requirements. We demonstrate our theoretical reflections by presenting two episodes of a case study of a child, whom we accompanied in her mathematical development over a period of about 3 years.

Introduction: A Story from Yellowstone—How the Wolves Help the Berries and the Bears[1]

Several years ago, the once extinguished wolves had been re-naturalized in the Yellowstone Park. The effects of it can be summarized as follows:

- The wolves decimate the moose.
- Fewer moose eat fewer young trees.
- The trees can then grow higher.
- That gives more room for berry bushes.
- Thus, the bushes produce more berries.
- That gives the bears the opportunity to eat more berries and hunt fewer animals.

[1] Taken from Süddeutsche Zeitung, Tuesday, July 30, 2013, number 174, p. 16.

G. Krummheuer (✉)
Goethe University, Frankfurt am Main, Germany
e-mail: goetzkrummheuer@mac.com

M. Schütte
Technical University Dresden, Dresden, Germany
e-mail: marcus.schuette@tu-dresden.de

T. Meaney et al. (eds.), *Mathematics Education in the Early Years*,
DOI 10.1007/978-3-319-23935-4_10

Without the wolves, the moose had a "perfect" niche and their population increased. The increased population of moose, however, produced worse conditions for the growth of trees and berries, and the bears had to hunt more animals. With the re-naturalization of wolves in the park, the niche for the moose became worse, but the niches for the trees and the berries improved and the food supply for the bears became more opulent and less strenuous to find.

Adaptability Within the Theoretical Framework of the Interactional Niche in the Development of Mathematical Thinking

Our research interest is the development of an empirically grounded theory concerning the generation of mathematical thinking of children between the ages of 3 and 8. Empirically, our longitudinal study concerns the observation of children over 3–4 years in situations of social interaction: in preschool, kindergarten, and elementary mathematics classes and partly also in families. For this particular research interest, we developed specifically designed mathematical problems (for the design of these problems, see Vogel 2013, 2014; Vogel and Huth 2010).

Theoretically, one major premise of our analyses is borrowed from Bruner's concept of Language Acquisition Support System (LASS). In his studies about language acquisition, Bruner rejects the idea of conceptualizing the child's acquisition of its mother language as a process that is entirely located within the child's cognition and its genetic predispositions. It is not only a genetically given capability of the human species to learn its mother tongue; it is also the cultural embeddedness of the children's development that provides a systematic backing for its development. Bruner speaks here of an acquisition support system which is mainly accomplished by certain characteristics of the adult–child interaction. What we adopt from his argument for our research is the approach that the child necessarily needs social support also for its development of mathematical thinking. For the support systems of language acquisition, Bruner reconstructs specific patterns of interaction within the mother–child dyad which he calls "formats." A format is a

> standardized, initially microcosmic interaction pattern between an adult and an infant that contains demarcated roles that eventually become reversible. (Bruner 1983, p. 120f)

Usually, neither the adult nor the child applies these formats intentionally. Commonly, they emerge spontaneously between adult and child in the course of interaction that is regulated by interaction moves, chained in the ongoing reiteration of the "adjacency pairs" of initiation and reply (Sacks 1998, pp. 521–541). This is the origin of patterns of interaction, which more or less functionally generate LASS. In most cases, the adults are the mother, father, or other close relatives, who regularly interact with the infant from birth in a familial context.

In a certain sense, one can understand the acquisition of the mother tongue as the prime example of a socio-constructivist conceptualization of any fundamental learning process in early childhood that progresses beyond this phase. By analogy, we attempt to reconstruct Mathematics Learning Support Systems (MLSS). For our theoretical reflections, we borrow from Bruner's theory the idea that it is a support system that is generated within the interaction system accomplished by the participants in a concrete social encounter. Such MLSS differs from LASS in that the emotional relationship between a nursery teacher and a child is not analogous to the adult–child attachment in the familial context (Bowlby 1969; Krummheuer et al. 2013). It also differs in terms of the content of "mathematics" that is under scrutiny in our studies which is not similar to the content of "language" in the studies of language acquisition.

One might tend to understand LASS as a kind of instruction or as a treatment in a psychological sense. My understanding of Bruner's concept of LASS is that it is a characteristic of the adult–child interaction. That means that the support for the child's language acquisition is no instruction of the adult and/or an application of her treatment. The support belongs in another system than that of the individual's actions, namely, into the system of social "interaction." Bruner describes this as a "microcosm." In recent terminology, one could say that LASS is concerned with the enabling and shaping of ways of the child's and the adult's participation in their joint social encounters. Therefore, one could speak of a concept of learning that is based on the incremental participation of the child within such a relatively stable interaction system that in the case of mathematics learning would be, in Sfard's (2008) terms, the "mathematical discourse." Thus, already in the 1980s, Bruner envisioned an alternative to a psychological view on learning that focuses on acquisition of knowledge.

This implies—with respect to the child's development of mathematical thinking—that one has to take into account a more complex "microcosm" than the adult–child dyad. Theoretically, this insight leads to the introduction of the concept of the interactional niche in the development of mathematical thinking (NMT) (Krummheuer 2012a, 2013a; Schütte and Krummheuer 2013). In general, NMT adopts Bruner's metaphor of a "microcosm" that characterizes this type of learning support interaction. Additionally, we also refer to the concept of "developmental niche" from Super and Harkness (1986):

> The developmental niche, [...], is a theoretical framework of studying cultural regulation of the micro-environment of the child, and it attempts to describe the environment from the point of view of the child in order to understand processes of development and acquisition of culture (p. 552).

Complementarily, we stress the aspect of the interactive local production of such processes in the "micro-environment of the child" and speak of NMT. It consists of the following:

- Provided "learning opportunities" of a group or society, which are specific to their culture and will be categorized as aspects of "allocation"
- Situationally emerging performance occurring in the process of negotiation of meaning which will be subsumed under the aspect of the "situation"
- Single child's activities and overt reflections, which will be concatenated under the aspect of the child's "contribution"

Table 1 NMT

NMT	Component: content	Component: cooperation	Component: pedagogy and education
Aspect of allocation	Mathematical domains; body of mathematics tasks	Institutions of education; settings of cooperation	Scientific theories of mathematics education
Aspect of situation	Interactive negotiation of the theme	"Leeway"[a] of participation	Folk theories of mathematics education
Aspect of a child's contribution	Actions of the child in the emerging interaction process	"Participation profile"[b] of the child	Competency theories

[a]Leeway taken here in the colloquial meaning of "room for freedom of action"; see Webster (1983, p. 1034), originally the notion of "Partizipationsspielraum" in the book by Brandt (2004)
[b]See again Brandt (2004)

These three aspects (allocation, situation, and child's contribution) are then subdivided into three components: "content," "cooperation," and "pedagogy and education"—a further development of the original components of the developmental niche of Super and Harkness especially taking into account the domain specificity of the mathematical content.[2]

Table 1 expresses the present version of the development of the concept of NMT which is empirically grounded in our research project. Acknowledging this to be a temporary artifact of this stage of research, in the following we further explicate the details of this table:[3]

Content: Children are confronted with topics from different domains of mathematics in their everyday lives. In the research project, mathematical topics are usually presented in the form of a sequence or body of tasks, which are adjusted to the assumed mathematical competencies of the children. On the situational level, the presentation of such tasks elicits processes of negotiation, which may not proceed either in accordance with the ascribed mathematical domain or with the activities that are expected in the tasks. With respect to the research interest in the individual development of mathematical thinking, we summarize in this subcategory the actions of the child in focus.

Cooperation: The children participate in culturally specific social settings which are variously structured as in peer interaction or small group interaction guided by a

[2]Super and Harkness (1986) defined "their" development niche by three components: "the physical and social settings in which the child lives," "culturally regulated customs of child care and rearing," and "the psychology of the caretakers" (ibid. p. 552). They conducted anthropological studies without focusing on the situational aspects of social interaction processes.

[3]In earlier publications, we worked with a preliminary version of this concept (Krummheuer 2011a, 2012, 2013a, 2014a).

nursery teacher, primary mathematics teacher, etc. These social settings do not exist automatically; in fact, they need to be accomplished in the joint interaction. Depending on each event, a different leeway of participation for the children will come forward. Over a complete episode, certain stabilities and/or characteristics of the child's participation might be reconstructed and is summarized as its "participation profile."

Pedagogy and education: The science of mathematics education develops theories and delineates—more or less stringently—learning paths and milestones for children's mathematical growth. In the concrete situation, however, it is rather the folk psychology and folk pedagogy of the participating adults and children that becomes operant. The concepts of folk psychology and folk pedagogy are defined by Bruner (1996):

> Folk psychologies reflect certain "wired-in" human tendencies (like seeing people normally as operating under their own control), but they also reflect some deeply ingrained cultural beliefs about "the mind." Not only is folk psychology preoccupied with how the mind works here and now, it is also equipped with notions about how the child's mind learns and even what makes it grow, just as we are steered in ordinary interaction by our folk psychology, so we are steered in the activity of helping children learn about the world by notions of *folk pedagogy*. (p. 45)

It cannot be assumed that these kinds of folk theories coincide with the theories of the science of mathematics education.

Competency theories refer to the diverse, applied theories of mathematics learning, the folk pedagogy of the educators, the leeway of participation, and the participation profile of the child in focus. One might be able to describe the change and/or progress of participation of the child and to theoretically reflect on this change and/or progress as an indicator of the child's (progress in its) development of mathematical thinking.

NMT refers to all the components in Table 1 and has to be accomplished in the process of interaction by the participants in the situation. However, in this article, we do not refer to all categories of the table of NMT.

In the "microcosm" (Bruner) or "micro-environment" (Super and Harkness) of NMT, we can reconstruct a twofold adaptability.

- The interaction system adapts itself to the possibilities of participation of the focus children in that it generates a kind of conversation that enables at least some children to contribute actively to this interaction. This adaptation can result in a pattern of interaction that one could characterize as a format. In general, the result of this adaption is the MLSS.
- Conversely, the individual adapts himself to such a pattern of interaction, if necessary, by making appropriate changes in his definition of the situation to a commonly shared interpretation. He then uses this patterned process of negotiation as his MLSS, and the changes in his definition of the situation are an expression of his cognitive achievement of adaptation. In the situation, the child can act with increasing autonomy in the evolving format, i.e., the child is learning mathematics.

In Table 1, these two forms of adaptability are represented as the interface between the aspects of:

- Allocation and situation
- Situation and contribution

With respect to the component "content,"[4] the interface between allocation and situation can be described as a change of subject matter in the process of negotiation of meaning. It brings the content to life as a working consensus. The substance of this consensus might differ from that of the intended one for the presented learning opportunity. Nevertheless, it might enable the child a more active and autonomous form of participation. We do not speak here of the "content" as it was intended to be introduced but of the "theme" that has been negotiated in the course of interaction. That means that the accomplished theme about the introduced content is principally open to change and leads, in an interactionist's terms, to the so-called taken as shared meaning (see, e.g., Krummheuer 2014b). One can understand this shift as the interactional modulation of the content toward the situational needs. This is the first dimension of the twofold concept of adaptability. The interface between situation and contribution refers to the achievements of a participant to contribute thematic input that suits or even promotes the momentary process of the negotiation of meaning. This is the second dimension of adaptability.

These processes of adaptation are not necessarily intentional actions of the participants. They rather are indebted to the interactional obligation of keeping the social encounter alive—as we likewise do not insinuate that the wolves deliberately decimate the moose in order to help the bear survive. The wolves' behavior solely serves to further their own survival.

In our analyses, we reconstruct the first kind of adaptation as a change in the way the participants talk and act about the presented and interactively negotiated problems. As in the two chosen episodes, there typically emerges a rather narratively structured exchange about a story that deals with the items, which the assisting adult introduced in the encounter. In the two examples, these are wooden animals, figures, and a toy train. Besides that, there also emerges a communicative exchange that is rather formalistic: The descriptions do not adhere to the peculiarities of the presented toys, and these items are rather taken as arbitrary objects of any discrete set. For the sake of easier discrimination, we speak of a "narrative discourse" and a "formal discourse."[5]

---

[4] In this chapter, we delineate the notion of adaptability as interface of two aspects of NMT only with regard to the component "content." Principally, it is also applicable to the components "cooperation" and "pedagogy/education."

[5] See here, too, Sfard's concept of "mathematical discourse" (Sfard 2008), Bauersfeld's notion of "language game" (Bauersfeld 1995) and/or Krummheuer's differentiation between "narrative argumentation" and "diagrammatic argumentation" (Krummheuer 2013a).

Two Episodes with Ayse

In the following two examples, with regard to this twofold notion of adaptability, we are concerned with the question about what the participants are challenged with while accomplishing a play situation that also offers the children the chance of learning a "bit" of mathematics. Due to the longitudinal design of our study, two episodes with the girl Ayse are chosen with a 2-year hiatus.

The two examples come from the longitudinal study "early Steps in Mathematics Learning" (erStMaL) in which we follow children in 12 daycare centers over a period of 4 years in which they are observed every 6 months. This research project is based in the center ***I****ndividual* ***De****velopment and* ***A****daptive Education of Children at Risk* (IDeA; www.idea-frankfurt.eu) and is concerned with the development of mathematical thinking in preschool, kindergarten, and early school years (for more details, see Acar Bayraktar et al. 2011). Several children are selected for deeper, long-term scrutiny. One of these is Ayse. Over the years, we observed her in different social settings during preschool, kindergarten, and the first years of school. The following two scenes stem from group work sessions, which are assisted by an adult. Our research interest is focused on the changes in her participation in these settings.

Ayse is the only daughter of Turkish parents who were born and went to school in Germany. Both parents work. The grandparents, who have immigrated to Germany, are caretakers of the child during the day. Ayse is 4.02 years old at the first time of observation. In accordance with the design of the erStMaL study, Ayse participated several times in varying settings of play, designed as discovery situations dealing with the content areas of number and operations, geometry and spatial thinking, measuring and size, and data and probability. In both analyzed episodes, the content areas can be classified as "data and probability" and "geometry and spatial thinking."

The First Example: The Polonaise of the Animals[6]

This example is the earliest episode from the data collection with Ayse. In this episode, Kai, 4.03 years old, is her partner. A further participant is the assisting adult B. All three participants sit on the floor on a round carpet. Their seating order can be seen in Fig. 1.

The intended play and discovery situation has to do with a question from combinatorics, which concerns the different order of three animal figures when they walk across a platform. The adult introduces the episode by talking about the platform from a circus. Without being asked, Ayse mentions that she has already seen a circus (possibly only in television). The adult B shows three, until now concealed,

[6] Melanie Huth and Rose Vogel conducted the first analysis of this episode (Vogel and Huth 2010).

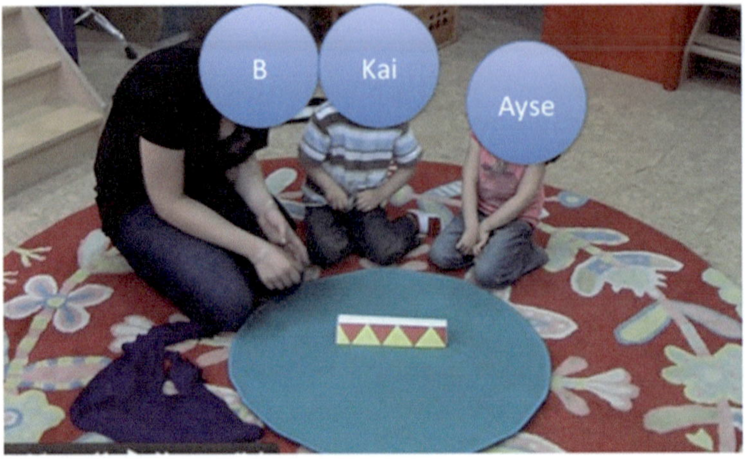

Fig. 1 Seating arrangement

animals: an elephant, a monkey, and a white tiger. Hardly had the elephant been placed on the carpet, Ayse picks it up and only gives it back after the adult intervenes. Perhaps to her own relief, Ayse offers Kai the monkey. This is also stopped by the adult. After the white tiger is placed on the carpet, B explains that it is a "baby tiger."

In this opening scene, Ayse is the more dominant of the two children: she mentions her circus experience; she grabs the elephant and attempts to talk Kai into taking the monkey. One could say that, for the time being, an NMT emerges that opens an active and dominant "participation option"[7] for Ayse.

Curiously, her leeway of participation changes when B says that the tiger is a baby. The motivation behind B's remark could have been that:

- The presence of a tiger could have aroused the association of the children with a predatory animal and that this "danger" could be calmed with the remark about a baby tiger.
- Baby animals are generally "cuter" than the grown specimens and B's explanation could produce a greater emotional relationship of the children to these animals.

This time Kai builds on this remark, declaring that all three animals are baby animals. In connection with B's following instruction for the play situation, which was to find out the different possible sequences of the three animals walking along the platform, the leeway of participation changes for Ayse. She is still and moves to the status of legitimate peripheral participation (Krummheuer 2011b; Lave and Wenger 1991).

[7] I encountered the concept "Partizipationsoption" in Höck (2015).

It appears as if the redefinition of the animals to baby animals introduces a change into a rather formal discourse. This discourse exists among others in that the specific qualities of the animals do not play a role anymore and that their sameness in the sense of arbitrary objects of a set is highlighted as a central theme: they are all babies and are thus not dangerous—just "mathematical objects." Additionally, B hints in distinct, nonverbal gesticulation that the animals should run across the platform.

In the following, four different orders of the animals are all produced by Kai, whereby Ayse is a little more active in the beginning. The specific qualities of the three animals do not play a role here. Kai even calls the tiger a lion at the end. Also B stops her exact, nonverbal remarks after the second permutation, and she, as well as Kai, change to a more formal way of articulating, as, for example, Kai's formulation: "the monkey right in the front, the elephant in the middle, and the lion must go here" <375–380>.

Along with the stronger formalistic words in connection with a definite reduction of nonverbal negotiation of meaning, that an NMT for Kai is produced that is beneficial for him. He seems to have a more fitting disposition for this more formal discourse and can occupy this niche more successfully.

Again, Ayse becomes more active toward the end of the episode, while Kai is no longer successful in producing permutations. Here she can broaden her leeway again in that she can act more strongly in the context of a circus and, for example, now attempts to push the animals under the platform.

We can describe two processes of adaption here:

- The process of negotiation of meaning adapts to a more formal discourse.
- Ayse and Kai adapt in more or less successful ways to the emerging leeway of participation that is defined by the participatory demands of such a formal discourse. Kai can act more successfully in such a formally structured discourse. For Ayse, this change of the discourse leads to a withdrawal to a more peripheral form of participation.

Thus, for the two children, their initial NMT changes, like it did for the moose, trees, and berries in Yellowstone by the impact of the wolves.

The Second Example: The Train

For this initiated play and discovery situation, four children, Ayse (6;04), Barbara (6;02), Elias (6;10), and Norbert (6;01), as well as an adult B are sitting on the floor. This play takes place about 2 years after the episode discussed above. The participants have the wooden train as play material in front of them. Also in this example, the children get the opportunity to become acquainted with the play material and to tell about other earlier experiences with it. Then, after a suggestion of B, mostly the two boys build an oval track as seen in Fig. 2.

B asks Ayse to place a red figure of a man inside the railroad system <03, 04>. Ayse refuses to try. Barbara offers to try and places the figure on top of the track so that it fits as best as it can between the two rails. B asks several questions which none of the children answer directly. Ayse takes the red figure and places it in the

Fig. 2 Seating arrangement

same way on the track. Elias turns away and starts to push the locomotive forward and backward on the track. In this way, he pushes over the red figure and it falls inside the oval circle.

Ayse places the figure immediately on the track. As the locomotive pushes over the figure, B says: "so now it lies on the track \ <43> I meant that the - <46> stands within this circle (so that) the train always runs around him\" <55, 56>. At the same time, B makes clear circular movements with his right hand above the oval circle.

B makes the connection for the first time in this scene between the train and the topological feature "within": "standing in the circle" can be recognized by the fact that the train can always run around it. This is not only expressed in relationship to the material in that he speaks of the oval track as a "circle" and no longer speaks in a more formal language of the "railroad system" but also in a flow of activity describing that the train always runs around the figure. Elias concretely demonstrates this flow of activity by moving the train around the track. Additionally, the red figure has fallen more or less intentionally inside the circle (see Fig. 3). Shortly afterward, Ayse places the red figure in the middle of the oval. B gives her positive feedback and gives Elias the problem: Elias, place the yellow one now so that it is standing outside of the circle <61–63>. He places the figure that was just given to him exactly on the same place where Ayse had placed the red figure at the beginning and returns immediately to the train <67–69>.

The report about this episode will be broken off here. We reconstruct the curious development of the process of interaction. In it, the solution that B intended is generated by the coincidence of three complementary actions:

1. Ayse places the red figure in front of herself on the track.
2. Elias pushes over the figure with the train and it falls into the inner area of the oval circle.

Fig. 3 The little *red* figure falls into the inner part of the circle

3. B reformulates his beginning question in reference to the materials, accompanying it with nonverbal gesticulation.

Ayse again places the figure back on the track and then places it clearly at a distance from the track in the middle of the oval. Evidently for Ayse, a discourse structure has been accomplished that makes it possible for her to become active. The structure is distinguished by a clear relationship to the material and a language that is supported by gestures which refer to the objects. One could call this a micro-narration: The formal diction of the "inner part of the track system" is transformed into a story related to the objects: the figure should stand so that the train can run around it. The formal description of "within" or "outside" a train system is transformed into a narrative.[8] By this kind of discourse, an NMT emerges for Ayse that broadens her leeway of participation. In analogy to the illustration from Yellowstone Park, the conditions of survival for the moose in the long run might improve, when on the average the trees grow higher which means that they might be better hidden from the wolves.

The Comparison of the Two Examples

There are 2 years between the first and the second episode. In the first one Ayse is aged 4 and in the second one 6. In both episodes, Ayse makes use of the opportunity to actively participate, as long as the discourse is oriented toward the options of

[8] The impact of narratives for children's argumentations is theoretically elaborated in Krummheuer (1999, 2000, 2009, 2013a).

actions that can be undertaken with the given material. In these parts of the episode, she acts spontaneously with a relatively high degree of autonomy. She takes her share in the production of a narrative. When the discourse is rather concerned with formal and mathematical attitudes, then she steps back and acts in the status of a legitimate peripheral participant. Her participation options seem to diminish.

The change in the interaction process between a narratively structured discourse and a rather formal one appears in the two episodes but in the reversed order. At the beginning of the first episode, the discourse begins with some narrative remarks, and B presents the mathematical problem in this vein. Most likely, the initiation is unintended by B and Kai, although their commonly accomplished declaration that the three animals are all baby animals sets a starting point of shifting the narratively structured talk into a rather formal one. For the boy, this swing in the interaction process opens new participation options, and thus for him, an NMT emerges in which he can participate in an active and relatively autonomous way: The NMT adapts toward Kai's participation strengths, and he adapts his thinking to the demands of this NMT by developing strategies for the systematic generation of permutations. For Ayse, however, this change of the type of discourse leads to a less supportive NMT: she only can participate in the status of a legitimate peripheral participant. In a certain sense, the interactively accomplished advancement of the interaction has left Ayse behind.

In the second episode, after a warming-up phase, B opens directly a formal discourse that obviously does not address the capabilities of all participating children. Necessarily, the type of discourse has to be modified, and by this it moves toward a micro-narrative with a strong relation to the play situation with a train toy, and in a quaint way, the expected answer is created by a train accident and B's simultaneous material-bounded and gesture-supported reformulation of the problem. This peculiar and interactively accomplished modification of the type of discourse appears to be advantageous for Ayse's participation options: immediately she can generate the desired solution. One can still doubt whether in mathematical terms she learned what the inner part of a railroad system is. For her, according to the produced micro-narrative, the little red figure is just located at a place where the train can circle around it.

For Ayse, an NMT emerges that provides successful participation opportunities for her. The interaction moves forward in this micro-narrative tone, and this time Ayse is not left behind; she is rather the trailblazer of the ongoing interaction process.

Theoretical Reflections

The two episodes focused on the participation options of Ayse are taken here as illustrations of the twofold process of adaptation within an NMT.

- First, the NMT alters or oscillates between narratively structured discourses and rather formal, "mathematical" ones. This movement is taken as the first kind of adaptation and is caused by the need for maintaining the process of interaction. These different kinds of discourses "define" what an appropriate kind of participation looks like.
- The second kind of adaptation is to be understood as the participant's attempt to find his/her way of partaking in the discourse that forms his/her changing NMT.

This can be a setback to the status of a legitimate peripheral participation. This is a less fortuitous condition for the participant's mathematical thinking in that it is solely based on inactive observation and a more or less passive making of meaning. This attempt can also lead to progress in the mathematical thinking. In this case, an MLSS comes into being that provides resources for the child in terms of developing an independent and autonomous participation status in the emerging mathematical discourse.

Applying this notion of a twofold adaptability to the two episodes, one somehow has the impression that Ayse has not experienced beneficial conditions for developing her mathematical thinking over the last 2 years. In both episodes, she is relatively active as long as the discourse is about a narrative. As soon as it shifts toward a rather formal and mathematical one, she is silent and observant. By analogy to the previous story about Yellowstone Park, one could take the emergence of a formal discourse as the re-naturalization of wolves and Ayse's regression into a quiet and observant child as the destiny of the moose.

However, one should be careful with such an assessment. First, one should consider that the mathematical domains are different: combinatorics versus elementary topology. The development of mathematical thinking might be domain specific, and Ayse's topological competencies might be less developed than those of her combinatory competencies. Secondly, the interactive settings are different: only one peer in the first episode and three peers in the second. Ayse's, as well as other children's, leeways of participation might be influenced by the specifics of the introduced social setting and how the participants cope with this learning situation in the concrete case. Thus, even within a constant mathematical domain, the emerging NMT might be structured in a different way through alternations of the social settings and might require different kinds of adaptations. Krummheuer (2013b) assumes that principally the child's development of mathematical thinking does not proceed in a direct, linear form but in highly situatively shaped, oscillating movements (p. 260). In terms of the analogy of the Yellowstone Park, not every year is identical; the trees might grow higher because of optimal climatic conditions, and that would have a positive effect on the living conditions of the berries and the bears, too.

Finally, applying the ecological concept of a niche for a theoretical conceptualization of the child's development of mathematical thinking casts a new light on the relations between teaching and the child's individual mathematical learning progress. At the first POEM conference, Norma Presmeg characterized this relationship as a "dance of instruction with construction" (2014, p. 11). In this metaphorization, she underlines the "dynamic way" (p. 11) of the interaction between teacher and pupils that leads to more profound reflection about mathematics. She also stresses the "canonical moves" (p. 11) in the interaction that lead to a coordinated exchange of ideas among the participants.[9]

As mentioned above, inspired by Bruner's insights into the supportive function of the adult–child interaction for the child's language acquisition, we can similarly

[9] See also Chen and McCray in this book (2014), who emphasize in more detail the "big ideas" of key mathematical concepts (p. 264) and the impact of "intentional teaching" (p. 269). This can be interpreted as a specific kind of the dance of instruction with construction.

conceptualize a mathematics-learning scenario in the early years that helps theoretically to more precisely describe this dance and thus also to recognize that the previous discussion involving the tension between instruction and construction was restricted to a merely psychological perspective on learning. We focus on those aspects of the child's development of mathematical thinking that take place without deliberate instructing activities, without theoretically reverting to a position that postulates that it is merely the challenge of a rich "natural" environment that enables the child to construct mathematics concepts and procedures by itself. The development of the child's mathematical thinking depends on the opportunities or chances of filling participatory leeways in NMTs in such a way that:

(a) The thematic movements within the niche emerge in a rather formalistic–mathematical discourse.
(b) The child finds a way to participate in this discourse in an increasingly autonomous manner.

We find the educational intention of regulating such a longitudinal process solely from an instructional perspective to be short-sighted and unproductive. However, a society decides how important and intentional mathematics learning should be taught to their children in the early years. The presented approach of NMT points at the rather "ungovernable" aspects of such learning processes by instruction. The NMT approach stresses more the situative processes of adaptability from both sides of instruction and construction rather than the possibility of didactically shaping forms of instruction.

Appendix: Transcripts

Polonaise of the Animals

0001	T01 143 144 000 000 000 000 090715 2 50301 3 0019-0450			
0002				*(B sitzt am linken Rand des runden Teppichs und*
0003				*Ayse etwa an der Mitte des Teppichs. Dazwischen*
0004				*sitzt Kai links von B. Zu Beginn der Situation*
0005				*hat B ein Podest quer in etwa in die Teppichmitte*
0006				*gestellt und fragte die Kinder, ob sie das schon*
0007				*einmal gesehen haben. Beide bejahen und Kai*
0008				*bezeichnet es als Zelt.)*

(continued)

(continued)

0009	00:19		B	*aha\ sieht aus wie ein Zelt\ ne/*
0010				*Berührt das Podest mit der rechten Hand und*
0011		<		*nimmt die Hand zu sich zurück. Blickt zu Ayse.*
0012		<	Ayse	*Deutet mit rechtem Zeigefinger auf das Podest.*
0013				*das hab ich schon mal bei Einsteins gekuckt*
0014				*Blickt zu B. Legt beide Hände zwischen die Beine.*
0015		<	Kai	*Blickt zum Podest und steckt beide Hände unter*
0016				*sein T-Shirt.*
0017			B	*aha\ im Fernsehen/*
0018			Ayse	*nickend ja\+ Blickt nach vorne.*
0019		>	B	*hmmh/ Berührt mit rechtem Zeigefinger das Podest*
0020				*von oben und fährt mit dem Zeigefinger von links*
0021				*nach rechts darüber. Nimmt die Hand zu sich*
0022	00:30			*zurück. das soll heute n Zirkuspodest für uns*
0023				*sein*
0024		>	Ayse	*Blickt zu B und danach zum Podest.*
0025		<	B	*Berührt mit linkem Zeigefinger die den Kindern*
0026				*zugewandte Seite des Podests und bewegt ihn am*
0027				*Podest entlang von links nach rechts. Nimmt dann*
0028				*die linke Hand zu sich zurück. Blickt zu Podest.*
0029				*guckt ma\ die sind auch ganz bunt gemustert hier*
0030				*neben*
0031		<	Kai	*Nimmt linke Hand aus dem T-Shirt und legt sie*
0032				*über seinen Mund. Steckt die Hand unter das*
0033				*T-Shirt zurück und blickt Richtung B.*
0034		>	Ayse	*Blickt zu B.*
0035		>	B	*Blickt zu den Kindern. im Zirkus da gibts ja*
0036				*immer Clowns und Tiere und*

(continued)

0037		<		Berührt das Podest von oben mehrmals abwechselnd
0038				und schnell mit Zeigefinger und Mittelfinger der
0039				rechten Hand von links nach rechts. Nimmt dann
0040				die Hand zu sich zurück.
0041				die machen dann da so Kunststücke auf dem Podest\
0042		<	Ayse	ich war schon mal n Clown\
0043			Kai	Nickt. Nimmt beide Hände aus dem T-Shirt. Beginnt
0044				mit beiden Händen am T-Shirt zu ziehen.
0045			B	ja/ Setzt sich seitlich nach rechts hin.
0046			Ayse	da warn auch Elefant\
0047			B	wow\
0048			Kai	und ich hab (des)#
0049			Ayse	#und Barbies\
0050			B	hmmh/
0051			Kai	und ' weißt du ich hab des schon mal . äm im im .
0052				beim Pettersson und Findus Spiel gesehn\ Blickt
0053				kurz zu Ayse und wieder zu B.
0054	01:00		B	hmmh\ mal kucken ob ich auch Tiere mitgebracht
0055		>		hab- Blickt nach rechts zu einem blauen Tuch und
0056				bewegt die rechte Hand unter das Tuch. hier
0057				unterm Tuch hab ich au noch glaub ich so n
0058				paar Tiere mitgebracht\ guck mal\ Zieht mit
0059				rechts einen Elefant unter dem Tuch hervor und
0060				bewegt ihn links neben das Podest. (Im weiteren
0061				Situationsverlauf werden alle Tiere stets mit
0062				Blick Richtung Kinder bzw. auf dem Podest in
0063				Laufrichtung nach rechts hingestellt.)
0064		>	Kai	Beobachtet Bs Hand.
0065		>	Ayse	Beobachtet Bs Hand. leise hää/+

(continued)

(continued)

0066			B	*Schiebt mit der rechten Hand das Podest etwa*
0067				*15 cm nach rechts und stellt es dabei schräg zur*
0068				*Kamera mit der linken Seite nach vorne. Stellt*
0069				*den Elefant etwa in die Teppichmitte.* kuckt doch
0070				mal da hin\ *Lässt den Elefant los und legt die*
0071				*rechte Hand rechts neben sich auf den Boden.* hm\
0072			Ayse	*Nimmt den Elefant mit der rechten Hand und bewegt*
0073				*ihn zu sich hoch auf Brusthöhe. Nimmt den Elefant*
0074				*mit beiden Händen in Brusthöhe. Blickt darauf.*
0075				*Lächelnd und leise* ein Elefant/
0076			Kai	*Beobachtet Ayse.*
0077			B	ja\ *Lächelt dabei und blickt zu Ayse.*
0078		>	Ayse	*Nimmt den Elefant mit links und bewegt beide*
0079				*Hände hinter ihren Rücken. Blickt dabei zu B.* ich
0080				hab den weg geklaut\
0081			Kai	*Blickt zu B und lächelt.*
0082		>	B	*Blickt zum Tuch und bewegt rechte Hand zum Tuch.*
0083				nein\ das darf mer aber nich klauen
0084		<		die wolln wir erst mal ankucken \ *Nimmt mit rechts*
0085				*einen Affen unter dem Tuch hervor und stellt*
0086				*ihn mit Blickrichtung zu Kai in die Teppichmitte.*
0087				*Blickt zu den Kindern.*
0088		<	Kai	*Beobachtet Bs rechte Hand.*
0089		<	Ayse	*Beobachtet Bs rechte Hand.*
0090			B	dann hab ich noch so eins\ *Lässt Affen los und*
0091				*bewegt rechte Hand zum Tuch.*
0092		<	Ayse	*Beugt sich nach vorne und nimmt den Affen mit*
0093				*der rechten Hand. Dreht die Handfläche mit Affen*

(continued)

0094				nach oben und bewegt die geöffnete Hand vor Kai.
0095				leise den kann doch (Kai habn)\ +Blickt zu Kai.
0096		<	Kai	Blickt zu B.
0097			B	Nimmt mit links den Affen aus Ayses Hand und
0098				stellt ihn in die Teppichmitte. und was ist das
0099				für ein Tier/ kennt ihr den auch/
0100		>	Kai	Blickt auf den Affen. ein Affe\
0101		>	Ayse	Blickt auf den Affen. ein Affe\ Bewegt die linke
0102				Hand mit Elefant vor sich. Nimmt mit rechts den
0103				Elefant und stellt ihn etwa5 cm schräg nach und hinten
0104				rechts versetzt zu dem Affen auf den Teppich.
0105		<		Lässt den Elefant los, berührt kurz ihn mit dem
0106				rechten Zeigefinger und lehnt sich nach hinten.
0107				Stützt sich mit beiden Händen hinter ihr auf den
0108				Boden. Bewegt dabei beide Knie etwa 20 cm nach
0109				oben. und das ist ein Elafant\
0110		<	B	aha\ Legt die linke Handfläche vor sich auf den
0111				Teppich und nimmt mit rechts einen weißen Tiger
0112				unter dem Tuch hervor. Blickt dabei kurz zum
0113				Tuch und dann wieder zur Teppichmitte. Stellt den
0114				Tiger etwa 2 cm schräg nach vorne und links
0115				versetzt neben den Affen auf den Teppich. und
0116	01:30			dann hab ich noch so einen\ Nimmt die Hand leer
0117				zu sich zurück.
0118			Ayse	eins\ Blickt zu B. äähm\ weiß ich nich\
0119				Blickt zur Teppichmitte. Bewegt mehrmals die
0120				Knie leicht nach oben und wieder nach unten.

(continued)

(continued)

0121				*Setzt sich dann gerade hin. Blickt zu den Tieren.*
0122			Kai	*weiß ich auch nicht\ Blickt dabei kurz zu B*
0123				*und dann zurück zur Teppichmitte.*
0124			B	*Legt die linke Hand auf ihr linkes Bein. Nimmt*
0125				*den Tiger mit der rechten Hand und dreht ihn*
0126				*mehrmals leicht von einer Seite auf die andere.*
0127				*Bewegt ihn dabei leicht nach links und rechts.*
0128				*Stellt ihn schließlich auf die vorherige Stelle*
0129				*und stützt sich dann mit rechts am Boden rechts*
0130				*neben sich ab.* weißt du nich- . mmm . also das
0131				ist jetzt noch n Baby\ Tiger\ is das\ *Blickt zu*
0132		>		*den Kindern. Berührt den Tiger kurz mit dem*
0133				*linken Zeigefinger und legt die Hand dann neben*
0134				*ihr linkes Knie.* n weißer Tiger\
0135		>	Ayse	*Beugt sich leicht nach vorne und stützt sich mit*
0136				*beiden Händen neben ihren Knien auf den Teppich.*
0137			Kai	*Blickt zu Ayse. lächelnd* hää/+
0138			B	hmmh/ *Blickt zu Ayse.*
0139			Kai	*Blickt auf die Tiere*
0140		<	Ayse	*Blickt zu B und dann auf die Tiere.* ein
0141				Babytiger\
0142		<	B	noch n Baby\ *Berührt mit linkem den Zeigefinger*
0143				*Affen.* und ich glaub der is auch noch n Baby\ *(Der*
0144				
0145				*Affe fällt auf den Rücken.) Nimmt mit der linken*
0146				*Hand den Elefant leicht hoch und stellt ihn etwa*
0147				*2 cm rechts und hinter den Affen.* und ich glaub

(continued)

0148				*der Elefant ist so klein- Stellt den Affen wieder*
0149		>		*auf. Schiebt den Elefant mit links direkt neben*
0150				*den Affen. Blickt zu Kai. Legt die linke Hand*
0151				*neben sich. der is bestimmt auch noch n Baby*
0152			Kai	*Lässt das T-Shirt mit beiden Händen los. Legt die*
0153				*linke Hand auf sein linkes Bein und die rechte*
0154		>		*Hand ans Gesicht. Deutet dann mit der rechten*
0155				*Hand Richtung Tiere und bewegt sie dabei von*
0156				*Tiger zu Elefant und wieder zurück. alle sind*
0157				*davon Babys- Blickt zu B. oder/*
0158			B	*ja*
0159		<	Ayse	*aber nis das hier\ Nimmt das Podest mit links*
0160				*bis etwa auf Brusthöhe hochkant hoch mit Boden*
0161				*Richtung B. Dreht den Podestboden zu sich und blickt*
0162				*darauf.*
0163		<	Kai	*Nimmt Hand zu sich zurück, blickt zum Podest.*
0164		>	B	*Nimmt mit links das Podest aus Ayses Hand. das*
0165				*ist das Podest*
0166		>	Ayse	*Blickt zu B und lehnt sich nach hinten. Stützt*
0167		<		*sich mit beiden Händen in ihrem Rücken ab. Blickt*
0168				*zur Teppichmitte.*
0169		<	B	*Nimmt das Podest mit rechts und stellt es an die*
0170				*vorherige Stelle zurück. Blickt in Teppichmitte.*
0171	02:00			*und das lassen wir mal so stehn\ okay/ Zieht die*
0172				*Hände auf dem Teppich zu sich. und jetzt im*
0173				*Zirkus- die Tiere wollen mal ein Kunststück*
0174				*machen auf dem Podest\ Nimmt den Tiger mit rechts*

(continued)

(continued)

0175				*und stellt ihn etwa2 cm nach rechts. Nimmt den*
0176				*Affen mit rechts und den Elefant mit links.*
0177				*Stellt sie etwa2 cm auseinander. Legt die linke*
0178				*Hand vor sich und zieht mit rechts das vordere*
0179				*Ende des Podests etwa 5 cm zu sich nahezu*
0180				*parallel vor die Tiere. Nimmt den Elefant und*
0181				*stellt ihn etwa 4 cm neben das linke Podestende.*
0182				*Nimmt mit rechts den Tiger und stellt ich etwa*
0183				*2 cm vor den Elefant. Stellt mit rechts den*
0184				*Elefant etwa 3 cm nach rechts und damit etwa 3 cm*
0185				*hinter das Podest. Legt die rechte Hand vor sich*
0186				*und blickt zu Ayse.* und zwar sollen die mal in
0187				einer Reihe da rüber laufen\ und da rüber
0188				balancieren\ wollt ihr mir mal helfen/
0189			Ayse	*Blickt zu B. Lächelnd* ja\+
0190			B	*Blickt zur Teppichmitte.* na stellt die mal aufs
0191				Podest dass die in einer Reihe rüber laufen
0192				können\ *Streicht sich mit links die Haare hinter*
0193		<		*ihr linkes Ohr und legt die Hand aufs linke Bein.*
0194		<	Kai	*Beugt sich nach vorne, stützt sich mit links auf*
0195				*den Teppich und nimmt mit rechts den Tiger. Zieht*
0196				*ihn etwa 5 cm zu sich.*
0197		<	Ayse	*Beugt sich nach vorne, stützt sich mit links auf*
0198				*den Teppich und bewegt rechte Hand zum Elefant.*
0199				*Schiebt den Elefant direkt neben das Podest und*

(continued)

0200				nimmt die Hand zu sich zurück.
0201			B	hm/ ein Tier muss ganz- Berührt mit linkem
0202				Zeigefinger das linke Podestende von oben. kann
0203				mer hier oben drauf stellen\ Tippt mit dem linken
0204				Zeigefinger etwa 2 cm weiter rechts und dann noch
0205				einmal etwa 4 cm weiter rechts auf das Podest.
0206				Legt die Hand aufs Bein. kuckt mal\ auf das weiße\
0207		<		ein Tier muss ganz vorne stehen/
0208				
0209		<	Ayse	Nimmt mit rechts den Elefant und stellt ihn
0210				auf das linke Podestende. Schiebt den Elefant
0211				etwa 2 cm nach rechts und wieder zurück.
0212		<	Kai	Bewegt den Tiger Richtung linkes Podestende.
0213				Stoppt mit der Bewegung über Ayses Hand.
0214		>	Ayse	Lässt den Elefant los und bewegt ihre Hand
0215				zu sich zurück.
0216		>	Kai	Stellt den Tiger auf das rechte Podestende und
0217				schiebt ihn dann auf dem Podest etwa 3 cm nach
0218				links. Lässt den Tiger los und legt die rechte
0219				Hand auf den Affen.
0220	02:30		Ayse	Nimmt den Tiger mit rechts und stellt ihn ganz
0221				rechts auf das Podest. Leise (muss hier) hin\+
0222		<		Blickt kurz zu B und wieder zum Podest.
0223			B	aha\ super\ und dann/ kommt/
0224			Kai	Stellt den Affen mit rechts direkt links neben
0225		<		den Tiger auf das Podest. der Affe/ Nimmt den
0226		>		Elefant und schiebt ihn etwa 4 cm auf dem Podest

(continued)

(continued)

0227				*entlang nach links.*
0228			Ayse	*Nimmt den Affen mit rechts und stellt ihn etwa*
0229		>		*auf die Mitte des Podests.*
0230			B	*und dann der Elefant*
0231			Kai	*Schiebt den Elefant etwa 3 cm nach rechts.*
0232			Ayse	*Lässt den Affen los und nimmt den Elefant.*
0233			Kai	*Lässt den Elefant los und bewegt die rechte Hand*
0234				*zu sich zurück. Setzt sich gerade hin.*
0235			Ayse	*Schiebt den Elefant mit rechts zum linken Ende*
0236				*des Podests zurück. so\\ Setzt sich gerade hin*
0237				*und stützt sich mit beiden Händen neben ihren*
0238				*Füßen auf den Boden.*
0239			B	*super\\ welches Tier steht denn jetzt ganz vorne*
0240			Ayse	*Deutet kurz mit links auf den Tiger. äm der*
0241				*Bewegt die Füße etwas nach außen und legt die*
0242				*Hände auf ihre Fußsohlen.*
0243			Kai	*Verschränkt die Arme vor der Brust. Reibt sich*
0244				*dann mit linkom Zoigofingor untor der Nase.*
0245			B	*aha\\ und wo ist hinten/*
0246		<	Ayse	*Beugt sich nach vorne, stützt sich mit links auf*
0247				*dem Boden ab und bewegt die rechte Hand bis etwa*
0248				*1 cm über den Elefant.*
0249		<	Kai	*da\\ Legt die linke Hand aufs Bein und deutet mit*
0250				*rechtem Zeigefinger auf den Elefant.*
0251		>		*Legt die rechte Hand auf sein rechtes Bein.*
0252		>	Ayse	*Setzt sich gerade hin und legt die Hände auf*
0253				*ihre Fußsohlen.*
0254		>	B	*da\\ aja\\ okay*

(continued)

0255				*Bewegt den rechten Zeigefinger von links nach*
0256				*rechts über das Podest und legt die Hand auf*
0257				*ihr rechtes Bein zurück.* ja die laufen ja auch
0258				nach da\ nö/
0259		>	Ayse	*Beugt den Oberkörper nach links und stützt sich*
0260				*mit links neben ihrem linken Bein auf den Boden.*
0261		>	B	*Nimmt mit rechts den Elefant und bewegt ihn mit*
0262				*kleinen hüpfenden Bewegungen nach rechts neben*
0263				*den Affen. Nimmt dann den Affen und bewegt ihn*
0264				*mit kleinen hüpfenden Bewegungen nach rechts*
0265				*neben den Tiger.* so na laufen die da alle drei
0266				so drüber\ nödülödüdün/ und dann können die am
0267				Ende/ *Nimmt den Tiger und stellt ihn etwa 3 cm*
0268				*rechts neben das Podest.* runter hüpfen- *Nimmt*
0269				*den Affen und bewegt ihn mit kleinen hüpfenden*
0270				*Bewegungen nach rechts über das Podest. Stellt*
0271				
0272				*den Affen vor den Tiger direkt rechts neben das*
0273				*Podest. (Affe fällt auf den Rücken.) Nimmt den*
0274				*Elefant und bewegt ihn mit etwas größeren*
0275				*hüpfenden Bewegungen nach rechts über das Podest*
0276	03:00			*und stellt ihn dann vor den Affen.* so\ und dann
0277				kommt noch der dicke Elefant/ wüm\ *Legt die*
0278				*rechte Hand auf das Tuch.* hm\
0279			Ayse	*Zeigt mit dem rechten Zeigefinger auf den Affen.*
0280				der Affe ist runter gefallen\ *Stellt den Affen*
0281				*mit rechts auf.*

(continued)

(continued)

0282			B	*Zeigt mit links kurz auf das Podest.* findet ihr
0283				noch ne andere Reihenfolge wie die drüber laufen
0284				können/
0285			Ayse	*Blickt kurz zu B, bewegt die rechte Hand dabei*
0286				*etwas nach links und lässt in der Bewegung den*
0287				*Affen los. (Der Affe fällt erneut um.) Setzt sich*
0288				*gerade hin, legt die Hände auf die Fußsohlen*
0289				*und blickt zur Kamera.* ja/ *Blickt Richtung*
0290				*Podest. Lehnt ihren Oberkörper nach hinten.*
0291			Kai	*Reibt mit linkem Zeigefinger unter der Nase.*
0292			B	*Nimmt alle drei Tiere mit rechts und legt sie*
0293				*links neben das Podest.* da müssen die sich wieder
0294				hinten anstellen\ glaub ich/ *Bewegt rechte*
0295				*Hand zu sich zurück.* könnt ihr die noch mal drauf
0296				stellen/ *Nimmt die Tiere kurz mit rechts und*
0297		<		*schiebt sie etwa 2 cm Richtung Kai. Lässt die*
0298				*Tiere los und legt die rechte Hand neben sich*
0299				*auf den Boden.* welches Tier soll denn jetzt
0300				ganz vorne stehen\
0301		<	Kai	*Beugt sich nach vorne und stützt sich mit links*
0302				*auf dem Boden ab. Legt den rechten Unterarm auf*
0303				*den Boden, nimmt mit rechts den Affen, bewegt*
0304				*die rechte Hand etwa 10 cm zu sich und stützt*
0305				*sich mit rechts auf den Boden.*
0306		<	Ayse	*Beugt sich nach vorne, stützt sich mit links vor*

(continued)

0307				dem Podest auf den Boden und nimmt mit rechts
0308				den Elefant. Bewegt den Elefant nach rechts
0309				über das Podest (die hinteren Elefantenbeine
0310				bleiben kurz am linken Podestende hängen) bis
0311				fast zum rechten Podestende. leise eins\
0312			B	hm/
0313			Ayse	Schiebt den Elefant mit rechts Stufenweise zur
0314				Mitte und dann zum linken Ende des Podests.
0315				Stellt ihn dann auf das rechte Podestende.
0316				die Elafant/
0317			B	ja\
0318		>	Kai	Bewegt die rechte Hand mit dem Affen etwa 10 cm
0319				nach oben.
0320		>	Ayse	Nimmt den Tiger mit rechts und stellt ihn auf
0321				das linke Podestende. und die Tiger hier/
0322		<		Nimmt den Affen von oben mit rechts und bewegt
0323				ihn Richtung Podestmitte.
0324		<	Kai	Bewegt die rechte Hand mit dem Affen hinter das
0325				Podest etwa in Podestmitte. Lässt den Affen los und legt
0326				den Unterarm auf den Boden. Bewegt dann die Hand
0327				zu sich zurück. Setzt sich gerade hin.
0328			Ayse	Stützt sich mit der rechten Hand (und dem Affen)
0329				auf den Boden hinter dem Podest in Podestmitte.
0330				und die Affen- Bewegt die rechte Hand etwa 10 cm
0331				nach oben ähm\ Stellt den Affen kurz auf die
0332				Podestmitte und dann hinter das Podest. in die
0333				mitten\ blickt zu B.

(continued)

(continued)

0334			B	*no\ aber alle Tiere sollen übers Podest laufen*
0335	03:30		Ayse	*Blickt zum Podest und stellt mit rechts den Affen*
0336				*etwa auf die Podestmitte (näher am Tiger).*
0337			B	*genau*
0338			Ayse	*Lässt den Affen los, nimmt die Hand zu sich*
0339				*zurück und setzt sich gerade hin.*
0340		>		*Beugt den Oberkörper nach links, nimmt mit rechts*
0341				*den Elefanten und bewegt in mit ausgestrecktem*
0342				*Arm nach oben über den Kopf. kchch\ Stellt*
0343				*den Elefant vor das rechte Ende des Podests.*
0344		>	B	*super\ dann steht ja der Affe schon wieder in*
0345				*der Mitte\ hm/*
0346			Ayse	*Blickt zu B. abgefrungen*
0347			B	*hm*
0348			Ayse	*Blickt zum Podest und nimmt den Affen mit rechts.*
0349				*Bewegt ihn mit drei kleinen hüpfenden Bewegungen*
0350				*zum rechten Podestende und hebt ihn mit fast*
0351				*ausgestrecktem Arm etwa auf Kopfhöhe. chchu*
0352				*Legt den Affen etwa 2 cm rechts neben das*
0353				*vordere rechte Podestende auf den Rücken.*
0354			Kai	*Nimmt den Tiger mit links, bewegt ihn vom Podest*
0355				*nach vorne herunter und dann mit hüpfenden*
0356				*Bewegungen vom linken zum rechten Podestende.*
0357		<		*Hebt ihn mit ausgestrecktem Arm über den Kopf und*
0358				*stellt ihn etwa 4 cm vor das rechte Podestende.*
0359				*(Berührt dabei den Elefant. Dieser fällt um.)*

(continued)

0360		<	Ayse	*Lässt den Affen los und setzt sich gerade hin.*
0361		<	B	*und finden wir noch ne Reihenfolge wie die*
0362				*drüber laufen können/*
0363			Ayse	*Nimmt die Hände hinter ihren Rücken. Bewegt den*
0364				*Oberkörper dabei schnell etwa 15 cm nach oben.*
0365				*Blickt zu B und lächelt. ja\ Bewegt den*
0366				*Oberkörper zurück und setzt sich dabei auf ihre*
0367				*Hände. Blickt zu den Tieren.*
0368			Kai	*ja\ Nimmt den Elefant mit links und stellt ihn*
0369				*etwa 2 cm schräg rechts vor das Podest. Nimmt den*
0370				*Affen mit links zu sich hoch.*
0371			B	*ich glaub ein Tier war noch gar nicht vorne*
0372				*Setzt sich anders auf ihre Beine.*
0373			Kai	*Nimmt den Affen mit der linken in die rechte Hand*
0374				*und stellt ihn auf das rechte Podestende.*
0375				*der und gan- der affe ganz vorne/ Nimmt den*
0376				*Elefant mit links zu sich hoch und dann mit*
0377				*rechts. Stellt ihn auf die Podestmitte. der af-*
0378				*und der Elefant in der Mitte- Nimmt mit rechts*
0379				*den Tiger und stellt ihn auf das linke*
0380				*Podestende. leiser werdend und der Löwe muss*
0381				*hier hin\+ Lehnt sich nach hinten und legt die*

The Train

001	T03 143 180 453 712 0002 00 111005 2 20202 3 0451-0638			
002	04:51		B	kannst du mal den roten so hinsetzen
				dass er innerhalb des Schienennetzes
003				steht/ hält Ayse mit der rechten Hand
004				das rote Männchen entgegen.
005		<	Ayse	hält die linke Hand vor den Mund, schaut
006				B mit großen Augen an, schüttelt den Kopf
007		<	Elias	setzt sich auf allen vieren vor B ich
008				hab keinen Norbert hat den blauen\
009		<	Norbert	hält das blaue Männchen in der rechten
010				Hand.
011			B	schaut für den Bruchteil einer Sekunde zu
012				Elias , schüttelt leicht den Kopf.
013				schaut zu Ayse wieso nicht/ zieht die
014				Hand zu sich zurück.
015			Elias	greift nach dem roten Männchen in B's
016		>		Hand unverständlich
017		>	Barbara	Ich kanns machen\
018		<	B	schaut zu Barbara du kannst das machen/
019				gibt Barbara das rote Männchen.
020		<	Elias	streckt die recht Hand aus und legt sie
021				auf B's Knie darf ich den gelben darf
022				ich jetzt den gelben/
023			B	wart mal ganz kurz\
024			Barbara	stellt das rote Männchen vor sich auf
025				die Schienen.
026			B	ok ahe\ hält ein gelbes Männchen vor
027				Sich.
028			Ayse	Nimmt das rote Männchen und stellt es vor
029				sich auf die Schienen\
030			B	Steht der jetzt in diesen Schienen
031				drin/
032			Elias	Ja\
033			Barbara	Ja\
034	5:21		B	nickt ahaj und können wir den auch mal
035				so stellen dass er- wedelt mit dem
036				rechten Zeigefinger hin und her in

037		>		diesem Kreis drin steht/
038		>	Elias	macht die an der Haltestelle stehende
039				Eisenbahn an und fährt mit ihr und lautem
040				Getöse über die Schienen.
041			Ayse	ok\ legt das rote Männchen um.
042		<	B	also liegt der ja jetzt auf den Schienen\
043		<	Elias	fährt mit der Eisenbahn das Männchen um
044				sodass es nun im Schienenkreis liegt.
045		>	B	ich meinte so dass der-
046			Ayse	stellt das rote Männchen wieder auf die
047				Schienen.
048			Barbara	ich weiß (sie)\ fasst das rote Männchen an
049				zieht die Hand wieder zurück, schaut zu
050				B.
051			Elias	stellt die Eisenbahn an der Haltestelle
052				ohne sie auszumachen ab das ist jetzt
053				mein Männchen unverständlich\
054			B	in diesem Kreis drin steht (so dass) die
055				Eisenbahn immer um ihn herum fährt\
056		<	Elias	darf ich jetzt mal den gelben/ streckt
057				seine linke Hand Richtung B aus.
058		<	Ayse	stellt das rote Männchen in die Mitte
059				des Schienenkreises.
060			B	Genauso\ schaut zu Elias stell du
061				doch mal den gelben jetzt so hin dass er
062				außerhalb des Kreises steht gibt Elias
063				das gelbe Männchen, welches größer ist
064				
065				als das rote.
066	5:51		Elias	Oha\ stellt das gelbe Männchen an die
067				selbe Stelle, auf der vorher das rote
068				Männchen stand.

References

Acar Bayraktar, E., Hümmer, A.-M., Huth, M., Münz, M., & Reimann, M. (2011). Forschungsmethodischer Rahmen der Projekte erStMaL und MaKreKi. In B. Brandt, R. Vogel, & G. Krummheuer (Eds.), *Mathematikdidaktische Forschung am "Center for Individual Development and Adaptive Education". Grundlagen und erste Ergebnisse der Projekte erSt-MaL und MaKreKi (Bd. 1)* (pp. 11–24). Münster: Waxmann.

Bauersfeld, H. (1995). "Language games" in the mathematics classroom: Their function and their effects. In P. Cobb & H. Bauersfeld (Eds.), *The emergence of mathematical meaning. Interaction in classroom cultures* (pp. 271–291). Hillsdale, NJ: Lawrence Erlbaum.

Bowlby, J. (1969). *Attachment. Attachment and loss* (Vol. 1). New York: Basic Books.

Brandt, B. (2004). *Kinder als Lernende. Partizipationsspielräume und -profile im Klassenzimmer.* Frankfurt a.m. Main: Peter Lang.

Bruner, J. (1983). *Child's talk. Learning to use language.* Oxford: Oxford University Press.

Bruner, J. (1996). *The culture of education.* Cambridge, MA: Harvard University Press.

Chen, J. -Q., & McCray, J. (2014): Intentional teaching. Integrating the processes of instruction and construction to promote quality early mathematics education. In U. Kortenkamp, B. Brandt, C. Benz, G. Krummheuer, S. Ladel, & R. Vogel (Eds.), *Early mathematics learning. Selected papers of the POEM 2012 conference* (pp. 257–274). New York: Springer.

Höck, B. (2015). *Ko-konstruktive Problemlöseprozesse im Mathematikunterricht der Grundschule. Eine mikrosoziologische Studie zum Zusammenspiel lernpartnerschaftlicher Ko-Konstruktion und individueller Partizipation.* Münster: Waxmann.

Krummheuer, G. (1999). *The narrative character of argumentative mathematics classroom interaction in primary education.* Paper presented at the European research in mathematics education I, Osnabrück, Germany. http://www.fmd.uni-osnabrueck.de/ebooks/erme/cerme1-proceedings/cerme1proceedings.html.

Krummheuer, G. (2000). Studies of argumentation in primary mathematics education. *Zentralblatt für Didaktik der Mathematik, 32*(5), 155–161.

Krummheuer, G. (2009). Inscription, narration and diagrammatically based argumentation. The narrative accounting practices in the primary school mathematics lesson. In W.-M. Roth (Ed.), *Mathematical representation at the interface of the body and culture* (pp. 219–243). Charlotte, NC: Information Age.

Krummheuer, G. (2011a). Die empirisch begründete Herleitung des Begriffs der "Interaktionalen Nische mathematischer Denkentwicklung" (NMD). In B. Brandt, R. Vogel, & G. Krummheuer (Eds.), *Mathematikdidaktische Forschung am "Center for Individual Development and Adaptive Education". Grundlagen und erste Ergebnisse der Projekte erStMaL und MaKreKi (Bd. 1)* (pp. 25–90). Münster: Waxmann.

Krummheuer, G. (2011b). Representation of the notion "learning-as-participation" in everyday situations of mathematics classes. *Zentralblatt für Didaktik der Mathematik (ZDM), 43*(1/2), 81–90.

Krummheuer, G. (2012). The "non-canonical" Solution and the "Improvisation" as Conditions for early Years Mathematics Learning Processes: The Concept of the "interactional Niche in the Development of mathematical Thinking" (NMT). *Journal für Mathematik-Didaktik, 33*(2), 317–338.

Krummheuer, G. (2013a). The relationship between diagrammatic argumentation and narrative argumentation in the context of the development of mathematical thinking in the early years. *Educational Studies in Mathematics, 84*(2), 249–265. doi:10.1007/s10649-10013-19471-10649.

Krummheuer, G. (2013b). Research on mathematics learning at the "Center of Individual Development and Adaptive Education" (IDeA)—An introduction. *Educational Studies in Mathematics, 84*(2), 177–181. doi:10.1007/s10649-10013-19502-10646.

Krummheuer, G. (2014a). The relationship between cultural expectation and the local realization of a mathematics learning environment. In U. Kortenkamp, B. Brandt, C. Benz, G. Krummheuer, S. Ladel, & R. Vogel (Eds.), *Early mathematics learning. Selected papers of the POEM 2012 conference* (pp. 71–84). New York: Springer.

Krummheuer, G. (2014b). Interactionist and ethnomethodological approaches in mathematics education. In S. Lerman (Ed.), *Encyclopedia of mathematics education* (pp. 313–316). Dodrecht: Springer.

Krummheuer, G., Leuzinger-Bohleber, M., Müller-Kirchof, M., Münz, M., & Vogel, R. (2013). Explaining the mathematical creativity of a young boy: An interdisciplinary venture between mathematics education and psychoanalysis. *Educational Studies in Mathematics, 84*(2), 183–199. doi:10.1007/s10649-013-9505-3.

Lave, W., & Wenger, E. (1991). *Situated learning. Legitimate peripheral participation*. Cambridge: Cambridge University Press.

Presmeg, N. (2014). The dance of instruction with construction in mathematics education. In U. Kortenkamp, B. Brandt, C. Benz, G. Krummheuer, S. Ladel, & R. Vogel (Eds.), *Early mathematics learning. Selected papers of the POEM 2012 conference* (pp. 9–17). New York: Springer.

Sacks, H. (1998). *Lectures on conversation*. Malden, MA: Blackwell. 3. Auflage.

Schütte, M., & Krummheuer, G. (2013). Changing mathematical content-related domains—A genuine mathematical action? In A. M. Lindmeyer & A. Heinze (Eds.), *Proceedings of the 37th conference of the International Group for the Psychology of Mathematics Education - Mathematics learning across the life span* (Vol. 4, pp. 185–192). Kiel: Institut für die Pädagogik der Naturwissenschaften (IPN).

Sfard, A. (2008). *Thinking as communicating. Human development, the growth of discourses, and mathematizing*. Cambridge: Cambridge University Press.

Super, C. M., & Harkness, S. (1986). The developmental niche: A conceptualization at the interface of child and culture. *International Journal of Behavioral Development, 9*, 545–569.

Vogel, R. (2013). Mathematical situations of play and exploration. *Educational Studies in Mathematics, 84*(2), 209–225.

Vogel, R. (2014). Mathematical situations of play and exploration as an empirical research instrument. In U. Kortenkamp, C. Benz, B. Brandt, G. Krummheuer, S. Ladel, & R. Vogel (Eds.), *Early mathematics learning—Selected papers of the POEM 2012 conference* (pp. 223–236). New York: Springer.

Vogel, R., & Huth, M. (2010). "…und der Elephant in die Mitte" - Rekonstruktion mathematischer Konzepte von Kindern in Gesprächssituationen. In B. Brandt, M. Fetzer, & M. Schütte (Eds.), *Auf den Spuren Interpretativer Unterrichtsforschung in der Mathematikdidaktik. Götz Krummheuer zum 60. Geburtstag* (pp. 177–207). Münster: Waxmann.

Webster, N. (1983). Webster's new twentieth century dictionary. Unabridged (2nd ed.). In J. L. McKechnie (Ed.). New York: Simon and Shuster.

"Similar and Equal…": Mathematically Creative Reflections About Solids of Children with Different Attachment Patterns

Melanie Beck

Abstract This chapter deals with mathematically creative processes in early childhood. The concept of the interactional niche in the development of mathematical creativity is introduced, which combines interactionistic theories of socio-constructivism, sociocultural theories, and a psychoanalytically based attachment theory in order to describe mathematically creative processes of children during early childhood development. Data are collected in the interdisciplinary project Mathematical Creativity of Children. Two empirical cases of children and their mathematically creative processes, from engaging in a task in a cooperative mathematical situation, are presented.

Introduction

Definitions of mathematical creativity differ on several aspects. On the one hand, creativity has been referred to as the individual ability of a person, such as in divergent thinking (Guilford 1967); the abilities to produce fluent, flexible, novel, and elaborated solutions to a given problem (Torrance 1974); or the ability to produce unexpected and original work, that is, adaptive (Sternberg and Lubart 2000). On the other hand, creativity is seen as embedded in a social process (e.g., Csikszentmihalyi 1997; Sriraman 2004; Vygotsky 2004), in which creativity is not solely located in a person's cognition but is also accomplished in the social interaction among members of the society.

My research interest is the examination of mathematical creativity from the specific perspective of early childhood development. In this contribution, I focus on the social and sociocultural approaches to creativity. The first section presents a theoretical approach that deals with the question in which forms of social interactions these early mathematically creative interactions of children are evoked and supported. Afterward, based on these theoretical assumptions, I define what I understand as a mathematical creative process in early childhood. Then I clarify which cultural

M. Beck (✉)
University of Frankfurt/IDeA Center, Frankfurt, Germany
e-mail: mbeck@math.uni-frankfurt.de

T. Meaney et al. (eds.), *Mathematics Education in the Early Years*,
DOI 10.1007/978-3-319-23935-4_11

and sociocultural dimensions should be considered in a theory of mathematical creativity in early childhood and in what way. After that, an overview of data collection and methods is given, followed by two empirical cases. The chapter finishes with a summary and some implications.

Theoretical Approach

In many fundamental works on children's creativity, play is regarded as a social situation in which creative actions arise and are fostered (e.g., Bateson and Martin 2013; Vygotsky 2004). Also from a psychoanalytical perspective, play is regarded as a location where children's creativity is formed (Winnicott 2012). According to Winnicott, play is neither part of the personal inner reality nor part of the actual external reality but of a third dimension, which he refers to as "potential space" (Winnicott 2012, p. 144). As soon as a child experiences his or her mother no longer being part of his or her own, a "playground" (Winnicott 2012, p. 64) emerges, which the child can use for creative activities. This complex process depends highly on a supportive mother, who is willing to participate and to reciprocate. Winnicott calls her the "good-enough mother" (p. 109), who is sensitive and reacts appropriately to her child's needs in opposition to the "not good-enough mother" (p. 109).

In developing a theory of mathematical creativity in the early years, one has also to consider the development of mathematical thinking. From a sociocultural perspective, children's play is considered as location for the development of mathematical thinking (e.g., Carruthers and Worthington 2011; van Oers 2002) and for this reason, the development of mathematical creativity, too. Mathematics comes into play through articulation by more knowledgeable people, "their companions in the cultural community" (van Oers 2002, p. 30). If play involves other more knowledgeable persons like children or adults, opportunities for scaffolding (Bruner 1986) or guided participation (Rogoff 2003) emerge. Lave and Wenger (1991) describe how learners become involved in a "community of practice" (p. 30), which embodies valued beliefs and behaviors. As the beginners or newcomers move from the periphery of a community to its center, they become more active and engaged within the culture and hence eventually take over the role of an expert or old-timer. They call this process "legitimate peripheral participation" to "full participation" (Lave and Wenger 1991, p. 34). According to "legitimate peripheral participation" (p. 29), newcomers become members of a community initially by participating in simple and low-risk tasks that are nonetheless productive and necessary and further the goals of the community. Through peripheral activities, novices become acquainted with the tasks, vocabulary, and organizing principles of the community.

Besides the aspect of children's play with competent partners, children are dealing with mathematics in their free self-initiated play as well as in play situation with peers (e.g., Carruthers and Worthington 2011).

Some examples of empirical attachment studies suggest that the development of (mathematical) creativity in children may also be influenced by their attachment pattern (e.g., Creasy and Jarvis 2003). The theory of attachment suggests that children come into the world biologically preprogrammed to form attachments with others (Bowlby 1969). The neonates develop special relationships and attachment figures between them and their parents, which influence the children's emotional as well as their cognitive and social development. In the first years of life, the children build up their own "internal world":

> a child is busy constructing working models of how the physical world may be expected to behave, how his mother and other significant persons may be expected to behave, how he himself may be expected to behave, and how each interacts with all the others. (Bowlby 1969, p. 354)

These internal "working models" (Bowlby 1969, p. 354) contain the early individual bonding experiences as well as the expectations, which a child has toward human relationships. An individual's primary attachment strategy can also include activation of mental representations, and "mental representations of attachment figures can become symbolic" (Mikulincer and Shaver 2007, p. 13). Thus, the introjected traits of security-providing figures become self-soothing in times of distress and vice versa for anxiety-producing attachment figures. When felt security is achieved, "the attachment system is deactivated and the individual can calmly and coherently return to nonattachment activities" (Mikulincer and Shaver 2007, p. 14). This triggers the exploratory system, which allows the individual to engage in nonattachment activities, which include new intellectual and cognitive pursuits, an increase of play, or more frequent social engagements. Simply put, if a child feels secure, it can activate its exploration system and explore its surroundings. If it perceives a danger, the attachment system is activated. The child interrupts its exploratory behavior and seeks safety by its parent according to the developed attachment pattern between them.

Mathematical Creative Processes in Early Childhood

In the interdisciplinary longitudinal study *Mathematical Creativity of Children* (MaKreKi), researchers from mathematics education and from psychoanalysis examine together the development of mathematical creativity in early years. In several publications of the project (e.g., Münz 2014), we show that "noncanonical" solving processes can be considered mathematically creative processes. Using the ideas of Krummheuer et al. (2013) and Sriraman (2004), mathematically creative processes are considered to include the following features:

Combinational Play

Under this aspect, the accomplishing of unusual combinations of insights and experiences and the sense of playfulness in the manipulation of procedures and its transfer to new areas are understood. With reference to Finke (1990), these activities are summarized with the "combinational play" (p. 3) of framing a mathematical situation.

Non-algorithmic Decision-Making

According to Ervynck (1991), mathematical creativity articulates itself when a unique and new way of problem solving emerges. For the age group of interest, processes of problem solving can be new, creative, and unique, although they may not be new for the mathematical community. Uniqueness can be seen as the "divergence from the canonical" (Bruner 1990, p. 19) way of solving a mathematical problem in early childhood, which adult mathematicians may not necessarily anticipate.

Adaptiveness

Sternberg and Lubart (2000) characterize creativity as the ability to create an unexpected and original result that is also adaptive to the given real situation. We have redefined this concept to our specific needs. In MaKreKi, adaptiveness describes the children's ability to accomplish (with sometimes unusual definitions of situations) new and adequate ways of answering questions, to solve problems, and additionally to convince their partners by their alternative framing of the situation. So a mathematical creative action has to be reasonable, which means there are arguments, why the chosen framing of the situation leads to a mathematically correct solution (e.g., Lithner 2008). Additionally, these arguments have to be somehow mathematically founded (e.g., Lithner 2008; Münz 2014).

Social and Sociocultural Dimensions

As already mentioned, a theory of mathematical creativity has to consider both social and sociocultural dimensions of creativity, because creative behavior can be seen as intertwined in a complex person–situation interaction. I introduce a framework that stresses these interactive structures and in which the emerging creative process between children is regarded as an aspect of an interactive process of negotiation of meaning between the involved participants.

For my research focus, the "concept of the *interactional niche in the development of mathematical thinking (NMT)*" of Krummheuer (2012, p. 317) seems appropriate as a theoretical framework for several reasons. For a start, the NMT describes the situational aspect in the development of mathematical thinking as a process of negotiation of meaning between participants. So it allows the description of this negotiation of meaning in a creative process and is able to capture insights about this creative process. Additionally, another procedural aspect is described in the form of the cooperation of the participants and the individual scope of action of a child. The NMT is a further development of the original components of the "developmental niche" introduced by Super and Harkness (1986), who describe it:

> as framework of studying cultural regulation of the micro-environment of the child, and it attempts to describe the environment from the point of view of the child in order to understand processes of development and acquisition of culture. (p. 552).

Table 1 The NMT of Krummheuer (Krummheuer and Schütte 2014)

	Component: content	Component: cooperation	Component: education/ pedagogy
Aspect of allocation	Mathematical domains; bodies of tasks	Institutions of education; settings of cooperation	Scientific theories of mathematics education
Aspect of situation	Interactive negotiation of the theme	Leeway of participation	Folk theories of mathematics education
Aspect of action	Individual contributions to actions	Individual profile of participation	Competency theories

Table 2 The NMC

	Content	Cooperation	Interpersonal relations
Aspect of allocation	Mathematical domains; bodies of tasks; mathematical potentials	Institutions of education; settings of cooperation	Attachment patterns of the involved persons
Aspect of situation	Interactive negotiation of the theme	Leeway of participation	Situational emerging of attachment patterns
Aspect of individual's creative action	Individual mathematically creative actions	Individual profile of participation	Acceptance in form of shared meaning or interim

Additionally, Krummheuer added the interactive, local production of such processes, which includes besides the aspect of allocation (under which the provided mathematical activities of a group are summarized; see Table 1) the aspect of situation (situationally emerging accomplishment occurring in the process of meaning making) and the aspect of action (which covers the individual contributions to the actions as well as the individual participation profile of a child). This approach can be adapted to the theory of mathematical creativity in early childhood so that it can examine the mathematical creative process. To describe a mathematical creative process in early childhood, the fourth line of Krummheuer's NMT "aspect of action" (Krummheuer and Schütte 2014) is renamed the aspect of individual's creative action. It highlights the mathematical concepts used by the individual child, which can be regarded as combinational play, divergence from canonical and adaptive, as well as the individual profile of participation of the child (see Table 2).

Furthermore, Krummheuer divides these aspects into three components: content, cooperation, and education/pedagogy (Krummheuer and Schütte 2014). I briefly summarize each of these components and emphasize their relevance to a theory of mathematical creativity in early childhood:

Content In the MaKreKi project at the level of allocation, mathematical topics are usually designed as mathematical situations of play and exploration (Vogel 2013), regarding the children's assumed mathematical competencies. They offer opportunities for children to demonstrate their mathematical creative potential. An assisting adult, who can be seen as a more knowledgeable person, presents the situations. At the situational level, this presentation generates processes of negotiation. The presentation of the mathematical situations of play and exploration and the processes of negotiations may lead to individual mathematically creative actions.

Cooperation Children participate in culturally specific social settings, which are variously structured as in peer interaction or small group interaction guided by a more knowledgeable person. These social settings do not succeed immediately. They need to be accomplished in the joint interaction. Depending on each event, a different "leeway of participation" (Brandt 2004, p. 47) of the children will arise. By embellishing these possibilities of participation, every child has an individual "profile of participation," which can be relatively stable over a given time (see Brandt 2004, p. 47).

For this chapter, I do not focus on the component education/pedagogy, which considers the influence of scientific as well as folk theories of mathematics education on the development of mathematical thinking (Krummheuer and Schütte 2014).

Psychoanalytical Dimensions in Mathematically Creative Processes

To integrate psychoanalytical insights about creativity in early childhood, I add a third column (see Table 2), the component interpersonal relations, that derives from Winnicott's concepts of the "good-enough mother" (Winnicott 2012, p. 109) as a requirement for the "potential space" (p. 64) as the origin of creativity in human life.

Interpersonal relations According to Winnicott, the initiation of playing is associated with the life experience of the baby who has come to trust the mother figure (Winnicott 2012), which develops when she reacts sensitively and warmly to the child's needs. In the first years of life, the child develops an inner "working model" through child–parent interactions (Bowlby 1969, p. 354). This model contains the early individual bonding experiences as well as the expectations, which a child has toward human relationships, derived from these experiences. They induce the child to interpret the behavior of the caregiver and to predict his or her behavior in certain situations. After the first year of life, this "working model" becomes more and more stable and turns into a so-called attachment pattern (p. 335). At this time, the child has developed mental representations in which the caregiver is seen as an independent, intentionally dealing object. The inner "working model" of the child encompasses their own motives and experiences in attachment relationships. The child does not yet regard caregivers as having their own plans, motives, experiences, and emotions, too, which may differ from the child's own.

The ability to recognize another's point of view develops from the age of three and leads to the possibility of child and caregiver developing a relationship, which Bowlby (1969) terms a "partnership" (p. 267). The beginning partnership appears when children are able to integrate the plans of the caregiver into their own plans and try to influence them. At the age of four, children display another behavior, which indicates the beginning of a new attachment relationship, called the "goal-corrected partnership" (e.g., Bowlby 1969; Benson and Haith 2009, p. 34). According to Bowlby (1969) and Marvin and Britner (1999), the basis of this new partnership is a cognitive and a communicative ability: the ability of the child to gain insights about the goals and emotions of the caregiver and to coordinate these on a representational level (perspective taking) and in case of conflict between plans of child and caregiver to negotiate a common plan with the caregiver. The children are now able to see two or more representations as components of a higher plan. In the relationship with the caregiver, they can represent their own plans as well as the plans of the caregivers simultaneously but separated, which allows them to compare both perspectives to see if they coincide or if they have to develop a common perspective.

The quality of the child–caregiver relationship in sense of the attachment patterns can be evaluated (Ainsworth et al. 1978). A rough distinction can be made between two types: the secure and the insecure attachment pattern. Children with a secure attachment pattern have, thanks to their sensitive mothers, a chance to build up secure relationships to their mothers in which the whole spectrum, of human feelings in the sense of communication with each other, can be perceived, experienced, and expressed. Children with insecure attachment patterns experience a mother who shows no intense affects and behaves in a distanced, controlled manner. Alternatively, a mother could sometimes react appropriately while at other times is rejecting and overprotective, thus being inconsistent in a way that is unpredictable for the child. Because of the antagonism between attachment and exploratory system mentioned previously, it seems plausible that children with a secure attachment pattern may develop great pleasure in exploration and have more chances to act creatively. Empirical observations of infant's exploratory behavior as well as children's play behavior highlight that children with a secure attachment pattern show more exploratory behavior and more positive affect and are more cooperative in their play than children with an insecure pattern (e.g., Creasy and Jarvis 2003). The quality of the play seems to depend on the attachment pattern, too. Significant differences between children with secure and insecure attachment patterns in play situations are also described in Grossmann's (1984) study. Securely attached children are more often initiators of the common play, and they seem to be rather extroverted instead of children with an insecure attachment pattern, who wait for instructions and seem to be rather introverted (Grossmann 1984). Crowell and Feldman (1988) observed that parents of children with insecure attachment pattern focus on basic task completion rather than on learning processes.

On the level of allocation of an NMC (see Table 2), this attachment pattern can be regarded as stable (Bowlby 1969); nevertheless, Bowlby suggests that this attachment pattern may change as the child begins to interact with other attachment figures, such as siblings, peers, and teachers. On the level of situation, a child with an insecure attachment pattern may meet other children or adults who show sensitivity to their needs in the sense of being a "good-enough" partner (Winnicott 2012,

p. 109) who enhances their potential for cooperation during the interactive process and creativity in the mathematical activities. The reverse conclusion is also conceivable. Regarding individual creative action, the "good-enough" partner may accept the contributions of the child to the mathematical process and understand its perspective in framing the mathematical situation. Therefore, the child also has to realize and accept the different perspective of their partner in the concrete situation and to use communicative strategies in the negotiation of a common perspective. The acceptance has not only to be understood as a shared meaning, it can also be seen as an interim, in which the participants have to match their framings of the situation and conclude that there is more than one possibility for that framing. If this is not the case, the creative process somehow fails in the concrete situation. Table 2 summarizes the additions to the concept of NMT, which I term as the "interactional niche of the development of mathematical creativity."

Empirical Approach and Methodology

The sample of participants for MaKreKi is based on the original samples from two larger projects that are conducted in the "Center for Individual Development and Adaptive Education of Children at Risk" (IDeA) in Frankfurt, Germany. One project is a study of the evaluation of two prevention programs with high-risk children in day care centers (EVA). It examines approximately 280 children. The second project is a study of early steps in mathematics learning (erStMaL). This project includes approximately 150 children. We asked the preschool teachers in the two original samples whether they knew children in their groups who show divergent and unusually sophisticated strategies while working on mathematical tasks. From their input, we identified 37 children, who seemed to work creatively on mathematical problems.

For the examination of the development of mathematical creativity in the selected children, we introduced mathematical situations of play and exploration constructed in the erStMaL project (Vogel 2013). The situations are designed so that the children can demonstrate their mathematical potential in the interactive exchange with the other participants. An assisting adult is supposed to present the material with limited verbal and gestural instructions or suggestions. To ensure that the implementations of these mathematical situations proceed in comparable ways, the mathematical situations of play and exploration are explicitly described (Vogel 2013). The children are observed every 6 months while they work on two mathematical situations of play and exploration. All these events are video recorded with two cameras.

For the diagnosis of the attachment pattern, we apply the Manchester Child Attachment Story Task, so-called MCAST (Green et al. 2000). This is a storytelling test that has good reliability and validity. A dollhouse and a child doll as well as a caregiver doll (usually the mother doll) are used. The child can choose his/her child and mother doll from several alternatives. The examiner provides four story beginnings one at a time, which are related to specific attachment stressors like a child has a nightmare, a stomach ache, and pain in their knee or gets lost in a mall.

After the first beginning is provided, the examiner interrupts and prompts the child to continue the story.

Two examples of analysis of interaction are described, which are based on the interactional theory of learning mathematics (Brandt and Krummheuer 2001). These examples illustrate how the complete data set was analyzed to identify mathematically creative moments. It focuses on the reconstruction of meaning and the structuring of the interaction process. Therefore, it is proper to describe and analyze topics with regard to the content and the negotiation of meaning in the course of interactional processes.

The applied analysis of interaction is derived from the ethnomethodologically based conversation analysis, in which it is stated that the partners co-constitute the rationality of their action in the interaction in an everyday situation. To do this, the partners constantly try to indicate the rationality of their actions and to produce a relevant consensus together. This is necessary for their own conviction about the purposes for acting as well as for convincing the other participants. This aspect of interaction is described with the term "accounting practice" (Lehmann 1988, p. 169). To analyze the "accounting practices" of children in mathematical situations, the reconstruction and analysis of argumentation of Toulmin (1969) have proved to be useful.

Victoria in the "Solid Figure Situation"

Information About the Girls and the "Solid Figure Situation"

Victoria (4 years and 4 months old) and Sina (4 years old), who both have secure attachment patterns, participate in the "Solid Figure Situation." They are close friends.

In the mathematical situation of play and exploration "Solid Figure," the attending children deal with two of each of the following geometrical solids: cube, rectangular prism, pyramid, triangular prism, cone, cylinder, and sphere. The material and the instructions for the assisting person provide the children with ways to engage with geometrically content by getting to know these solids and their properties. To enable the children to focus their attention more easily on the geometric figures, a little bag in which the children feel them is used.

"…Because This Is a Gyroscope"

At first, the assisting person calls on both girls to handle the cone and describe what they have touched. Afterward, she puts a red cube, a red pyramid, and a blue cylinder on the table. The children grab and label each solid without being asked: The cone has been designated as "castle" by Victoria and "hat" by Sina, and the cube as "cube," the cylinder as "gyroscope," and the pyramid has been identified as a "cornflake" by the children.

Fig. 1 Towers

In this context, Sina inquires if they have to build a castle and begins to put some solids on top of each other. The assisting adult says, "No, because I have something in this bag," and presents a little bag, in which she has put a cylinder, which the children have not seen. She invites the girls to find out which solid is located in this bag only by touching and gives the clue that the solid in the bag is also on the table. Each girl feels the solid in the little bag. Victoria starts and says: "A gyroscope." Next, Sina follows and says: "A gyroscope, too." After validating their conclusion, the girls put together some solids to build towers. First, they put two cylinders on top of each other and place the pyramid on the cube to see which one is the tallest tower. Then Victoria asks, "Should I fetch the yellow one?", and Sina answers, "Okay. Good. Come on." Victoria continues: "Then we can look which one is bigger." Victoria places the cone and the pyramid side by side. Immediately thereafter, Sina puts the pyramid on top of the cylinders and the cone on the cube and then interchanges the cone and pyramid (Fig. 1).

The assisting adult asks: "Which one was the biggest Victoria? How did you see it so fast?" Victoria puts her right hand on the pyramid and says: "This one." Whereupon the adult responds: "Put them down again [referring to the pyramid and the cone]. Then you can look again which one is bigger." But Victoria looks at Sina's towers and says by grabbing the cube–pyramid tower: "Or Sina no. Do you know what? These ones belong to these ones." And by touching the cone on top of the cylinders, she notes: "And these to these ones." Sina moves the double cylinder–cone tower toward the cube–pyramid tower and tells: "Bigger." The assisting adult looks at Victoria and wants to know why these solids belong together whereupon Victoria shrugs her shoulders and Sina responses, "This is red and red," by touching the pyramid and the cube. Victoria continues, "Yes because and look and this belongs to the gyroscope because this is a gyroscope," she points at the double cylinder–cone tower. Then, she holds both hands at the bottom of the cube–pyramid tower, "because it," and turns her hands and shows her palms.

Analysis of the Episode of Victoria and Sina in the "Solid Figure Situation"

Regarding the component **content** on the level of *allocation*, the assisting adult provides geometrically content for the children with regard to solids and their properties. The task is to find identical solids. On the *situational* level, this content is extended to build castles or towers (Sina), to find out which solid is the biggest one by direct height comparisons (Victoria) and to find similar solids, which belong together (Victoria). These extensions lead to the *mathematical creative action* of Victoria by grouping pyramid and cube as well as cylinder and cone together.

Her *combinational play* in matching cylinder and cone is interpreted in order to match solids, the cylinders, from their properties. This similarity is expressed in Victoria's terming of both solids as "gyroscope" (in German "kreisel"). The German word "kreisel" includes the word "kreis" which is translated in English "circle." Cylinder and cone have both circular base areas. In an analogous manner, Victoria matches cube and pyramid, which she does by showing her palms, which can be considered a nonverbal addressing of the base areas of the solids, because like the cylinder and cone, the cube and pyramid also have same base areas. Victoria's description of the solids can be regarded as a conclusion, *divergent from the canonical*, because the intended task of the assisting adult was to find identical solids. Her conclusion seems *adaptive* to the group, because no one disagrees. However, the adult invites her to explain her findings, which she does by using a plausible and mathematical underpinned warrant in emphasizing the same base areas of the solids, which has also a plausible and mathematical backing (solids with one equal property can be grouped together).

Regarding the topic **cooperation**, Victoria and Sina are in a dyad, together with an assisting person in their day care center. They have to work on a task consecutively, because the assisting adult requests Victoria and then Sina to feel the unknown solid. The adult has the role of an *initiator* of tasks and *evaluator* of solutions, while the girls have to process that task as *processors*. The assisting adult has rejected their first attempt to build a castle. The polyadic changes to a more dyadic interaction structure between the assisting adult and one girl, whereas Sina's referring to Victoria's solution "A gyroscope, too" focuses on maintaining the polyadic interaction. The adult's initiation of an evaluation of the solution addresses both girls. After completing the task set to them by the assisting adult, the roles of the girls change. They refocus on Sina's idea of building towers, and they *initiate* new tasks such as the comparison of heights.

They engage in a dyadic interaction between each other. The adult is more reserved, as her role shifts to be a *facilitator*, who inquires. This could be seen when she invites Victoria to say which solid is the biggest. It seems somehow as if she is one step behind the girls. Victoria is able to conduct two dyadic discourses simultaneously between her and the assisting adult and between her and Sina. In the second dyad, Sina puts the cone and cylinder together as well as cube and pyramid. Victoria comments on that these solids are belonging together. At this moment, both girls are not working on instructions of the assisting adult, but have made their own choices and developed their own tasks. The adult requests further information by asking Victoria for an explanation of the groupings. Sina's expression "Bigger" may refer to the adult's first question concerning the size comparison, and her explanation of the grouping "This is red and red" answers the adult's second question, so again Sina focuses on maintaining the polaydic interaction. In the polyadic discourse, Victoria extends Sina's explanation with the evidence of the same base areas of the matched solids. In this way, she can be seen as the *initiator* of the explanation about why the matched solids belong together. The girls' role changes from a *processor* to an *initiator*, and the dyadic interaction structure becomes a polyadic one.

Concerning the component **interpersonal relations**, both girls have secure attachment patterns, which mean that they show high exploratory behavior and cooperative strategies in their play (see Creasy and Jarvis 2003). In the situation, Sina displays efforts aimed at maintaining a polyadic interaction. Furthermore, she supports Victoria in her idea of comparison of sizes ("Okay. Good. Come on."). Also, Victoria supports Sina's idea of building castles or towers, even though the assisting adult has rejected it. The scene shows how Victoria and Sina cherish their idea of building castles. They first respond on the adult's question and afterward immediately ask their questions and build castles. Victoria is able to perceive the two (competitive) perspectives of Sina and the assisting adult and furthermore to integrate both in her activities by turning first toward the adult and secondly turning to Sina.

It seems that Victoria has in Sina a "good-enough" partner (Winnicott 2012, p. 109). Later on, the assisting adult can also be seen as a "good-enough" partner, because she shows interest in Victoria's conclusion and invites her to explain her findings. In the polyadic interaction, a mathematical "playground" (Winnicott 2012, p. 64) emerges, which enables Victoria to engage in a mathematically creative process. The conclusion from grouping the pyramid and the cube as well as the cone and the cylinder together can be interpreted as producing a shared meaning between the three participants. But both girls have different explanations: Sina argues by referring to the same color, whereas Victoria focuses on the same base area. Victoria situates her explanation as an extension of Sina's. She starts with "Yes because and ...." So both explanations are not mutually exclusive, but have equal rights. Both perspectives of framing the (mathematical) situation are legitimate in the group.

Jared in the "Solid Figure Situation"

Jared (3 years and 10 months old) and Maria (3 years and 8 months old) participate in the "Solid Figure Situation," too. Jared has an insecure-avoidant attachment pattern. Both children attend the same group in their day care center. In this day care center, only Jared participates in the MaKreKi project, so we have only measured his attachment pattern. Marie is participating in the erStMaL project, so she interacts with Jared in the mathematical situations of play and exploration.

"These Are Equally..."

At first, the female assisting adult presents a little bag with a cone in it and asks the children to touch the bag and to describe what they have touched. Afterward, she calls on Maria and then on Jared to put her/his hand in the bag and redescribe what she/he has touched. Maria answers that she does not know what she has touched, and Jared describes his touch as "funny." Then the assisting adult also puts her hand into the bag and states that she touches "a top, a circle, and an abrasive surface." Afterward, she fetches the cone from the bag and says: "I think this looks like a an ice-cream cone." After a short conversation about eating ice cream, Jared begins to examine the cone by touching and pivoting it on its apex. Twenty seconds later, Maria insists that she also wants to have the cone, and the assisting adult answers "soon." Another 10 s later while Jared continues pivoting the cone, Maria tells that it is her turn. Another 15 s later, Maria holds both hands in front of Jared and insists he gives her the cone. Finally, the assisting adult looks for another cone, but Jared has already given the cone to Maria. In the next 15 s, each child is handling a cone. The assisting adult asks if they have seen something like this before. Jared answers that he has seen it at home and that this was much bigger. Maria gives her cone back to the adult and says that she is finished. The assisting adult replies: "Oh no, let's stay with this because I have a great idea." Then she tells both children that she will fill the little bag again and that they have to find out what is in the bag. Both children close their eyes while she is putting a cube in the bag. When they open their eyes, the assisting adult puts a cube and a sphere in the middle of the table and presents the little bag again and says: "One of these two is in the bag. Touch it." She holds the bag in front of Maria, "How does it feel?," and Maria answers, "Funny" and "this one" by holding the cube, which she gives to Jared. The assisting adult replies: "What do you think? Give it to Jared so that he can touch it, too." Jared takes the bag and points his finger at the cube: "This is this one, too." The assisting adult responds, "Do you think so too?," which Jared confirms. Then the adult suggests to open the bag and to check if they are right. When Jared opens the bag, the assisting adult asks if Jared was right, which he affirms again. The adult comments this with a "wow, look at it." Jared gets the cube out of the bag and puts both cubes next to each other, so that two surface areas of one side lie on top of each other. He says: "Equal."

Maria bowls the sphere along the table and Jared lifts both cubes. The assisting adult looks at Maria and says, "Look they are equal" by pointing her finger at the two cubes. Then Maria looks at both cubes and Jared says "equal" again. He gives one cube to Maria and begins to throw the other one like a die. Maria follows Jared and throws her cube like a die, too. The assisting adult asks if the children have seen something like this before. Jared says "yes" and Maria "no." And the adult continues: "What does this look like? With what can you do the same?" Maria gets up from her chair and says that she is finished, but the adult asks her to remain and so she stays. Then the assisting adult continues, "It's like a cube, or?" and threw Maria's cube as if it were a die.

Analysis of the Episode of Jared and Maria in the "Solid Figure Situation"

Regarding the component **content** on the level of *allocations*, the assisting adult provides geometrical content for the children about solids and their properties. The task is first to describe properties of an unknown solid and second to find identical solids by touching an unknown solid in a bag and comparing it with two other solids on the table. The assisting adult prompts both children to describe what they have touched. In this way, she focuses on geometrical properties like the apex or the circular base area of the cone. Jared seizes on this idea when he shows that both cubes are equal by putting them next to each other so that both sides align (see Fig. 2). This parity refers to the fact that both sides of the cubes have the same shapes as well as they are of equal area. Nevertheless, Jared presents another reason why the chosen cubes are equal: because both cubes can be thrown like dice. The orientation toward what the cube could be used for seems to relate to Wittgenstein's conclusion that the meaning of a concept is its use in language (Wittgenstein 1977). Jared's *mathematical creative action* emerges by *combining* geometrical reasons with pragmatically *linguistic* (Brandom 2000) reasons about why both cubes can be seen as equal. So on the *situational* level, Jared expands the mathematical content about geometrical solids and what they can do. His choice is *divergent from the canonical*, because the intended task was (only) to find out which solid is hidden in the bag. His explanation seems *adaptive* to the group, because no one disagrees. Nevertheless, he makes sure that the group can follow him by explaining his findings with mathematically and linguistically underpinned warrants.

Regarding the topic **cooperation**, Jared and Maria are paired in a dyad together with an assisting person in their day care center. They have to work on the task consecutively because the assisting adult first requests Maria and secondly Jared to touch the bag, which shows a more dyadic orientated interaction structure between the

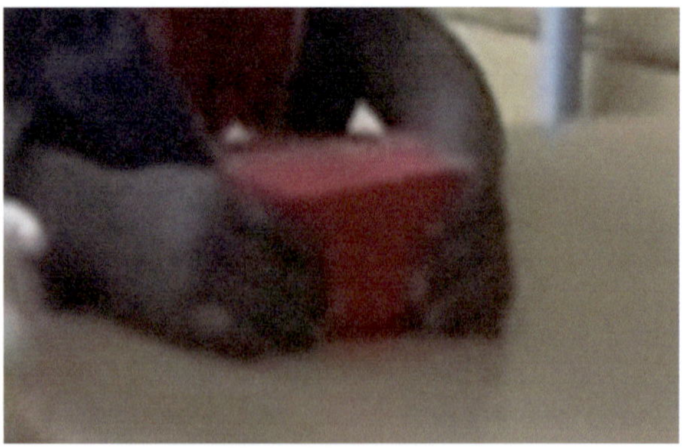

Fig. 2 Two cubes next to each other

adult and one child. Nevertheless, she also addresses both children simultaneously sometimes, e.g., when she asks if they have seen something like this before and Jared gives credit to Maria's answer, when he says: "This is this one, too," where one can see a more polyadic interaction structure. There are also two times when both children focus a dyadic interaction between each other: (1) Maria wants Jared to give her the cone and (2) Jared shows Maria the two cubes and tells her that they are equal. Jared's profile participation first seems to be like that of a "legitimate peripheral participation" (Lave and Wenger 1991, p. 29). He works very closely on the instructions of the assisting adult and answers all her questions, when he first touches the bag, describes what he feels, or tells about solids he has seen before. There are only 30 s when he explores the cone on his own. But his profile of participation changes to a more independent one when he presents the idea that both cubes are equal. He demonstrates and explains his conclusion, even though neither the assisting adult nor Maria objects or prompts him to explain his findings. Additionally, the adult has already given credit to him when she says: "wow, look at it." He follows a "practice of a reflexive argumentation" (Krummheuer 1997, p. 3) in which he endeavors to bring arguments for his conclusions, which have to exhibit the rationality of his acting.

Regarding the aspect **interpersonal relations**, Jared has an insecure-avoidant attachment pattern. Attachment theory suggests that as a child with an insecure attachment pattern, he would show a less exploratory behavior and less cooperative strategies in his play. But as the example shows, this is not the case. Concerning the first aspect, exploratory behavior of Jared can be seen when he examines the cone. So on the *situational level*, the assisting adult and Maria seem to be "good-enough" partners (Winnicott 2012, p. 109) for Jared to enhance his exploratory behavior by leaving him some space for his examination of the cone. Nevertheless, he seems to work closely with the instructions of the assisting adult, which is typical for a child with an insecure attachment pattern. At the moment when Jared's profile of participation changes from a "legitimate peripheral participation" (Lave and Wenger 1991, p. 29) to a more independent one, he shows great interactionistic endeavor to state the rationality of his acting and to integrate Maria into his dialogue with the assisting adult by giving her one cube. In the analyses, this behavior is assigned to the category of *adaptiveness* according to the three fundamental aspects of mathematical creativity, mentioned previously. With respect to the interactional function of this *adaptiveness* in the situation, namely, to convince his partner by means of argumentation, one can see his wish to keep himself as an accepted common partner in his group, although his mathematical inventions are far from being mutual. As a child with an insecure-avoidant attachment pattern, he is used to experience a caregiver who behaves in a distanced and controlled way. In the relationship between child and caregiver, there is always a risk of becoming detached. So Jared's *adaptiveness* in the "Solid Figure Situation" may be linked to his attachment pattern and inner "working model" (Bowlby 1969, p. 354) because he wants to avoid and minimize this risk of being detached from his interlocutors.

Summary and Implications

Understanding mathematical creativity in early childhood as a cooperative process, which emerges in the situational negotiations of meanings in social interactions, the concept of the niche of the development of mathematical creativity has been introduced and used with two empirical cases. It highlights the *allocative* and *situational* terms of a *mathematically creative process* as well as the *individual mathematically creative action*.

As the first example shows, from the socio-constructivist view, a *mathematically creative process* arises in the concrete situation once a not anticipated interpretation of a mathematical situation occurs in the interaction between the involved interlocutors. Although both girls follow the instruction of the assisting adult, which differs from their own interpretation of the situation, they do not forget their own plans in building castles. So first they respond to the adult's invitation and her expectation in answering which solid is hidden in the bag. Then, they reorganize the social order of the interactional process by raising their own questions and building castles. These reorganizations are necessary for the ongoing interactional process in which Victoria accomplishes her mathematically creative action. As the second example shows, a *mathematically creative process* arises once a child combines different perspectives to explain a mathematical issue. Jared uses mathematically and linguistically underpinned warrants to justify why both cubes are equal.

From a psychoanalytic point of view, the profile of participation of the children in the mathematical situations of play and explorations seems to be linked to their attachment patterns. Victoria shows more exploratory behavior than Jared, whose profile of participation is to be akin to the profile of a "legitimate peripheral participation" (Lave and Wenger 1991, p. 29). Jared works more closely with the instructions given by the assisting adult than Victoria, who quickly develops her own interpretations of the situation and her own questions, which she wants to answer. Therefore, in promoting children's creative mathematical potential in early years, it seems to be important to use different models of instruction. Children with an insecure attachment pattern need a situation in which they first have the chance to embellish their leeway of participation in form of a "legitimate peripheral participation," while children with a secure attachment pattern need a situation where they can participate in a more autonomous way right from the beginning.

Both examples show that children independent of their attachment pattern show great interactionistic endeavor and communicative abilities in convincing their interlocutors of their creative framing of the mathematical situation. In the analyses, these endeavors are summarized under the term *adaptiveness*. However, the *adaptiveness* of Victoria and Jared differs in how it emerges. In the case of Victoria, the assisting adult has to request an explanation for why the chosen solids belong together, whereas Jared provides arguments, which exhibit the rationality of his acting in the sense of a "practice of a reflexive argumentation" (Krummheuer 1997, p. 3) unrequested. From a psychoanalytic point of view, these differences may also be found in the different attachment patterns of the children. Victoria as a child with a secure attachment pattern has a secure basis from which she explores the world.

So for her, it is not always necessary to explain her conclusions, because she knows that she is accepted and loved by her caregiver. Jared as a child with an insecure-avoidant attachment pattern is used to explain his views to make sure that he does not lose the connection between him and his caregiver.

The example of Jared points out that the assumption that only children with a secure attachment pattern show exploratory behavior and mathematically creative activities during the engagement in cooperative mathematical situations seems not to be true. In fact, most children in the MaKreKi project have an insecure attachment pattern ($n=24$). This could be for several reasons. The first emphasizes the "good-enough" partners or environment in the sense of Winnicott's (2012) "good-enough mother," which could open possibilities for all children to enhance their potential for mathematically creative activities as well as for cooperative strategies. The second is the use of the inner "working model" (Bowlby 1969, p. 354) and the experience of children with an insecure attachment of losing the connection to their partners. Maybe these experiences are accountable for the early development of some kind of sensitivity, perspective taking, and the need to explain the own perspective, seen in these children, so that an impending disconnection can be avoided. The third one comes from resilience studies and therapeutic experiences, which state that children use a kind of special talent, e.g., mathematical talent to compensate for an emotional weakness or an early traumatization. Runco and Richards (1997) claim that children can become more flexible in emotional and intellectual ways when they are susceptible to difficult and unfavorable situations. So for Jared, the engagement in mathematical themes may be some kind of compensation. The last one is that the attachment theory may not be a strong predictor for creative and exploratory behavior of children at preschool age.

While the current cases offer some initial insights regarding the niche of the development of mathematical creativity in the early years, further research is needed. The observations concerning the link between attachment pattern and leeway of participation have to be examined with additional cases of children with secure and insecure attachment pattern. Additional analysis of scientific and folk pedagogy's concepts about mathematical creativity and mathematics education arising in the social setting may give further insights about the conditions of the development of mathematically creative processes.

Acknowledgments The preparation of this chapter was funded by the federal state government of Hessen (LOEWE initiative).

References

Ainsworth, M. D., Blehar, M. C., Waters, E., & Wall, S. (1978). *Patterns of attachment*. Hillsdale, NJ: Erlbaum.

Bateson, P., & Martin, P. (2013). *Play, playfulness, creativity and innovation*. Cambridge: Cambridge University Press.

Benson, J. B., & Haith, M. M. (2009). *Social and emotional development in infancy and early childhood*. Oxford, UK: Academic Press.

Bowlby, J. (1969). *Attachment. Attachment and loss* (Vol. 1). New York: Basic Books.

Brandom, R. B. (2000). *Begründen und Begreifen [Articulating reasons]*. Frankfurt a.M: Suhrkamp.

Brandt, B. (2004). *Kinder als Lernende: Partizipationsspielräume und -profile im Klassenzimmer [Children as learners. Leeway of participation and profile of participation in classroom]*. Frankfurt a. M.: Peter Lang.

Brandt, B., & Krummheuer, G. (2001). *Paraphrase und Traduktion: Parti-zipationstheoretische Elemente einer Interaktionstheorie des Mathematiklernens in der Grundschule [Paraphrase and traduction]*. Weinheim: Beltz.

Bruner, J. (1986). *Actual minds: Possible worlds*. Cambridge, MA: Harvard University Press.

Bruner, J. (1990). *Acts of meaning*. Cambridge, MA: Harvard University Press.

Carruthers, E., & Worthington, M. (2011). *Understanding children's mathematical graphics beginnings in play*. New York: Open University Press.

Creasy, G., & Jarvis, P. (2003). Play in children. An attachment perspective. In O. N. Saracho & B. Spodek (Eds.), *Contemporary perspectives on play in early childhood education* (pp. 133–153). Charlotte, NC: Information Age.

Crowell, J., & Feldman, S. (1988). Mother's internal models of relationships and children's behaviour and developmental status: A study of mother-infant-interaction. *Child Development, 59*, 1273–1285.

Csikszentmihalyi, M. (1997). *Creativity: Flow and the psychology of discovery and invention*. New York: Harper Perennial.

Ervynck, G. (1991). Mathematical creativity. In D. Tall (Ed.), *Advanced mathematical thinking* (pp. 42–53). Dordrecht: Kluwer.

Finke, R. (1990). *Creative imagery. Discoveries and inventions in visualization*. Hillsdale, NJ: Lawrence Erlbaum.

Green, J., Stanley, C., Smith, V., & Goldwyn, R. (2000). A new method of evaluating attachment representations in the young school age children: The Manchester child attachment story task (MCAST). *Attachment and Human Development, 2*, 48–70.

Grossmann, K. (1984). *Zweijährige Kinder im Zusammenspiel mit ihren Müttern, Vätern, einer fremden Erwachsenen und in einer Überaschungssituation. Beobachtungen aus bindungs- und kompetenztheoretischer Sicht [Two year olds in joint play with their mothers, fathers, a stranger and in a surprise situation: Observations from the attachment and competence viewpoint]*. Regensburg: Universität Regensburg.

Guilford, J. P. (1967). *The nature of human intelligence*. New York: McGraw-Hill.

Krummheuer, G. (1997, March 24–28). *Reflexive arguing in elementary school classes: Opportunities for learning*. Paper presented at the annual meeting of the American Research Association, Chicago, IL. http://files.eric.ed.gov/fulltext/ED409109.pdf. Accessed 19 June 2015.

Krummheuer, G. (2012). The "non-canonical" solution and the "improvisation" as conditions for early years mathematics learning processes: The concept of the "interactional niche in the development of mathematical thinking" (NMT). *Journal für Mathematik-Didaktik, 33*(2), 317–338. doi:10.1007/s13138-012-0040-z.

Krummheuer, G., Leuzinger-Bohleber, M., Müller-Kirchof, M., Münz, M., & Vogel, R. (2013). Explaining the mathematical creativity of a young boy: An interdisciplinary venture between mathematics education and psychoanalysis. *Educational Studies in Mathematics, 84*(2), 183–200. doi:10.1007/s10649-013-9505-3.

Krummheuer, G., & Schütte, M. (2014). Das Wechseln zwischen mathematischen Inhaltsbereichen—Eine Kompetenz, die nicht in den Bildungsstandards steht [The changing between mathematical domains—A competence that is not mentioned in the educational standards]. *Zeitschrift für Grundschulforschung, 7*, 126–138.

Lave, J., & Wenger, E. (1991). *Situated learning: Legitimate peripheral participation*. Cambridge, UK: Cambridge University Press.

Lehmann, B. E. (1988). *Rationalität im Alltag? Zur Konstitution sinnhaften Handelns in der Perspektive interpretativer Soziologie [Rationality in everyday life? Construction of meaningful acting in the perspective of interpretative sociology]*. Münster: Waxmann.

Lithner, J. (2008). A research framework for creative and imitative reasoning. *Educational Studies in Mathematics, 67*(3), 255–276. doi:10.1007/s10649-007-9104-2.
Marvin, R. S., & Britner, P. A. (1999). Normative development: Ontogeny of attachment. In J. Cassidy & P. Shaver (Eds.), *Handbook of attachment: Theory and research* (pp. 44–67). New York: Guilford Press.
Mikulincer, M., & Shaver, P. (2007). *Attachment in adulthood: Structure, dynamics, and change*. New York: Guilford Press.
Münz, M. (2014). Non-canonical solutions in children-adult interactions: A case study of the emergence of mathematical creativity. In C. Benz, B. Brandt, U. Kortenkamp, G. Krummheuer, S. Ladel, & R. Vogel (Eds.), *Early mathematics learning* (pp. 125–146). New York: Springer.
Rogoff, B. (2003). *The cultural nature of human development*. Oxford: Oxford University Press.
Runco, M., & Richards, R. (1997). *Eminent and everyday creativity*. London: Ablex.
Sriraman, B. (2004). The characteristics of mathematical creativity. *The Mathematics Educator, 14*(1), 19–34.
Sternberg, R. J., & Lubart, T. I. (2000). The concept of creativity: Prospects and paradigms. In R. J. Sternberg (Ed.), *Handbook of creativity* (pp. 3–15). Cambridge, UK: Cambridge University Press.
Super, C. M., & Harkness, S. (1986). The developmental niche: A conceptualization at the interface of child and culture. *International Journal of Behavioural Development, 9*, 545–569.
Torrance, E. P. (1974). *Torrance tests of creative thinking*. Bensenville, IL: Scholastic Testing Service.
Toulmin, S. E. (1969). *The uses of argument*. Cambridge, UK: Cambridge University Press.
Van Oers, B. (2002). The mathematization of young children's language. In K. Gravenmeijer, R. Lehrer, B. van Oers, & L. Verschaffel (Eds.), *Symbolizing, modeling and tool use in mathematics education* (pp. 29–58). Dordrecht: Kluwer.
Vogel, R. (2013). Mathematical situations of play and exploration. *Educational Studies in Mathematics, 84*(2), 209–226. doi:10.1007/s10649-013-9504-4.
Vygotsky, L. (2004). Imagination and creativity in childhood. *Journal of Russian and East European Psychology, 42*(1), 7–97.
Winnicott, D. W. (2012). *Playing and reality*. Retrieved from http://www.eblib.com. Accessed 19 June 2015.
Wittgenstein, L. (1977). *Philosophische Untersuchungen [Philosophical investigations]*. Frankfurt a.M.: Suhrkamp.

Children's Play as a Starting Point for Teaching Shapes and Patterns in the Preschool

Kerstin Bäckman

Abstract This chapter contributes to knowledge about the teaching and learning of mathematical content through play in preschool. The study focuses on the potential for teaching and learning mathematical content with children's play as the starting point. The question is important because according to research, children's mathematical encounters in play activities are educational experiences. The understanding of children's mathematical encounters in play and teachers' teaching is presented in terms of learnable and teachable moments in "here-and-now" situations. The data consists of video-recorded observations of young children's play in four Swedish preschools. Two 14-min-long excerpts from the recordings illustrate the potential of children's play for the teaching and learning of shapes and patterns. The results show that a teacher's questions and didactical choices in play can support children's explorations if the teacher observes and recognizes the mathematical content. The results also indicate the potential for teachable and learnable moments and dilemmas when play is the starting point.

The conclusion is that "here-and-now" situations provide teachable and learnable moments. There are also dilemmas, in that teachers have to observe and discern the mathematics in children's play and direct the child's attention towards this.

Introduction

In this chapter, children's early learning of a mathematical content in everyday activities like play is regarded as essential. Research on early mathematics highlights the importance of education and early mathematical skills (Clements and Sarama 2009; Claessens and Engel 2013; Ginsburg and Amit 2008). According to Claessens and Engel (2013), early mathematical knowledge and skills predict the learning of other content areas like language. One way of learning mathematics in the preschool is to provide children with rich and varied mathematical encounters

K. Bäckman (✉)
University of Gävle, Gävle, Sweden
e-mail: kbn@hig.se

T. Meaney et al. (eds.), *Mathematics Education in the Early Years*,
DOI 10.1007/978-3-319-23935-4_12

and experiences (Clements and Sarama 2009). Research also shows that children learn mathematics in daily interactions with peers (e.g. Bäckman 2015), in the environment (e.g. Carruthers and Worthington 2006; Ginsburg 2006) and in culture (e.g. Starkey and Klein 2008; van Oers 2010).

Two strong discourses on education and teaching/learning in research concerning early childhood can be identified. The first discourse is based on play and children's experiences of the mathematics encountered in their activities (e.g. Carruthers and Worthington 2006; Ginsburg 2006). In this discourse, the idea is that children learn mathematics through play. This leads to informal teaching, in which the teacher follows the children's interests and strives to discern the mathematical content in everyday situations.

In the second and more formal discourse, the teacher chooses the content in advance and plans the teaching situations. Researchers maintain that in this discourse, teachers' instructions are important (e.g. Clements and Sarama 2009; Claessens and Engel 2013; Starkey and Klein 2008). Both discourses can support children's mathematical learning. With a play-based approach, as in this study, the teacher has to observe mathematics in the children's activities and identify potential teachable moments (Hyun and Marshall 2003). This means that teachers have to quickly decide which solutions are appropriate at any given moment. In the second more "teacher-directed approach", the teacher is able to plan what he or she wants the children to learn and design specific teaching situations. The planning can also include deciding which questions and instructions to use to promote the learning of certain content.

In this chapter, I explore the potential for teaching and learning mathematical content in two "here-and-now" situations, with children's play as the starting point. "Here-and-now" situations (Bäckman 2015) mean that the preschool teacher is not tied to any specific situation, but tries to target the learning situations that spontaneously arise in everyday activities, especially in children's play. In these situations, preschool teachers try to ensure that children are able to distinguish mathematical content, for example, shapes and repeat patterns in their activities. "Here-and-now" situations are closely connected with play and have a temporal aspect, which means that the teacher has to make quick decisions on the spur of the moment.

Children's encounters with mathematics in activities provide them with experiences that form a basis for education. In this chapter, the mathematical content consists of geometry, that is, shapes, geometric figures and patterns. Ginsburg (2006) argues that when children play, objects provide opportunities for mathematical thinking. Children construct their knowledge of, for example, shape by playing with shapes.

According to Clements and Sarama (2009), teachers can and should challenge children's experiences of different shapes and create environments with rich and various shapes and thereby promote learning. In Sweden, most children spend a great deal of their early childhood in preschool.[1] Play is an important part of their

[1] 83 % of Swedish children attend preschool between the ages of 1 and 5 years (National Agency for Education 2011).

daily lives, and the preschool curriculum suggests that education should be playful (National Agency for Education 2011). If preschool teachers observe and discern mathematics in children's play, their observations can form the basis for discussion and reflection together with children. In this way, education in preschool can be designed according to children's perspectives.

The aim of this chapter is to explore the potential of children's play as a starting point for teachers' teaching and children's learning of shapes, geometric figures and patterns. Here, *teachable* and *learnable moments* and *dilemmas* in play activities are problematized. According to Hyun and Marshall (2003), teachable moments occur when teachers observe, discern and interpret children's spontaneous interest in play. Teachers can create learnable moments and take advantage of the teachable moments that arise (Cheeseman 2015). A moment is learnable when a child discerns the mathematical content and when the situation promotes learning. A teachable moment can also be a learnable moment, especially if a child has the same focus as the teacher and is receptive to the teacher's questions. If at the same time the teacher is able to discern children's earlier experiences and reflect with the children and challenge their existing thoughts, this could be both a learnable and a teachable moment. One dilemma is that children's own intentions in play can make it difficult for the teacher to visualize the mathematical content that is appropriate for them. It can be problematic for the teacher to direct the child's attention to a mathematical content when the child's attention is directed towards a play content. Another dilemma is whether the teacher is able to recognize the mathematical content in children's play. Teachers need mathematical knowledge and knowledge of relevant issues that can challenge the child's thinking in the moment.

Teaching and Learning Through Play

Children's learning occurs in meaningful and social contexts when they learn something new that builds on their earlier experiences (Vygotsky 1978). In the preschool context, social interactions with peers, adults and the environment all offer children mathematical experiences. Vygotsky (1978) also emphasizes play as the most important part of a child's learning. For researchers to be able to say that learning occurs, knowledge is often required about children's understanding before and after a learning situation. In this study, the interpretations of children's learning are based on whether the moment is learnable or not. According to Bruner (2002), teachers can create interesting environments by making use of such interactions and by using play materials like building blocks and geometrical figures to "scaffold" children's learning.

Play is an important aspect of teachers teaching and children's learning in preschool (Ginsburg 2006; Ginsburg and Amit 2008; Munn and Kleinberg 2003; Pramling Samuelsson and Fleer 2008; Wager and Parks 2014; Wood and Attfield 2005). Ginsburg (2006) describes play and learning as two sides of the same coin, i.e. children play and learn at the same time. The author argues that play motivates and enhances children's cognitive and socio-emotional development.

According to Wood and Attfield (2005), the potential of play in teaching has great importance because it is integrated into the learning process. In play, children can develop skills such as language, mathematics, communication and social skills. Children have opportunities to think hypothetically and follow rules. In play, children are able to guess, estimate or predict what might happen. They can also explore shapes, geometric figures and patterns, dimensions and positions and develop their reasoning about different aspects. Wager and Parks (2014) discuss children's opportunities to learn mathematics in play and how teachers can support that learning in both informal and formal settings. They argue that it is important to observe children's play at school, at home and in the community in order to understand how play facilitates mathematical learning and what the children learn.

Different play contexts provide meaningful opportunities for children to use and develop mathematical skills, such as problem-solving situations in which children can think, experiment, draw and say what they are thinking (Ahlberg 1998). Furthermore, Wood and Attfield (2005) argue that education should give children the opportunity to use flexible and creative ways of thinking and acting and that various play contexts can offer rich opportunities for this.

Preschool children's learning and spatial thinking in geometry imply an understanding of space and, for example, shapes and pattern. Van den Heuvel-Panhuizen and Buys (2005) highlight the importance of geometry and spatial thinking. Children's understanding and meaning making in geometry influence their thinking and spatial development. Van Hiele (1959) and later Tirosh et al. (2011) describe the different levels of children's geometric thinking describing the first level as the visual level. This means that children have an early experience of an object when they see its structure or form. At this level of their spatial thinking, children assess figures belonging to the same category. For example, a rectangle could be a door or a table. The second level is the descriptive level. At this level, children examine the properties of shapes, rather than their appearance. Children can verbally describe that triangles have three corners and three sides and that a circle is round. This means that at this descriptive level, language is important. The third level is the deductive level and means that children are able to formulate definitions for shapes like triangles and rectangles. When children explore various items and look at them and touch them, they have a visual and tactile experience which can support them to discern similarities and differences that will form the basis for future experiences of shapes.

Spatial thinking and spatial orientation are important for children's exploration of the world, because they indicate where things are located and placed and the distance between them. Van den Heuvel-Panhuizen and Buys (2005) describe the spatial ability and orientation that is important for children's spatial development and their discernment of shapes and patterns. This is in line with Ginsburg and Ertle's (2008) suggestions of describing spatial relations and different kinds of patterns, e.g. alternating patterns with or without repetition and growing patterns. A pattern refers to an underlying rule, such as the repetition of circular shapes. Clements and Sarama (2009) argue that children develop their geometric thinking through play that this can be carefully planned by using materials like mosaics and puzzles, but can also occur in spontaneous play.

In order to experience and learn about geometric content such as shapes, figures and patterns, environments are required that offer a variety of geometric shapes and figures. According to Clements and Sarama (2009), this environment should give children an opportunity to explore and discern the similarities and differences in the artefacts used and give rise to discussions about different kinds of shapes and their properties and the kind of geometric tasks and challenges that arise. The authors highlight four guiding features in the environment that can provide education about shapes. The first feature is that preschools should give children opportunities to experience a lot of different kinds of shapes. This includes varied examples of the characteristics of shapes and opportunities to discern the similarities and differences among them.

The second feature is that preschool teachers should encourage and challenge children's descriptions in order to enrich their language. For example, children should have opportunities to explain why a shape belongs or does not belong to a certain category.

The third feature includes the environment. Preschools should offer different classes of shape such as various sizes and orientation of circles, triangles, squares and rectangles, as well as different colours and materials. Clements and Sarama (2009) argue that this includes showing children that squares are examples of rectangles.

The fourth feature is to stimulate children by providing a wide range of interesting activities and tasks, including reflection and discussion, so that children can compare, identify and explore the different shapes and figures that are important in geometry.

Claessens and Engel (2013) suggest that when children are able to focus on pattern recognition, measurement and advanced numbers in the early years of schooling, it will benefit their learning later in school. The authors highlight that teachers' instructions are necessary for children's outcomes in mathematics. Bruner (2002) highlights scaffolding which consists of teachers' feedback and the use of different strategies, such as active listening, questions, affirmation and mathematics-related talk (Bruner 2002).

Play gives teachers the opportunity to observe children's expressions and their mathematization (actions and reflections on mathematical content, articulation of concepts and features) of shapes (Carruthers and Worthington 2006; Van Oers 2010). Play also helps children to use their imagination and creativity by, for example, considering how an object works, how a ball rolls and what they can do with a ball. Teachers and children can discuss and reflect on the characteristics of objects, e.g. whether shapes are round or curved, which forms have corners and how many corners there are (Clements and Sarama 2009). When children play with blocks and build constructions, preschool teachers have an opportunity to reason with them about different classifications and attributes and to offer opportunities that will develop their spatial abilities, such as body and spatial awareness, and knowledge about measurement. Carruthers and Worthington (2006) and Van Oers (2010) argue that there is a mathematical content in children's play and that it is up to the teacher to discern when mathematics occurs in different play contexts.

Methodology

The aim of this study is to explore the potential of children's play as a starting point for teachers' teaching of shapes and patterns. Children's actions and intentions are in focus, specifically what they direct their attention towards in their mathematical encounters during play, as these have the potential to become learnable moments. The research also focuses on teachers' approaches to and teaching in teachable moments.

This study is a part of a larger study (Bäckman 2015). Thirty-five 4-year-old children from four Swedish preschools took part in the study relating to children's experiences of mathematics in everyday situations. Video observations were used to focus on children's activities with a mathematical content. Children's and preschool teachers' formations of mathematics in the preschool constitute the study's research object and include an analysis and interpretation of children's actions and mathematical encounters. It also includes interactions with teachers.

For this chapter, two 14-min excerpts from the study (18 h of video observations) are used. These excerpts—one observation with shapes and another with geometric figures and patterns—have been chosen because they reflect common situations found in the preschool and highlight two dimensions of the teaching and learning potential of play.

The video observations make the mathematical content in children's play and children's actions visible. The observations show verbal and non-verbal language, glances, gestures, nods, smiles, the artefacts the children are using and how they use them and whether they use them on their own or together with peers/adults. The observations also highlight teachers' actions and their interactions with one child.

The research is directed towards teachers' teaching and preschool children's learning, which entails a particular responsibility to comply with applicable ethical considerations. As video observations are used to observe different situations in the preschool, it is necessary to protect the participants' identities and integrity. Both parents and teachers gave their written consent for the children's participation. In the video-recorded situations, the participating children's oral consent was obtained. The teachers also gave their consent to participate in the study.

Results

The results demonstrate children's experiences with building blocks and geometric figures and their creation of patterns. One result is the identification of the potential of learnable and teachable moments in play. Children often play on their own, with material that the teacher provides. The boy in the first excerpt focuses on the different attributes of the shapes from a stable and durable perspective. He seems to have set goals in his building and tries to put different shapes on top of each other.

The observation started early one morning in a Swedish preschool when Erik, aged 4, was playing with building blocks in the hall. One of the preschool teachers was standing beside him talking to a parent.

Erik does not seem to be paying any attention to the adults' conversation. He is using different shaped blocks to build a high stack of eight blocks. The blocks consist of seven cubes in three different sizes, as well as a pyramid. Every second block is a small cube, and the alternative block is a larger cube. A pyramid is placed on the top of each stack. Erik then started to make a shorter construction consisting of six half cylinders. He placed half a cylinder with the short edge towards the floor on one side of the stack and another half cylinder opposite the first one with the long surfaces against each other. He looked at a third half cylinder, twisted and turned the block, looked at the construction and then put the half cylinder between the first two.

Erik is totally focused on the construction. He seems to want to build both horizontally and vertically. He twists and turns the blocks in an attempt to find a stable and durable way of placing them on top of each other. At the same time, there also seems to be a desire to make his building work symmetrical.

He picks up a fourth half cylinder and places it on the other side of the stack. He then creates a similar construction as the one on the opposite side of the stack with identical half cylinders on each side of the half cylinder in the middle. When the blocks are in place and everything is stable, he places a pyramid on the top of both constructions.

Erik explores the various geometric shapes by placing them on top of and next to each other in different ways. He distinguishes the various qualities of the blocks as he twists and turns and builds with them. The observation shows the potential of learning about the critical aspects of geometric shapes in the play context. Erik seems to have an idea or intention for the construction and experiences how these differently shaped blocks can or cannot be stacked. Erik is focused on his construction, although he does look up from time to time. He seems to have specific goals in mind with his building and is not disturbed by the adults and children talking to each other next to him. Erik continues to create the construction:

Erik points to two cubes in the high stack in the middle and says,

E: "It's over and it's over".

Then he points to the two top blocks and says,

E: "Those should be removed".

He takes the top two blocks from the highest tower and places them on the floor, a cube with a pyramid top. He then takes a pyramid lying on the floor and places it on top of the tall tower. He does this while holding a little blue car in his hand. He looks at the high tower and says,

E: "There you go".

Erik's building seems to give him experiences of the similarities and differences between the various blocks. He does not talk about the shapes or the features, but is totally engaged in the construction. The teacher talks to one parent in the hall and later on walks through the hall passing Erik and his construction. She stops and says:

Teacher: "What a nice building. Very tall!"

Erik looks up when the teacher is talking to him but doesn't say anything. The teacher then leaves the room.

Erik is engaged in his construction work and the teacher observes, and at a later moment, the teacher did make use of these experiences in discussions with Erik about different shapes.

Another excerpt from the data shows a 4-year-old girl, Meg, playing with geometric figures. The figures in the stack on the table in front of her have different shapes and colours, such as circles, triangles, squares, pentagons and polygons.

Meg has started to twist and turn the geometric figures into different shapes. She then selected only the red figures of different shapes (circles, triangles, squares, pentagons and polygons) from the stack in front of her. She created a red pattern with the figures and after that she selected a blue circle from the stack with the geometric figures. She put the blue circle on the table in front of her, then a yellow circle, an empty space and then a yellow circle.

The teacher, who is sitting at another table observing Meg's designs, asks Meg,

Teacher: "What should be put in the empty space in your pattern now?"

Meg looks at the geometric figures and picks up a blue circle, which she places in the empty space.

Teacher: "What is next in your pattern?"

Meg looks at the teacher and smiles as she picks up the last yellow circle.

Teacher: "What colour is the circle?"

Meg: "Yellow", she says as she puts down the shape. Meg then chooses a blue circle.

Teacher: "Blue".

Meg laughs and adds the blue circle to the yellow one in the pattern.

The teacher makes Meg's pattern visible to her by verbally supporting and confirming the girl's actions. The use of questions and colours are strategies employed in the feedback process to make the pattern visible. Meg is able to think abstractly and reflect on the pattern. The interaction between the teacher and the child highlights colour as a criterion for the circles in the pattern. It seems that Meg has not noticed that the coloured circles make a potential repeat pattern, but the teacher does and draws this to her attention. In this case, the teacher scaffolds and gives feedback using questions and by giving the features names. In the beginning of the observation, Meg explores different geometric figures and has an opportunity to discern the similarities and differences between them. She starts by focusing on the differently shaped red figures and then starts to create a potential repeat pattern. Maybe it is the teacher's comments that make Meg think what kind of circle is appropriate.

The observations show the kind of experiences children engage in. The observation of Erik illustrates a common situation in the preschool. Teachers talk to parents and other adults at the same time as they take care of many children. Despite this they want to give the children some kind of feedback. Erik seems to study the characteristics of the blocks before he puts them together, which may make him wonder about the various features and if the construction is stable. Erik's experiences in this play setting provide valuable opportunities to explore different shapes, which makes the situation learnable. The play also has the potential to be teachable if the teacher stays and reflects with the child. The provided material and the teacher's observation can be used again in a new teachable moment. The teacher did make use of Erik's experiences at a later occasion when they reflected on different shapes.

The example with Meg highlights a play situation in which a child explores coloured circles and puts some of them together in what appears to be a repeat pattern. Like Erik, Meg is engaged in the exploration of shapes and patterns. Here, the teacher draws Meg's attention to the possibility of repeating the colours of the circles to make a pattern and uses this as a teachable moment to guide Meg into recognizing a repeat pattern. The teacher does this by asking questions such as "what is in your pattern now?" The girl picks a blue figure and the teacher says "blue!" The situation can also be said to be learnable in that the girl seems to observe the possibility of repeating the colours and perhaps also discerning the pattern.

These two observations take account of both the child's perspective and the teacher's perspective. Erik initiates the play himself and he seems to have set goals in sight. The teacher gives him brief feedback about the height of his construction and says that it looks nice. In Erik's case, the teacher has an opportunity to really pay attention to the mathematical aspects of his construction. She observes Erik's play but does not ask any questions and only briefly comments on his construction. However, she does have an opportunity to elaborate on Erik's construction at a later date. The teacher in Erik's case could have asked him how many sides and how many corners the different shapes had and about his choices of different shapes. Here, Erik is only exposed to the qualities of shapes when he works and plays with them.

Meg does not seem to be as goal-oriented as Erik. She looks at the different shapes and places them on the table. The teacher recognizes a possible pattern in the coloured shapes. She draws the child's attention to the possibility of repeating the colours by asking questions and stating the fact that there is a pattern, i.e. both whole and in part. The girl's actions display that she seems to be aware of the patterns she made. In Meg's case, the teacher's intention is to support the girl's discernment of patterns by asking her about the colours of the shapes. From the child's perspective, the teacher's interest, questions and statements serve as positive feedback and are perhaps enough for Meg's discernment at that particular moment.

Discussion

In this section, I reflect on and discuss the potential for teachable and learnable moments in two "here-and-now" situations with the children's play as the starting point. This includes children's experiences and the teachers' scaffolding. The designed environment with interesting play material like building blocks and geometrical figures can be part of a teacher's scaffold (Bruner 2002).

Ginsburg's (2006) suggestion that play gives children an opportunity to explore shapes, geometric figures and patterns, dimensions and positions seems to fit these situations. When the teacher observes Meg's play and the spontaneously emerging situation, she is able to exploit this and turn it into a teachable moment. Guided by the teacher's questions, children can be challenged, stimulated and acquire new experiences, which makes the situation learnable. In the examples with Erik and Meg, the teacher's attention and reaction to the child's actions are aspects of the guiding and feedback process. This agrees with Clements and Sarama's (2009) research, which points out that preschool teachers' use of questions, feedback and the provision of rich environments can lead to deeper understanding and learning. Both children in the described observations experience shapes in their different constructions. Such experiences can form the basis for children's learning. Children's cognitive processes in activities like these affect their learning, as does observing and participating with others in play. Children are not passive recipients, but are active in the processes in which they are involved. The two observations also highlight some of the dilemmas that can arise with teachable and learnable moments in "here-and-now" situations. One such dilemma is time like the teacher in the example with Erik's construction and another is teachers' knowledge. Ginsburg and Amit (2008) argue that teaching mathematics to young children is almost the same as teaching mathematics to older children. They maintain that a preschool teacher must know what the content is and how this can be made visible to the children. They also stress that the teacher must have pedagogical content knowledge to know how to teach the content and in this study to preschool children. The teacher in the example with Erik observed a mathematical content in the boy's construction and did make use of the moment. She was able to reconnect to the boy's experiences at a later time. Wager and Parks (2014) highlight the importance of observing children's play in order to understand how play can facilitate mathematical learning.

The teachers in the excerpts could have given more feedback by asking the children explain their thinking and actions. In the example with Erik, the teacher could have asked him about his thoughts and suggestions and could also have provided specific information about the different shapes he was playing with. They could have talked together about the different features of the shapes, but in this situation, they did not. In Meg's case, the teacher pointed to the possible repeat pattern as a way of scaffolding. Teachable moments can provide learning experiences for the child, although as Hyun and Marshall (2003) argue, it can be difficult for the teacher to respond to the teachable moments that arise, especially if there are a lot of children in the group or parents like in the example with Erik. This is a dilemma for the teacher, and in the example with Erik, the teacher talked to a parent which means that she could not respond to him directly.

Bäckman (2015) has highlighted and provided insights into the importance of continuing to raise awareness among preschool teachers regarding preschool children's mathematical experiences in everyday life. The most important thing to note is that mathematical content is present in a variety of situations in the preschool. It may not always be the mathematical content that is focused on by the children, but the activity itself. It is the teacher who can direct the child's intention to the mathematics and make the situation teachable and learnable. This can also be a dilemma, because children like Erik and Meg have their own intentions in play, and it can be difficult for the child to have the same focus as the teacher.

"Here-and-now" situations like the situations with Erik and Meg can be both teachable and learnable moments depending on what children express in their actions and what opportunities the preschool teacher have to exchange thoughts and reasoning around the object's various features. When children are at this visual level of their spatial thinking, shapes that look similar belong to the same category (Clements and Sarama 2009). Both Erik and Meg show by their actions that they reflect on differences and similarities. According to Carruthers and Worthington (2006) and Van Oers (2010), it is up to the teachers to observe the mathematical aspects of children's play and support them by providing the relevant material and giving appropriate feedback. Sometimes, like in Erik's case, the material gives feedback in the moment, and the teacher observes and gives feedback later on.

The language that teachers use when talking to children about shapes is important, because it helps them to make the necessary connections. Teachers can also ask children to describe and reflect on the various features of the shapes they are playing with in the moment or afterwards like in Erik's case. In this context, the teacher's questions and feedback may be more important than instructions. The use of questions like in Meg's case can direct the child's attention to the similarities and differences among the coloured circle shapes, but the teacher lost the opportunity to ask about the specific characteristics and attributes of shapes and figures.

Clements and Sarama (2009) suggest that a carefully designed learning environment in preschool supports children's spatial development and provides opportunities for children like Erik and Meg to mathematize in play. Various play materials in the learning environment that offer reflection about similarities and differences among shapes are important for children's mathematical thinking (Ginsburg 2006). Teachers can help them by using different didactic strategies and making didactic choices even if they did not in these two examples. The presence of teachers offers the opportunity for teachable and learnable moments in "here-and-now" situations such as play like in the example with Meg. In the other example, the teacher observed Erik's play with blocks, and she had the opportunity later on to reflect with him about his choices. Maybe it could have been more mathematically useful if the teacher talked to him in the moment about how different shapes can be used to ensure that the construction is tall and stable. The didactic choices include questions directed at both the mathematical content and children's perceptions of the specific content. Here, flexibility around the mathematical content in children's activities "in the moment" and responsiveness to what the children direct their attention to are important aspects of teaching.

Conclusion

The study shows that in Erik's and Meg's exploration of shapes, possible patterns and so on, children create teachable and learnable moments in "here-and-now" situations, with play as the starting point. Teachers can observe and reflect on the experiences that children have in play and thereby provide teachable and learnable moments. Teacher's use of feedback strategies like attention, questioning and statement is also of importance. It could be teachers' questions together with various play materials that support children's mathematizing and learning in preschool rather than instructions. Play is a valuable part of children's everyday lives and can give teachers opportunities to encounter and reflect on mathematics from a child's perspective. Children have their own intentions in play, and teachers need to be attentive to these intentions and the child's experiences in the teaching situation.

References

Ahlberg, A. (1998). *Meeting mathematics: Educational studies with young children*. Göteborg: Acta Universitatis Gothoburgensis.

Bäckman, K. (2015). *Mathematical formation in the preschool*. Åbo: Åbo akademis förlag.

Bruner, J. S. (2002). *Making stories: Law, literature, life*. New York: Farrar, Straus, and Giroux.

Carruthers, E., & Worthington, M. (2006). *Children's mathematics: Making marks, making meaning* (2nd ed.). London: Sage.

Cheeseman, J. (2015). Mathematical conversations that challenge children's thinking. In B. Perry et al. (Eds.), *Mathematics and transition to school* (Early mathematics learning and development, pp. 273–293). Singapore: Springer.

Claessens, A., & Engel, M. (2013). How important is where you start? Early mathematics knowledge and later school success. *Teachers College Record, 115*(6), 1–29. Columbia: Columbia University.

Clements, D. H., & Sarama, J. (2009). *Learning and teaching early math: The learning trajectories approach*. New York: Routledge.

Ginsburg, H. P. (2006). Mathematical play and playful mathematics: A guide for early education. In D. G. Singer, R. M. Golinkoff, & K. Hirsh-Pasek (Eds.), *Play=learning: How play motivates and enhances children's cognitive and social-emotional growth* (pp. 145–165). New York: Oxford University Press.

Ginsburg, H. P., & Amit, M. (2008). What is teaching mathematics to young children? A theoretical perspective and case study. *Journal of Applied Development Psychology, 29*(4), 274–285.

Ginsburg, H. P., & Ertle, B. (2008). Knowing the mathematics in early childhood mathematics. In O. N. Saracho & B. Spodek (Eds.), *Contemporary perspectives on mathematics in early childhood education* (pp. 45–66). Charlotte, NC: Information Age.

Hyun, E., & Marshall, J. D. (2003). Teachable-moment-oriented curriculum practice in early childhood education. *Journal of Curriculum Studies, 35*(1), 111–127.

Munn, P., & Kleinberg, S. (2003). Describing good practice in the early years—A response to the 'third way'. *Education 3–13, 31*(2), 50–53.

National Agency for Education (2011). Curriculum for the Preschool Lpfö 98: Revised 2011. Stockholm: National Agency for Education.

Pramling Samuelsson, I., & Fleer, M. (Eds.). (2008). *Play and learning in early childhood settings*. Dordrecht: Springer.

Starkey, P., & Klein, A. (2008). Sociocultural influences on young children's mathematical knowledge. In O. N. Saracho & B. Spodek (Eds.), *Contemporary perspectives on mathematics in early childhood education* (pp. 253–276). Charlotte, NC: IAP-Information Age.

Tirosh, D., Tsamir, P., Levenson, E., & Tabach, M. (2011). From preschool teachers' professional development to children's knowledge: Comparing sets. *Journal of Mathematics Teacher Education, 14*(2), 113–131.

Van den Heuvel- Panhuizen, M., & Buys, K. (2005). *Young Children Learn Measurement and Geometry. A Learning Teaching Trajectory with Intermediate Attainment Targets for Lower Grades in Primary school*, Freudenthal Institute, Utrecht University, Utrecht.

Van Hiele, P. M. (1959/2004). The child's thought and geometry. In T. P. Carpenter, J. A. Dossey, & J. L. Koehler, (Eds.), *Classics in mathematics education research*. Reston, VA: National Council of Teachers of Mathematics.

Van Oers, B. (2010). Emergent mathematical thinking in the context of play. *Educational Studies in Mathematics, 74*(1), 23–37.

Vygotsky, L. S. (1978). *Mind in society: The development of higher psychological processes*. Cambridge, MA: Harvard University Press.

Wager, A. A., & Parks, A. N. (2014). Learning mathematics through play. In S. Edwards, M. Blaise, & L. Brooker (Eds.), *The Sage handbook of play and learning in early childhood*. London: SAGE. eBook.

Wood, E., & Attfield, J. (2005). *Play, learning and the early childhood curriculum* (2nd ed.). London: Paul Chapman.

Preschool Children Learning Mathematical Thinking on Interactive Tables

Dorota Lembrér and Tamsin Meaney

Abstract In many countries around the world, young children use different kinds of information and communication technologies (ICT) on a daily basis. In this chapter, the use of games or apps on these technologies is explored in relationship to children's learning of mathematical thinking. The work of Biesta on education and socialisation is combined with that of Radford on subjectification and objectification to theorise young children's learning of mathematical thinking. Two Swedish preschool children's interactions with a balance game on an interactive table are analysed to consider the value of this theory and what it says about the affordances of the game.

Introduction

In this chapter, we explore how young children could learn mathematical thinking through information and communication technologies (ICT). Our aim is to develop criteria for evaluating games and apps purported to support young children learning mathematics, something that we consider is missing in current mathematics education research about ICT. Such an exploration is needed because currently a large amount of money is being spent on putting different kinds of technologies into preschools, including in Sweden.

When new games are produced for ICTs, often there is limited evaluation of them from the perspective of children's mathematical thinking. In regard to commercial games for preschoolers, Lange and Meaney (2013) investigated one child's interactions with a variety of applications on a tablet. They found that mathematical concepts appeared in the apps with some features appearing to be more likely to

D. Lembrér (✉)
Malmö University, Malmö, Sweden
e-mail: dorota.lembrer@mah.se

T. Meaney
Bergen University College, Bergen, Norway
e-mail: tamsin.meaney@hib.no

T. Meaney et al. (eds.), *Mathematics Education in the Early Years*,
DOI 10.1007/978-3-319-23935-4_13

support discussions about mathematics. However, they did not investigate what the child learnt from playing the games. Similarly, Palmér and Ebbelind (2013) investigated the potential of using tablets when teaching mathematics, by elaborating on Bernstein's (1971) notions of framing and classification, when classifying applications. By studying how the designs of different applications could influence the dialogues between teachers and children, the authors suggested that the more control the children had over the interactions with the games, the more possibilities for mathematics to be available for discussion. Although providing useful information about the designs of apps and games, this research also did not provide information about what children learnt.

We begin by discussing mathematical thinking, focusing on how it is learnt, before describing possible affordances of ICT for developing this thinking. In order to trial out these criteria, we analyse how two Swedish preschool children play a game using a virtual balance on an interactive table and the mathematical thinking they engage in.

Mathematical Thinking

The background document (Utbildningsdepartementet 2010) to the Swedish preschool curriculum (Skolverket 2011) stated that there are four goals which are related to mathematics. One is connected to content, but the other three goals, listed below, require preschools to provide opportunities for children to:

- Develop their ability to use mathematics to investigate, reflect over and test different solutions to problems raised by themselves and others
- Develop their ability to distinguish, express, examine and use mathematical concepts and their interrelationships
- Develop their mathematical skill in putting forward and following reasoning (Skolverket 2011, p. 10)

Although not labelled as such, all of these would contribute to children engaging in what Pimm (1995) described as mathematical thinking—'ways of thinking developed to work on mathematical forms and entities' (p. 167). Nevertheless, definitions of the mathematical thinking of young children are generally rare. Mathematics education researchers working in the early childhood area tend to mention it in passing (see Highfield and Mulligan 2007; Edo et al. 2009; Dockett and Perry 2010) or simply link it to children completing activities using recognisable mathematical content, such as number (see Hunting and Mousley 2009; Tudge and Doucet 2004).

Working with school students, Henningsen and Stein (1997) suggested that there were different kinds of thinking processes connected to mathematics. These 'can range from memorization to the use of procedures and algorithms (with or without attention to concepts, understanding, or meaning) to complex thinking and reasoning strategies that would be typical of "doing mathematics" (e.g., conjecturing, justifying, or interpreting)' (p. 529). The aims of the preschool curriculum seem more in alignment with the complex thinking and reasoning strategies. Thus, it is possible to consider mathematical thinking to involve conjecturing, justifying and interpreting.

Still, there is a question about whether young children's immaturity means that they are unable to engage in this sort of mathematical thinking. It is, after all, described as complex. In relationship to number understandings, Starkey et al. (2004) stated that Piaget with his colleagues considered that 'young children's mathematical thinking is *informal*, because it depends upon the actual presence or mental representation of concrete entities (e.g., sets of concrete objects) and the transformation of those entities' (p. 100). Although children could conjecture, justify and interpret using concrete materials or their representations, this quote suggests that young children are unable to think mathematically about ideas that are abstract in nature.

An alternative view of mathematical thinking comes from research that draws on Vygotsky. While also recognising the importance of representations, Dijk et al. (2004) stated:

> Mathematical thinking is conceptualized as a form of thinking about quantitative and spatial relationships with the help of symbolic means. The construction of symbolic forms (like schemes, diagrams, drawings) and the reflection on the interrelationships between these forms and their meanings is essential for mathematical thinking. (p. 73)

This is in alignment with Greenes et al. (2004) who considered 'mathematical metacognition—learning to think about and express one's thinking—is of critical importance in the context of mathematics learning' (p. 161). However, in regard to children under the age of 3, van Oers (2010) stated, 'as long as these actions are not intentionally and reflectively carried out, we cannot say that children perform mathematical actions' (p. 28).

This view of the mathematical thinking capabilities of young children depends on adults identifying whether or not young children have carried out their actions intentionally and reflectively. As seems to be the case with van Oers' (2010) comment that very young children cannot be classified as acting mathematically, Lee (2001) proposed that a young child's age affects adults' perceptions of them being able to have opinions and desires. In contrast, there have been calls to consider young children as social actors in their own right, rather than in comparison to adults (Ebrahim 2011). Consequently, an alternative perspective is to consider what mathematical thinking is from what children can do, rather than from what they cannot. If they are able to conjecture, justify and interpret about concrete or abstract mathematical ideas, then they should also be considered as having done so intentionally and reflectively.

Perry and Dockett (1998) described a justification given by a preschooler in which she argued that she should be the mother in a play situation, by holding onto a child to indicate that she had a family, unlike another participant who was also vying to be a mother. This example which highlights the resources that the child utilised, such as her action of holding onto the other child, also shows her understanding that evidence is required for a justification to be considered valid. Thus, the child's intentionality and reflectivity about her mathematical thinking are visible in her actions as well as her words. Therefore, children should not be expected to show their intentionality and reflectivity through verbal fluency alone as their actions and gestures can provide evidence that they can think about a mathematical idea differently (see Johansson et al. 2014).

Drawing on Henningsen and Stein's (1997) description of complex mathematical thinking, we define mathematical thinking as conjecturing, justifying and interpreting. However, rather than just relating these to quantitative and spatial relationships, we consider that these processes can be connected to the mathematical activities identified by Bishop (1988). These activities are playing, explaining, designing, locating, measuring and counting. In particular, we consider that explaining is closely aligned with mathematical thinking. Using such a definition of mathematical thinking supports us in identifying how the affordances of some ICT games or apps contribute to it.

Learning

Consequently, learning to think mathematically means learning to conjecture, justify and interpret. As opposed to traditional views of learning as a cognitive activity (Gifford 2004), Radford (2008), building on Vygotsky's work, considered learning as becoming progressively conversant with the collectively and culturally constituted forms of reflection. Learning is 'not just about knowing something but also about becoming someone' (Radford 2008, p. 215). In this way, the object of learning is not only within the awareness of the learner, but the learner himself/herself is part of what is to be appropriated in the learning process. Learning to conjecture, justify and interpret will involve a child in learning about themselves and their relationship to mathematics and mathematical thinking. Learning embeds the child within the historically developed societal context around mathematical thinking. Reflection of this kind contributes to the development of what Radford calls the *communal self*, an individual who is part of the social and historical environment in which knowledge and thus the individual's possibilities for becoming have been developed. Reflection is the necessary tool for learning to occur, but as individuals reflect in combination with others, their reflective capabilities also develop so that they mirror the forms of reflection valued by the societies in which they live.

Although not his focus, Radford (2008) noted that learning can produce new understandings. In contrast, Biesta (2007) made a distinction between socialisation and education to highlight the importance of learning for adapting to new situations. What Radford (2008) described as learning, Biesta (2007) considered to be socialisation—"insertion of 'newcomers' into existing cultural and socio-political settings" (p. 26). Biesta described how education as defined by Kant was about the self-education needed to achieve rational autonomy in order to become fully human. Thus, education from this perspective is a form of socialisation because it sets up what the end product of self-education had to be—rational autonomy.

Individuals must take on the attributes of existing members of a society but without necessarily recognising the role of the community in the process (Radford 2008). For Biesta (2007), this philosophy suggests that those who did not have or did not gain the appropriate attributes were unable to be considered human, including young children.

Consequently, Biesta (2007) postulated that education would be better deemed as a preparation for an uncertain future, where he stated freedom 'needs to be realised again and again' (p. 32). This is because adulthood in the twenty-first century is less stable, uncertain and thus predictable. Stabilities, such as having the same job, are not expected to last for a person's lifetime. Lee (2001) suggested that to reflect this reality, conceptualisations of childhood need to accept and respond to this uncertainty. In so doing, new definitions of what it means to be human can be produced (Biesta 2007). In regard to young children's learning of conjecturing, justifying and interpreting, there is a need to recognise that young children should not merely reproduce established forms but explore them so new forms can arise. The reflection from this exploration is likely to contribute to an understanding about how culturally valuable ways of doing mathematics, such as mathematical thinking, are socially constructed and adapted to meet new situations.

Learning about mathematical thinking requires reflection, in our case about conjecturing, justifying and interpreting. This reflection contributes to a child's developing subjectification and also their metalevel understanding about the role that reflection itself plays in learning. Learning needs to open up possibilities for new kinds of conjecturing, justifying and interpreting by valuing not just current culturally constituted forms but by valuing possibilities for exploration to identify when alternative forms may be needed and what these might be. Such a view of learning positions all children as human, whether or not they can articulate their intentions and reflections.

While acknowledging the contribution that artefacts make to the learning process, Radford (2008) stated:

> Objects cannot make clear the historical intelligence that is imbedded in them. This requires that they be used in activities as well as in contact with other people who know how to—read this intelligence and help us to acquire it. (p. 224)

Perceptions of ICT games and apps for learning conjecturing, justifying and interpreting are discussed in the next section.

ICT

The importance of being familiar with ICT has been highlighted in reports and policies for some time. For example, the Commission of the European Communities (2000) stated that all 'school-leavers must be digitally literate in order to be prepared for a knowledge-based economy' (p. 8). Directive such as this has resulted in preschools in many Western countries being flooded by various forms of ICT.

After reviewing the literature on children using ICT, Sarama and Clements (2009) suggested that 'compared with their physical counterparts, computer representations may be more manageable, flexible, extensible, and "clean" (i.e., free of potentially distracting features)' (p. 147). This suggests that the affordances of computers could make them more advantageous for developing mathematical thinking than physical objects. Although noting weak empirical evidence for some ideas, Sarama and Clements (2009) suggested seven affordances that computer manipulatives provide to learners. We discuss each one of these in relationship to learning conjecturing, justifying and interpreting through ICT-based activities. However, the research that we draw on mostly comes from small-scale studies because studies are rare of how ICT could be used to develop preschool children's learning about mathematical thinking. Unlike schools, there is much more variety in the format for preschools found around the world, making large-scale studies that could be generalised to a range of circumstances difficult:

1. Bringing mathematical ideas and processes to conscious awareness
   In our discussion of learning, the role of reflection was highlighted. Conscious awareness of conjecturing, justifying or interpreting would seem to be a necessary component of reflection. Sarama and Clements (2009) stated that it was the way that this affordance was used within a task or through teacher guidance that supported children to become consciously aware of the mathematics. Consequently, it is only when designers actively incorporate this affordance into ICT-based activities that conscious awareness of the need for conjecturing, justifying and interpreting would be likely to occur.
2. Encouraging and facilitating complete, precise explanations
   Complete and precise explanations are components of mathematical conjecturing and justifying. Computer software can contribute to this in a number of ways. For example, Clements (2002) suggested that having children write commands for the logo turtle required them to think more carefully about the characteristics of a geometric shape. Explorations of this nature are not so easy to achieve when children use paper and pencil. Although young children's developing hand-eye coordination would seem to be facilitated by computers (Highfield and Mulligan 2007), the relationship between this and precise explanations is not clear.

   Much of the research that looks at children's use of ICT highlights the adult's role suggesting that it is the interactions with adults that allow for precise explanations to appear (Perry and Dockett 2007; Lee and Ginsburg 2009). On the other hand, Ginsburg (2006) provided an example of a preschool child correcting another's language so that it became more precise. As well, Pareto et al. (2012) found that children in Year 3 who played an interactive game in pairs increased their confidence in giving mathematical explanations. Lamberty (2007) in designing her software program DigiQuilt for school children built in feedback about the mathematical concepts, such as fractions. Therefore, the software contributed to the precision of the explanations. Nevertheless, it remains unclear how the design of games and apps can contribute to children providing precise mathematical explanations.

3. Supporting mental 'actions on objects'
   Doing mental actions on objects could contribute to young children's interpretations of mathematical representations and so could be connected to learning mathematical thinking. Clements and Sarama (2007) showed that young children gained significant mathematical understandings from using computer manipulatives to construct and deconstruct geometric shapes. However, Highfield and Mulligan (2007) suggested that more research is needed on how computer manipulatives support visualisation, especially with preschool children.
4. Changing the very nature of the manipulative
   In their small study, Highfield and Mulligan (2007) showed how preschool children could use computer manipulatives to engage with mathematical ideas, such as scaling and shearing, which are more difficult to do outside a virtual environment. Therefore, there seems to be affordances for the altering aspects through computer manipulatives. However, it is not clear how this affordance relates to learning about conjecturing, justifying and interpreting.
5. Symbolising mathematical concepts
   The work of Ladel and Kortenkamp (2012, 2013) on multitouch tables suggests that this type of technology may allow children to move their understandings about amounts using their fingers to symbolic forms. Although conjecturing, justifying and interpreting could involve symbolising mathematical concepts, this is not a necessary component of these processes.
6. Linking the concrete and symbolic with feedback
   In Lamberty's (2007) work, students used virtual quilt blocks to design a 16-block quilt. Lamberty made a conscious decision to include feedback on the fractional amount of a block that was filled by one specific colour. However, like the examples that Sarama and Clements (2009) provide, the symbolic feedback about fraction and angle size is not intuitively understandable and thus may be difficult for young children to interpret. As well, symbolic representation may not be necessary for children learning about conjecturing, justifying and interpreting.
7. Recording and replaying students' actions
   Riesbeck (2013) showed how Swedish Year 2 students were able to use the recording features of a tablet to produce videos of themselves explaining mathematical properties. The reflections that the children engaged in during the recording allowed them to refine their explanations. Most forms of technology allow for recordings to be made and replayed, and it may be possible for children to refine their conjecturing, justifying and interpreting skills using these features.

In our analysis of a video of two Swedish children using a virtual balance game on an interactive table, we consider which of these affordances are present and how they contribute to children learning mathematical thinking. This is discussed in more detail in the next section.

Methodology

The data analysed in this study comes from the trialling stage of the project which followed the development of games for an interactive table. Interactive tables are similar to interactive whiteboards but allow young children to stand beside them as they use them. Research with older children, such as Davidsen and Georgsen's (2010), indicated that interactive tables, such as the SmartTable™, have the potential to foster children's mathematical engagement while also developing their language abilities, including their possibilities for conjecturing, justifying and interpreting. We considered that collaboration between young children might be facilitated by the layout of the interactive table and thus had the possibility to support children's learning in a way that tablets or computers do not. However, this technology has not been researched in any detail in regard to preschool children, and few software applications are available for it. Ladel and Kortenkamp's (2012, 2013) work is the only research that we found in mathematics education. This meant that many decisions had to be tested during the design process (see Lembrér et al. 2014).

As we could not find appropriate mathematical games or apps for preschool children, four games were designed as pilot tasks. The wider project followed this development to identify the components of Bower's (2008) affordance analysis of e-learning design methodology. The task designers, first-year university students enrolled in a gaming design course, decided on and developed the actual games, which were memory, shapes, balance and cubes.

In this study, two children's use of the balance game is explored (see Fig. 1) to see which of Sarama and Clements' (2009) affordances were available in the game and seemed to contribute to children developing their mathematical thinking.

As can be seen in Fig. 1, the game consisted of a virtual set of scales, with different combinations of cubes available at the bottom of the screen which could be dragged to the scale pans. The cubes have different sizes and colours, and weight

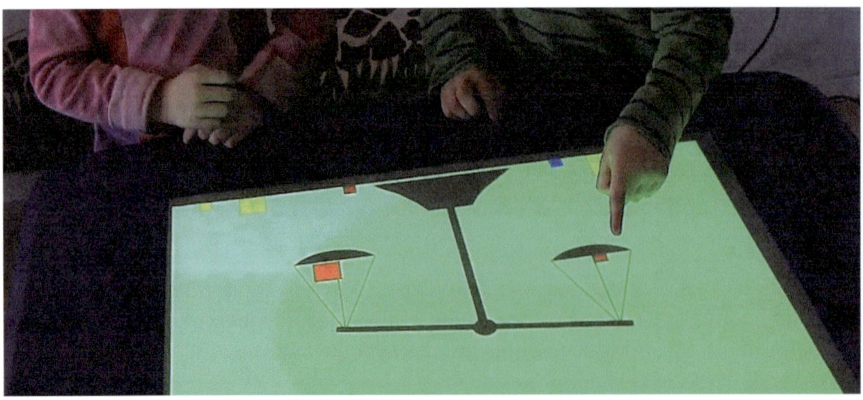

Fig. 1 The balance game

could be inferred from the movement of the balance arms. The balance bar tips to suggest that one side is heavier than another. The children had previously had some experiences with a real balance. When the pans are considered to be equal in weight, a light in the middle of the balance bar glowed green. In Fig. 1, a small cube can be seen in the left pan. The larger cube in the right pan has lines drawn on it to indicate visually that it is equivalent to four of the smaller cubes.

In most runs of the game, the amount of blocks needed to be the same, so that although the discussion was about heavier and lighter, often the evidence the children used was the amount of small cubes on each side. However, on one occasion equivalence was not achieved with the same amount of weights. This is discussed in section "Justification".

Our data consisted of a 6-and-a-half-minute video recording of two Swedish preschool children, aged between 4 and 5 years, using a balance game on an interactive table and responding to the teacher's questions. The teacher held the camera and is in front of the children. This video was one of several recorded during the trialling of the games with different groups of children at one preschool. We chose it because it seemed typical of the kinds of interactions that this game invoked and thus seemed valuable in providing information about whether Sarama and Clements' (2009) list of affordances provided insights into children's use of mathematical thinking.

The video was first analysed to determine when the children seemed to be using conjecturing, justifying and/or interpreting. Conjecturing was considered to involve making a suggestion about the outcome of a potential action. This suggestion had to be more than a guess and involve some sort of evidence base or reflection. Justifying was judged to be when the children provided evidence for a statement about the balance. Interpreting was when the children described in words or action what they saw happening using mathematical concepts from Bishop's six activities, mostly measuring and counting.

These episodes were then reanalysed to identify if and how the children were learning about conjecturing, justifying and interpreting. This was done by determining if there were changes in these processes which came from the children watching each other's action, listening to each other's points or responding to the teacher's questions or suggestions. We then considered whether the changes seemed to show a movement towards mathematically valued forms of conjecturing, justifying or interpreting or whether the changes explored alternatives. We then determined how the game itself supported the changes and how these were related to Sarama and Clements' (2009) list of affordances.

In the next sections, we discuss the situations in which children seemed to be learning about conjecturing, justifying and interpreting and how the game and the interactive table seemed to contribute to this learning. The children's names have been changed in the transcripts. Each transcript includes the original Swedish as well as an English translation. It is worth noting that these young children's fluency in Swedish was developing and they did not always use complete or clear sentences. This made translations into English somewhat difficult. Time indications are provided with each transcript.

Learning Mathematical Thinking

There were several ways in which the game contributed to learning through culturally constructed forms of reflection (Radford 2008), often by prompting the teacher to ask questions. In some interactions, the children's learning about conjecturing, justifying and interpreting was clearly integrated. This is to be expected in that it is hard to conjecture without having interpreted the situation and having a justification for considering the conjecture to be valid. However, justifications were often intrinsic and rarely would the children provide an explicit one without being prompted. This means that in regard to conjecturing and interpreting as well as justifying, the children's intentionality had to be determined from their actions.

Conjecturing

The game encouraged the children to provide conjectures about how to make the balance even. After the first runs of the game, they realised that the green light in the centre of the balance beam indicated when they had been successful. The following episode which occurred with Fig. 1 shows how the teacher's interactions contributed to the children verbalising their conjectures.

(0:00:05.9)Läraren: Vilken sida är tyngst nu?	Teacher: Which side is heavier now?
Albin och Anna tittar upp mot läraren och ler.	Albin and Anna look up at the teacher, smiling.
Anna: Min	Anna: Mine
Anna svänger med kroppen och ler.	Anna swinging her body while smiling.
Läraren: Din. Hur ska vi göra för att det ska vara lika?	Teacher: Yours. What shall we do for it to be equal?
Albin: Ta lika många på varje sida.	Albin: Take an equal number on each side.

Although it can be presumed that these children already had some sense of what it meant to make appropriate conjectures in regard to the game, the teacher's questions highlighted that she valued verbalising what they planned to do before they carried out the action. Over 2 min later (see Fig. 2), Albin indicated to Anna how she could put two single cubes onto her pan to equal the two-cube block in his pan.

(0:02:19.9) Nytt spel. Albin tittar mot Annas sida vågen och drar upp en röd kub2. Han pekar på två gula kuber på Annas sida.	New game. Albin looks at the pan on Anna's side and pulls up a red 2 cube. He points to two yellow cubes on Anna's side.
Albin: Då kan du ta de två.	Albin: Then you can take the two.
Anna drar upp två gula kuber i vågskålen.	Anna pulls up two yellow cubes in the balance.
Läraren: Kommer det bli samma då?	Teacher: Will it be the same then?
Anna: Ja.	Anna: Yes.

Fig. 2 Two single blocks are the same as a two-cube block

Anna lägger armarna i kort och ser nöjd ut.	Anna crosses her arms and looks pleased.
Läraren: Hur vet ni det?	Teacher. How do you know?
Albin: En, två. En, två.	Albin: One, two. One, two.
Albin pekar på vågskålarna och kuberna samtidigt som han förklarar med ord.	Albin points on the scales and the cubes while he explains with words.
Läraren: En, två. En, två	Teacher: One, two. One, two.

In this episode, Albin can be considered to have accepted the need to verbalise his conjecture that the two single cubes are the same as one two-cube block. In addition, the set-up of the game in which each child was responsible for one of the pans meant that he needed Anna to carry out this action, making the verbalisation essential. Up till this point, Anna had seemed uncertain about the relationship between larger blocks of cubes being the same as several smaller blocks of cubes. For the first half of the video, she put smaller blocks as close together as possible so that they looked the same as the larger blocks. This possibility within the game allowed her to form conjectures later, as can be seen in section "Interpretation".

Justification

As can be seen in the last episode, quantifying the amounts on each pan was a common way for the children to justify why the pans were equal. Like the conjectures, justifications were provided almost always as a response to a prompt by the teacher. However, the children's control of when to make a new game often frustrated the teacher's attempts to have the children reflect on why the balance was even. In the following episode (see Fig. 3), Albin uses groups of three, two on each side, to justify why the balance beam is level. This was in contrast to Anna's justification which was based on the positioning of her two 1×3 blocks of cubes.

Fig. 3 Justifying sameness

Läraren: Hur vet ni att det är lika nu då?	Teacher: How do you know that it is same now?
Anna: För att jag satte den bredvid där.	Anna: Because I put it next to there.
Anna pekar på sina kubstaplar som står bredvid varandra i vågskålen.	Anna points to their cube groups standing next to each other in the pan.
Albin: (Sjunger nästan) Det är tre och de är tre och det och det är tre.	Albin: (Almost singing) There are three and they are three and there and that's three.
Albin visar på staplarna med sitt pekfinger.	Albin points at the cube groups with his index finger.
(0:02:07.1) Albin byter spel mitt i den pekande rörelsen	Albin changes the game in the middle of the pointing movement.
Läraren: Ahaaaa. Det är många treor.	Teacher: Ahaaaa. There are many threes.

The way the game is set up allows the children to use different kinds of justifications. At this point in Anna's development of mathematical understanding, the sameness of the shapes on the two pans is what makes them equivalent. For Albin, it is that he can see equivalent number of groups of three-cube blocks. This suggests that shape is still a determining factor in his understanding of sameness, but he can describe this using amounts for his justification. The game does not value any particular justification as it only indicates when equivalence is achieved. However, the teacher's repetition of Albin's focus on the threes could have contributed to the children sensing what the teacher considered to be an appropriate justification.

In a later episode, the teacher reinforced the value of knowing the amount of cubes to justify the difference between equivalence as determined by the balance and sameness of amounts (see Fig. 4).

Fig. 4 Sameness of amounts is different to equivalence on the balance

Vågen visar tyngre på Albins sida.	Balance shows that Albin's side is heavier.
(0:05:06.5) När Albin tar bort en gul stapelkub2 × 3 flimrar punkten grönt en kort stund. Albin placerar gul stapelkuben2 × 3 på foten av vågen. Nu lyser punkten grönt med ett långsamt pulserande.	When Albin removes a yellow 2 × 3 cube group, the balance point flickers green briefly. Albin puts yellow 2 × 3 cube group at the foot of the balance. Now the balance point shines green with a slow pulse.
Lärare: (drar in andan som av spänning) Kolla, hur blev detta nu?	Teacher: (draws breath as with tension) Look, how was this now?
Anna: Han tog av en kloss.	Anna: He took a block.
Läraren: Hur många rutor har du Albin, hur är din sida?	Teacher: How many cubes have you Albin, how is your side?
Albin räknar kuberna på sin sida han pekar på kuberna.	As Albin counts the cubes on his side, he points to them.
Albin: 1 2 3 4 5 6 7 8 9 10 10!	Albin: 1, 2, 3, 4, 5, 6, 7, 8, 9, 10. 10!
Läraren: Och hur många har Anna?	Teacher: And how many has Anna?
Anna räknar kuberna på sin sida hon pekar på kuberna.	As Anna counts the cubes on her side, she points to them.
Anna: 1 2 3 4 5 6 7 8 9 10 11 12 16	Anna: 1, 2, 3, 4, 5, 6, 7, 8, 9, 10, 11, 12, 16.
Läraren: Var där 16? Albin du hjälper oss och räkna en gång så vi höra så det blir riktigt.	Teacher: Were there 16? Albin help us and count again so we hear that it becomes correct.
Albin: 13!	Albin: 13!
Albin klappar i hop händerna.	Albin hops, claps his hands.
Läraren: Är där det?	Teacher: Is it there?
Albin: Mmm	Albin: Mmm
Läraren: 13… Och du hade 10 och ändå står det på lika.	The teacher: 13 … And you were 10 and it still shows it as the same.

Albin tittar på läraren och slår händerna mot varandra.	Albin looks at the teacher and claps his hands together.
Läraren: Hur kan detta nu hända?	Teacher: How can this happen now?
(0:05:49.1) Albin byter spel.	Albin change games.

This run of the game produced an unexpected result in that the amounts of cubes were different, but the balance beam indicated that they were equivalent. The justification that the teacher was prompting the children to make was about what was different and what was the same. The children showed initial interest in this problem. However, after counting the blocks to implicitly justify that there was a difference, the teacher is unable to extend the conversation as Albin changed the game.

The game contributed to bringing up a problem which allowed the teacher to support the children to provide a justification. At the same time, the children's ability to change the game meant that the teacher was unable to prompt for further reflections which could have led to conjectures about possible reasons for such a result.

Interpretation

In the first episode in "Conjecturing" section, the teacher used the term 'heavier' to indicate to the children that she interpreted what was happening in the game as modelling a real set of scales. In the next episode, Anna used the term 'same weight' to show that she also interpreted the situation as modelling a real balance.

When an unexpected event occurred in the game, the children needed to make sense of it. In the following episode, although Albin had taken out some of the blocks of cubes from his pan, the balance did not move (see Fig. 5).

Programmet har inte reagerat på att Albin har flyttat sina kuber. Det tror att kuberna är kvar i vågskålen.	The program has not responded to Albin moving his cubes. It considers that the cubes are still in the balance.
Albin försöker flytta röd stapelkub2 × 3	Albin tries to move the red 2 × 3 cube block.
Läraren: Vad har hänt där nu?	Teacher: What's happening there now?
Anna: Vi har lika tungt.	Anna: We have the same weight.
Läraren: Har ni det lika tungt?	Teacher: Do you have the same weight?
Albin: Mhm	Albin: Mhm.
Läraren: Det ser ju konstigt ut på vågen. Emils sida är mycket längre ner än din.	Teacher: It looks a bit strange on the scale. Albin's side is far lower than yours.
(0:04:13.2)Vågen visar att Albins sida är tyngre, det ligger lika många kuber på varje sida.	The balance shows that Albin's side is heavier, but there is an equal number of cubes on each side.
Albin flyttar på gul stapelkub2 × 3 och då rör sig vågen.	Albin moves a yellow 2 × 3 group and then the balance moves.
Albin: Det är bara för att den var fast där.	Albin: It's just because it was stuck there.
När Albin förklarar visar han samtigt vilken gul stapelkub2 × 3 han menar.	As Albin explains, he shows simultaneously which yellow 2 × 3 cube group he means.

Fig. 5 When the same amount is not equivalent on the scales

Fig. 6 Interpreting the *green* button

In this episode, it seems that Anna interpreted the equal number of cubes of each side to mean that the pans had the same weight as this had been the appropriate interpretation in early runs of the game. However, the teacher queried this interpretation by highlighting that the pans were not level. In this way, she indicated that there were alternative ways for Anna to interpret what they were seeing. It is worth noting that the teacher did not tell Anna that she was wrong; rather, she indicated that her interpretation that they had the 'same weight' did not match how she, the teacher, interpreted how the pans were sitting.

The next example shows how the children interpreted the green button as showing that they had the correct answer (see Fig. 6). Although the teacher prompted for a verbal justification, neither child seemed to consider that this was necessary; instead, they pointed to the green button.

Albin drar upp en röd stapelkub2 × 3 Anna tittar på vad Albin gör, tittar sedan på sina egna kuber och drar upp en röd stapelkub2 × 3.	Albin pulls up a red 2 × 3 cube block, Anna looks at what Albin does, then looks at her own cubes and pulls up a red 2 × 3 cube block.
Läraren: Nämen, näee kolla! (uppmuntrande).	Teacher: Why, wait, check! (encouraging)
Vågen blir grön.	Balance turns green.
(0:02:14.8) Albin pekar på den gröna punkten. Anna vänder sig mot läraren med armarna i kors och tittar rakt på läraren och ler. Samtidigt tittar Albin ner i spelet och byter till nästa spel direkt.	Albin points to the green dot. Anna turns to the teacher with arms folded and looks straight at the teacher, smiling. While Albin looks at the game, he changes the game.

As noted in Lembrér et al. (2014), games for young children need to be intuitive as they cannot read instructions and might struggle if they are explained orally. Therefore, features such as the green button need to be interpreted by the children if the game is to be played successfully. The green light was mentioned early by the teacher, and the use of green as a positive colour in Western society could support the children to learn to make a similar appropriate interpretation. Simultaneously the interpretation of green to mean success in this game is also likely to contribute to children interpreting other situations which used green to also mean they were successful. Learning to interpret is more than learning to listen, especially in mathematics where visual displays and symbols carry much information.

Affordances of the Balance Game

In the video, the children showed intentionality and reflectivity (Radford 2008) in resolving the problems in the balance game. We also suggest that in doing so, they were engaged in learning about conjecturing, justifying and interpreting as well as exhibiting the knowledge of these that they already had. In this section, we discuss the features of the balance game and the interactive table which seemed to contribute to the children's learning and how they related to Sarama and Clements' (2009) list of affordances of computer manipulatives.

As was noted by Sarama and Clements, we acknowledge the importance of the teacher. The balance game influenced the mathematics that was possible for the children to learn, but it was the teacher who focused their attention on the mathematical aspects. In the balance game, the children had a range of options for making the balance even, and this prompted the teacher to ask several times, 'how do you know?' Although she also asked closed questions, to which she knew the answer, the children's possible choices seemed to support her to find out about their thinking. This result is similar to that found by Palmér and Ebbelind (2013), and the teacher's role and the type of questions they asked were important for supporting

German preschool children to externalise their understandings using multitouch table (Ladel and Kortenkamp 2012).

Of the seven affordances nominated by Sarama and Clements (2009), we suggest that the ones that would most likely support learning about conjecturing, justifying and interpreting were:

1. Bringing mathematical ideas and processes to conscious awareness
   By bringing mathematical thinking into the children's conscious awareness, the teacher also supported them to learn what were the valuable features of conjecturing, justifying and interpreting. For example, in section "Justification" where Anna used evidence about the shapes and Albin used numbers to indicate the same amounts, the teacher only repeated Albin's justification. This valuing of amounts as numbers was reinforced later when the teacher had the children count the amounts in their pans to form a justification about the differences.
2. Encouraging and facilitating complete, precise explanations
   None of the children's conjectures, justifications or interpretations could be consider as complete, although they might be considered precise. Yet, within the game the children's contributions appeared to be understood by each other and the teacher. As with the previous affordance, it is a game which allows for multiple possibilities and so supports a teacher to ask questions which encourage children to give explanations of their conjectures, justifications and interpretations. However, the children's control of when to change the game often frustrated the teacher's attempts to have the children reflect more on their mathematical thinking. Increasing children's agency through giving them control of the game means that a need for explanations needs to be built in (Lange and Meaney 2013). If, as was the case with the balance game, the teacher could not contribute to the children developing reflective understandings about mathematical thinking, there would be a limited development of their communal selves (Radford 2008).
3. Supporting mental 'actions on objects'
   Conjecturing requires manipulation of mental actions on objects. At the start of the video, the children's actions in pulling up different blocks to put them on the pans suggested that they were most comfortable with working with small blocks of cubes. Initially Anna needed to place blocks of cubes together to accept that they were the same as the larger blocks on Albin's pans. The children's explorations and the teacher's prompting meant that by the end of the video, they seemed more comfortable to work with large blocks, using those on both their own and their partner's side to ensure that the scales became equal. This meant that they were manipulating different possibilities to ensure that they minimised their actions. In learning mathematical thinking, children need to use their current knowledge and skills to reflect on the knowledge and skills they are acquiring. This is what seemed to be occurring as children became more familiar with the game, suggesting that mental actions on objects require familiarity with what kind of actions are possible.

4. Recording and replaying students' actions
   The game did not make use of any record or replay options. Given the difficulties with having children reflect before they changed the game, it might have been useful to have at least been able to record a particular situation before the game was changed. This might have supported the teacher to have a session away from the interactive table which focused solely on reflections. In Lange and Meaney's (2013) research, one app allowed the child to take a 'photo' of their drawing which was stored in the tablet's gallery. If such a possibility was included in the game, then it would be possible for the teacher to show it to the children the next day and have another reflective discussion about the children's mathematical thinking.

However, it also seemed that Sarama and Clements' (2009) list did not include all the features that contributed to children learning about conjecturing, justifying and interpreting. The game utilised the layout of the interactive table so that it was possible for the two children to have their own pans but need to interact in order for the scales to be equal. As others have found with older children, interactive tables have the possibilities to support collaboration (Harris et al. 2009). Some features of the game which supported children's intuitive interactions, such as the green button, made use of cultural knowledge. Understanding how cultural knowledge can be integrated into mathematical thinking is an important part of developing Radford's (2008) communal self. By playing with the balance game, children may have constructed new knowledge about what it meant to engage in mathematical thinking and so participated in a process of cultural reproduction. Becoming aware of this knowledge can be seen as an active process of understanding and interpreting through which the children showed that they were capable of encountering new experiences and producing new knowledge. Consequently, learning about mathematical thinking is a process of reflection, and opportunities for reflection need to be built into games for ICT.

References

Bernstein, B. B. (1971). On the classification and framing of educational knowledge. In M. F. D. Young (Ed.), *Knowledge and control* (pp. 47–69). London: Collier-Macmillan.

Biesta, G. (2007). The education-socialisation conundrum or 'Who is afraid of education?'. *Utbildning & Demokrati, 16*(3), 25–36.

Bishop, A. J. (1988). Mathematics education in its cultural context. *Educational Studies in Mathematics, 19*, 179–191.

Bower, M. (2008). Affordance analysis—Matching learning tasks with learning technologies. *Educational Media International, 45*(1), 3–15.

Clements, D. (2002). Computers in early childhood mathematics. *Contemporary Issues in Early Childhood, 3*(2), 160–181.

Clements, D. H., Sarama, J. (2007). Early childhood mathematics learning. In F. K. Lester (Ed.), *Second handbook of research in mathematics teaching and learning* (pp. 461–555). Charlotte, NC: Information Age.

Commission of the European Communities. (2000). *eEurope 2002: An information society for all.* Brussels: European Commission.

Davidsen, J., & Georgsen, M. (2010). ICT as a tool for collaboration in the classroom—Challenges and lessons learned. *Designs for learning, 3*(1-2), 54–69.

Dijk, E. F., van Oers, B., & Terwel, J. (2004). Schematising in early childhood mathematics education: Why, when and how? *European Early Childhood Education Research Journal, 12*(1), 71–83.

Dockett, S., & Perry, B. (2010). Playing with mathematics: Play in early childhood as a context for mathematical learning. In L. Sparrow, B. Kissane, & C. Hurst (Eds.), *Shaping the future of mathematics education.* Proceedings of the 33th annual conference of the Mathematics Education Research Group of Australia (pp. 715–718). Freemantle, Australia: MERGA.

Ebrahim, H. (2011). Children as agents in early childhood education. *Education as Change, 15*(1), 121–131.

Edo, M., Planas, N., & Badillo, E. (2009). Mathematical learning in a context of play. *European Early Childhood Education Research Journal, 17*(3), 325–341.

Gifford, S. (2004). A new mathematics pedagogy for the early years: In search of principles for practice. *International Journal of Early Years Education, 12*(2), 99–115.

Ginsburg, H. P. (2006). Mathematical play and playful mathematics: A guide for early education. In D. Singer, R. M. Golinkoff, & K. Hirsh-Pasek (Eds.), *Play = Learning: How play motivates and enhances children's cognitive and social-emotional growth* (pp. 145–165). New York: Oxford University Press.

Greenes, C., Ginsburg, H. P., & Balfanz, R. (2004). Big math for little people. *Early Childhood Research Quarterly, 19*, 159–166.

Harris, A., Rick, J., Bonnett, V., Yuill, N., Fleck, R., Marshall, P., et al. (2009, June). Around the table: Are multiple-touch surfaces better than single-touch for children's collaborative interactions? In *Proceedings of the 9th international conference on computer supported collaborative learning—volume 1* (pp. 335–344). International Society of the Learning Sciences.

Henningsen, M., & Stein, M. K. (1997). Mathematical tasks and student cognition: Classroom-based factors that support and inhibit high-level mathematical thinking and reasoning. *Journal for Research in Mathematics Education, 28*(5), 524–549.

Highfield, K., & Mulligan, J. (2007). The role of dynamic interactive technological tools in preschoolers' mathematical patterning. In J. M. Watson & K. Beswick (Eds.), *Mathematics: Essential research, essential practice: Mathematics: Essential research, essential practice.* Proceedings of 30th Mathematics Education Research Group of Australasia, Hobart (pp. 372–381). Adelaide: MERGA. Retrieved from http://www.merga.net.au/.

Hunting, R., & Mousley, J. (2009). How early childhood practitioners view young children's mathematical thinking. In M. Tzekaki, M. Kaldrimidou, & C. Sakonidis (Eds.), *Proceedings of the 33rd conference of the International Group for the Psychology of Mathematics Education* (Vol. 3, pp. 201–208). Thessaloniki: PME.

Johansson, M. L., Lange, T., Meaney, T., Riesbeck, E., & Wernberg, A. (2014). Young children's multimodal mathematical explanations. *ZDM, 46*(6), 895–909.

Ladel, S., & Kortenkamp, U. (2012). Early maths with multi-touch—An activity theoretic approach. In M. Pytlak, T. Rowland, & E. Swoboda (Eds.), *Proceedings of the seventh congress of the European society for research in mathematics education* (pp. 1792–1801). Rzeszów: University of Rzeszów.

Ladel, S., & Kortenkamp, U. (2013). An activity-theoretic approach to multi-touch tools in early mathematics learning. *International Journal of Technology in Mathematics Education, 20*(1), 1–6.

Lamberty, K. K. (2007). *Getting and keeping children engaged with constructionist design tool for craft and mathematics.* PhD dissertation, Georgia Institute of Technology, Atlanta, GA. Available from http://smartech.gatech.edu/handle/1853/14589.

Lange, T., & Meaney, T. (2013). iPads and mathematical play: A new kind of sandpit for young children. In B. Ubuz, Ç. Haser, & M. A. Mariotti (Eds.), *Proceedings of the eighth congress of European research in mathematics education (CERME 8)* (pp. 2138–2147). Ankara: Middle East Technical University.

Lee, N. (2001). *Childhood and society: Growing up in an age of uncertainty*. Maidenhead: Open University.

Lee, J. S., & Ginsburg, H. P. (2009). Early childhood teachers' misconceptions about mathematics education for young children in the United States. *Australasian Journal of Early Childhood, 34*(4), 37–45. Available from http://www.earlychildhoodaustralia.org.au/australian_journal_of_early_childhood/ajec_index_abstracts/ajec_vol_34_no_4_december_2009.html.

Lembrér, D., Johansson, M. L., Meaney, T., Wernberg, A., & Lange, T. (2014, August 18–20). *Assessing the design of collaborative mathematical activities for preschool children using interactive tables*. Poster presented at the biennial meeting of the EARLI Special Interest Group 20 Computer-Supported Inquiry Learning, Malmö University, Malmö.

Palmér, H., & Ebbelind, A. (2013). What is possible to learn? Using iPads in teaching maths in preschool. In A. M. Lindmeier & A. Heinze (Eds.), *Proceedings of the 37th conference of the International Group for the Psychology of Mathematics Education (PME37)* (pp. 425–432). Keil: PME.

Pareto, L., Haake, M., Lindström, P., Sjöjén, B., & Gulz, A. (2012). A teachable-agent-based game affording collaboration and competition: Evaluating math comprehension and motivation. *Educational Technology Research and Development, 60*, 723–751.

Perry, B., & Dockett, S. (1998). Play, argumentation and social constructivism. *Early Child Development and Care, 140*(1), 5–15.

Perry, B., & Dockett, S. (2007). *Play and mathematics*. Adelaide: Australian Association of Mathematics Teachers. Retrieved November 3, 2012, from http://www.aamt.edu.au/Professional-reading/Early-Childhood.

Pimm, D. (1995). *Symbols and meanings in school mathematics*. London: Routledge.

Radford, L. (2008). The ethics of being and knowing: Towards a cultural theory of learning. In L. Radford, G. Schubring, & F. Seeger (Eds.), *Semiotics in mathematics education: Epistemology, history, classroom and culture* (pp. 215–234). Rotterdam: Sense.

Riesbeck, E. (2013). The use of ICT to support children's reflective language. In B. Ubuz, Ç. Haser, & M. A. Mariotti (Eds.), *Proceedings of the eighth congress of European research in mathematics education (CERME 8)* (pp. 1566–1575). Ankara: Middle East Technical University. Available from http://www.mathematik.uni-dortmund.de/~erme/.

Sarama, J., & Clements, D. H. (2009). "Concrete" computer manipulatives in mathematics education. *Child Development Perspectives, 3*(3), 145–150.

Skolverket. (2011). *Curriculum for the preschool lpfö 98: Revised 2010*. Stockholm: Fritzes.

Starkey, P., Klein, A., & Wakely, A. (2004). Enhancing young children's mathematical knowledge through a pre-kindergarten mathematics intervention. *Early Childhood Research Quarterly, 19*(1), 99–120.

Tudge, J. R. H., & Doucet, F. (2004). Early mathematical experiences: Observing young Black and White children's everyday activities. *Early Childhood Research Quarterly, 19*(1), 21–39.

Utbildningsdepartementet. (2010). *Förskola i utveckling: Bakgrund till ändringar i förskolans läroplan [Preschools in development: Background to revisions of preschool curriculum]*. Stockholm: Åtta 45.

van Oers, B. (2010). Emergent mathematical thinking in the context of play. *Educational Studies in Mathematics, 74*(1), 23–37.

What Is the Difference? Young Children Learning Mathematics Through Problem Solving

Hanna Palmér

Abstract This chapter reports on a design research study where young children were taught mathematics through problem solving. In the study, a sequence of five problem-solving lessons was conducted in two Swedish preschool classes with children, aged approximately 6 years. This chapter utilizes interviews made with the children after the problem-solving lessons. The lessons in the study were categorized as work by the children, though with some differences from how they usually worked. The differences emphasized by the children were that they had worked together with their classmates, thought a lot, and used manipulatives rather than textbooks. Further, they emphasized that they were surprised by the outcomes many times. At the same time as these answers say something about the problem-solving lessons, they also say something about the ordinary mathematics teaching in these preschool classes. The results emphasize the need to discuss how much and, especially, in which way young children are to be taught mathematics.

Introduction

Young children spontaneously engage in problem-solving activities outside formal schooling (English 2004a), and throughout the early years of school, children's problem-solving experiences should include a wide range of problem-solving activities (English 2004b). This chapter reports on a Swedish design research study implementing problem-solving lessons in two Swedish preschool classes with children approximately 6 years old. One aim of the study is to develop knowledge of how to implement mathematics problem solving with young children who may not always know how read or write. Another aim is to develop knowledge of how young students learn and perceive problem solving. Because research on problem solving

H. Palmér (✉)
Linnaeus University, Växjö, Sweden
e-mail: Hanna.Palmer@lnu.se

T. Meaney et al. (eds.), *Mathematics Education in the Early Years*,
DOI 10.1007/978-3-319-23935-4_14

has been criticized for seldom influencing school practice (Lesh and Zawojewski 2007), design research was used to increase the likelihood that the research would influence the practice.

The study shows that the children did learn a lot from the intervention. However, to claim the intervention as being successful, it is important to know how the children perceived the intervention. This is the main focus of this chapter. Based on interviews with children after they had experienced a sequence of problem-solving lessons, this chapter focuses on how the children themselves perceived the intervention they had been involved in.

Problem Solving

When teachers are teaching mathematics, the tasks given to the students are an important part of the teaching design. However, a mathematics task does not necessarily have to imply a written task in a textbook. According to Stein and Smith (1998), a mathematics task is an activity in the context of mathematics teaching which aims to develop a specific mathematical idea. It can involve one single or several mathematical problems and/or extended work. A mathematics problem-solving task is a task to be solved where the method or methods for solving it are not known beforehand:

> A task, or goal-directed activity, becomes a problem (or problematic) when the 'problem solver' […] needs to develop a more productive way of thinking about the given situation. (Lesh and Zawojewski 2007, p. 782)

When working with mathematics problem-solving tasks, the students have to investigate, be creative, and try out different strategies to solve the task. From this perspective, problem solving is what the problem solver does to reach a solution to a mathematics problem-solving task (Wyndhamn et al. 2000). As such, problem solving is not just about using mathematical skills appropriately, but also about interpreting, describing, and explaining situations mathematically (Lesh and Zawojewski 2007).

Problem solving in mathematics has a significant role in the syllabi in many countries (Lesh and Zawojewski 2007). Nevertheless, teaching mathematics through problem solving has not been substantially implemented in classrooms (Cai 2010; Lesh and Zawojewski 2007; Lester and Lambdin 2007). In Sweden, problem solving is part of the curriculum for both preschool and primary school. Problem solving is not new in these educational systems; however, the emphasis regarding how and why students are to be taught problem solving has changed throughout the years. The emphasis has shifted slowly from a view where students first need to learn mathematics in order to become problem solvers to a view where problem solving is to be taught as content itself toward today's view that problem solving is a strategy for acquiring new mathematical knowledge (Wyndhamn et al. 2000). The basic idea is that "students will learn important mathematics more effectively if they encounter the concepts and techniques of the subject through carefully organized collaborative investigations of mathematically rich problems" (Harris et al. 2001, p. 310). Similarly positions are adopted in other countries, such as in the NCTM (2000) Principles and Standards for School Mathematics.

Preschool Class and Problem Solving

The social and cultural context within which children learn mathematics influences what they think mathematics is, how they think about mathematics learning, and what they learn (Cross et al. 2009; Perry and Dockett 2008). As such, social and cultural contexts must be accounted for in addition to the specific problem-solving activities. This study is conducted in two preschool classes. These were instigated in Sweden in 1998 to make a smooth transition between preschool and primary school and to prepare children for further education. The preschool class is part of the education system and is located at primary schools. Even though it is included within the curriculum of obligatory schooling, it is voluntary and there are no regulations or goals around the teaching of mathematics. Its working methods and pedagogy are not supposed to be like school (with a tradition of learning) or preschool (with a tradition of play) but a combination of the two (The Swedish National Agency for Education 2014). This leads to uncertainty regarding how much and in which way mathematics is to be taught. Consequently mathematics teaching in Swedish preschool classes differs considerably across Sweden, thereby creating differences in students' experiences of mathematics and how it is taught (Agency for School Improvement 2004).

Because preschool class is supposed to support a smooth transition between preschool and primary school and to prepare children for further education, the mathematical content of the curriculum for both primary school and for preschool is starting points for its mathematics teaching (The Swedish National Agency for Education 2014). One goal in Swedish preschools is that every child should develop his or her "ability to use mathematics to investigate, reflect over and test different solutions to problems raised by themselves and others" (The Swedish National Agency for Education 2010, p. 10). According to the curriculum for primary school, students are to develop the skills to formulate and solve mathematical problems, as well as to evaluate the strategies and methods to be used (The Swedish National Agency for Education 2011). This emphasis in the curriculum can be understood as a reaction to a national inspection of mathematics teaching made in 2009, which showed that mathematics teaching in Sweden was dominated by individual calculating, with limited possibilities for students to develop their ability to solve problems (The Swedish Schools Inspectorate 2009).

The majority of the teachers working in Swedish preschool classes are educated as preschool teachers. This means that they are qualified to work in preschool and preschool classes but not in primary school. For many years the Swedish preschool teacher education, which is a university education, did not include courses in mathematics education, leaving many preschool teachers in Sweden without a teacher education in mathematics (SOU 2004:97). However, research has shown that it is essential for teachers, including preschool teachers (Tsamir et al. 2014), to be knowledgeable about children's ways of thinking about mathematics as well as about how they learn and perceive mathematical conceptions (Ball et al. 2008; Björklund 2013; SOU 2004:97). To work successfully with mathematics problem solving, teachers need a deep understanding of the mathematics embedded in the problem-solving tasks. Further, the teachers must act as facilitators, asking questions that do not direct students toward the solutions, but instead help them to solve the problems on their own (Harris et al. 2001). However, research has shown that many preschool teachers have limited knowledge of early years' mathematics as well as limited knowledge regarding where this mathematics might lead (Björklund 2013; Ginsburg 2009; Perry and Dockett 2008; Tsamir et al. 2014). Further, many teachers of young children have negative experiences in learning mathematics when they were at school and so often avoid teaching mathematics (Ginsburg 2009). Taken together, limited knowledge and negative experiences result in "many early childhood settings [that] do not provide adequate learning experiences in mathematics" (Cross et al. 2009, p. 2).

The Study

In order to support preschool class teachers' understandings about teaching mathematics, education design research was the basis for the intervention. Design research is a cyclic process of designing and testing interventions situated within an educational context. The intention of the methodology is to enable the impact and transfer of research into school practice by building theories that "guide, inform, and improve both practice and research" (Anderson and Shattuck 2012, p. 16):

> Design-based research, by grounding itself in the needs, constraints, and interactions of local practice, can provide a lens for understanding how theoretical claims about teaching and learning can be transformed into effective learning in educational settings. (The Design-Based Research Collective 2003, p. 8)

There is however no single "fixed method" (p. 3) of design research but instead a genre of research conducted with rich variations (McKenney and Reeves 2012). Design research often starts with a particular teaching or learning problem. In this study, the starting point was how much and in which way is mathematics taught in Swedish preschool classes. The experience of the researcher was that the mathematics teaching in preschool class seldom included problem solving but instead used quite formal instruction techniques focused on numbers and counting. Based on these experiences, the researcher wanted to initiate, explore, and investigate possibilities for and barriers to teaching mathematics through problem solving to children in preschool classes. Is it possible to teach mathematics through problem solving in preschool class, and if it is, how can the problems be designed to take into consideration the children's perceptions and learning, the regulations for preschool class, and the mathematical content? As such, this study is design research on an intervention striving to generate theoretical understanding of its functions and characteristics, understanding which can be used to underpin the design of similar interventions (McKenney and Reeves 2012).

The main data analyzed in this chapter come from interviews made with the children after the first design cycle, in which a sequence of five problem-solving lessons was conducted in two preschool classes. These preschool classes were selected based on the preschool class teachers' interest in being involved. The four preschool teachers had, according to their own descriptions, limited knowledge of problem solving and so had not been teaching mathematics through problem solving. During the design cycle, one lesson was held in the preschool classes each week for 5 weeks. The guardians of the children were given written information about the study and had approved their children's participation. Forty-nine children were included in the study.

In order to understand the context for the children's interviews, a short presentation of the problem-solving lessons is needed. The lessons were conducted in the children's ordinary classrooms, and the researcher was the teacher of half of the class, while the other half of the class worked with their usual teacher. The researcher was known to the children before the first lesson because the researcher had spent some time in the classes before the intervention. Each lesson focused on one problem-solving task. Mostly the lessons followed the same pattern. First, the children worked alone; after that they worked together with a classmate, and finally the lesson ended with a whole class discussion led by the researcher/teacher. Each lesson lasted for about 30–40 min. The problem-solving tasks did not require the children to read or write, but instead used pictures and manipulatives to support their understandings. Before starting the first lesson, the children were told that the aim of the lessons was for the researcher to learn about how students solve and perceive mathematical problems. They were told that they were to teach the researcher, not vice versa. The aim of this was to make the children comfortable in the situation (Alderson and Morrow 2011).

An example of a problem-solving lesson from the first lesson in the sequence began with handing the children a picture of a tower (Fig. 1).

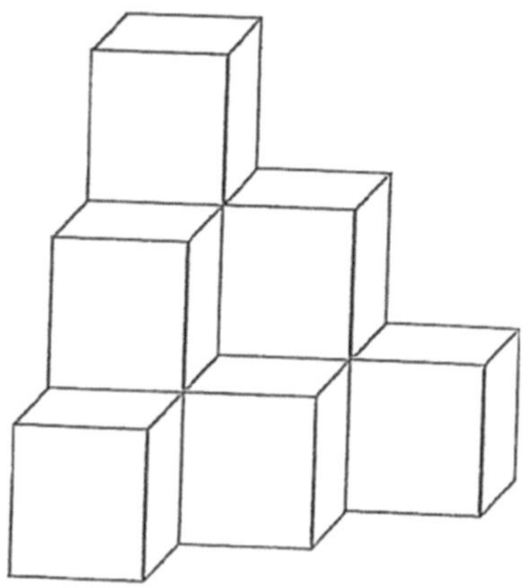

Fig. 1 The picture of a tower given to the children in the first lesson (the task is from http://ncm.gu.se/kangaru)

The question put to the children is how many blocks are needed to build the tower. One child initially said that he would need ten blocks to build the tower. Three other children said that they would need eight blocks to build the tower. The remaining 40 children (only 44 of the 49 children were present this day) said that they would need six blocks to build the tower. After working on their own and after discussing their solutions with a classmate, the children, in pairs, were to build the tower. First they built with blocks, and after that they built in a virtual building program on a tablet computer. From doing this, the children found out that they needed (at least) ten blocks, and discussions were held about which blocks were "hidden" in the first picture as well as the connections between the different representations.

The children were interviewed before and after the sequence of problem-solving lessons. This chapter will focus on the follow-up interview. The purpose with this interview was to develop knowledge of how young students perceived problem solving after they had been involved in doing it. Such knowledge is important in the evaluation of the design cycle because it enables understanding of how the students themselves perceive the intervention. The interviews were held 1 or 2 weeks after the last problem-solving lesson, which meant 6 or 7 weeks after the first problem-solving lesson. The students were interviewed in pairs to equalize the power imbalance between the researcher and the children (Alderson and Morrow 2011). In the interview, the children were first asked what they remembered from the lessons. After that, they were shown the five problem-solving tasks they had worked with and were asked to comment on what they remembered from working with each problem. Finally, they were asked if they had learned something from the lessons and if there was any difference between these lessons and what they otherwise did when working with mathematics in preschool class.

Results

The majority of the children remembered a lot from the five mathematical problem-solving tasks they had worked with, even though the first one was conducted almost 2 months before. For example[1]:

> *Vi skulle bygga ett konstigt torn. Och så fick vi bygga samma torn med iPaden.*
> We were supposed to build a strange tower. And then we were to build the same tower with the iPad.

> *Att man skulle bygga på en pyramid med Lego.*
> You were to build on a pyramid with Lego.

[1] The quotations are from different children. To support understanding of the quotations, only those related to the tower task are presented. However, of course, the children did also talk about the other problem-solving tasks in the interviews. The Swedish transcript is provided, followed by the English translation.

Vi skulle se med klossarna. Man kunde bygga med klossarna. Och vi försökte också lista ut på hur många sätt bilarna kunde parkera.

We were supposed to see with the blocks. You could build with the blocks. And we also tried to figure out how many ways the cars could park.

Vi gissade med klossarna.

We guessed with the blocks.

When shown each problem-solving task, several children remembered their own solutions. They also talked about whether they had experienced the tasks as easy or difficult or fun or boring. Their most common evaluation of the tasks was that they had been difficult at first but at the same time fun to work with. Below are some examples of the children's reflections on the tower task presented in the previous section. As mentioned, only one of the 44 children initially said that he would need ten blocks to build the tower. The others found out the need for ten blocks when trying to build the tower with blocks:

Det var väldigt svårt men det var roligt. Jag kommer ihåg att det var tio. (Räknar till tio samtidigt som han pekar på både synliga och "osynliga" klossar på bilden)

It was very hard but it was fun. I remember that it was ten. *(The child counts to ten at the same time as he points at both the visible and 'invisible' blocks on the picture.)*

Adam klurade ut det först. Det var lite svårt. Man kan se det. De kan ju inte flyga i luften. Det måste vara tre där bakom.

Adam[2] figured it out first. It was a little difficult. You can see it. They can't fly in the air. There have to be three behind.

Det var ganska svårt men jag tyckte det var roligt. Det var konstigt att det var några bakom.

It was pretty hard but I thought it was fun. It was strange that some were behind.

Jag tyckte det var svårt. Man kunde se sex, men det var tio.

I thought it was hard. You could see six, but it was ten.

Det är bara att göra såhär. Den är tre. (Pekar på den bakre stapeln på tornet på bilden) En, två, tre, fyra, fem, sex, sju, åtta, nio, tio. (Pekar på både de synliga och "osynliga" klossarna på bilden samtidigt som han räknar.) Jättelätt. Rätt så roligt också.

You just have to do like this. That one is three. *(Points at the back part of the tower on the picture.)* One, two, three, four, five, six, seven, eight, nine ten. *(Points at both the visible and 'invisible' blocks on the picture at the same time as he counts.)* Really easy. And quite fun too.

The two most common answers when asked if they had learned something from the problem-solving lessons were that they had "learned maths" and that they had "learned to think":

Det känns som om jag har lärt mig nya saker varje gång du har varit här.

It feels like I learned new things every time you were here.

Med dig har vi lärt oss lite mer att tänka. Vi lär oss kanske att tänka med dom [förskoleklasslärarna] också, men vi tänker mer hos dig.

---

[2] This name is changed.

With you we learned to think a little more. Maybe we learn to think with them [their preschool class teachers] also, but we think more when we're with you.

Vi har lärt oss att tänka bättre.
We learned to think better.

However, there were also children who answered that they had learned nothing (even though their previous answers showed that they probably had learned something). Several children also gave answers that included solutions from one or several of the mathematical problems they had been working with:

Innan visste jag inte att det skulle vara tre klossar under. (Pekar på bilden av tornet)
Before I didn't know that there should be three blocks under. *(Pointing at the picture of the tower)*

Jag trodde det var sex men det var tio!(Pekar på bilden av tornet)
I thought it was six but it was ten! *(Pointing at the picture of the tower)*

On the question of whether there were any differences between the lessons within the study and what they otherwise did when working with mathematics in preschool class, it became visible that the children saw a difference between work and play as preschool class activities. The problem-solving lessons in the study were categorized as work by the children, but with some differences from how they usually worked with mathematics. The differences emphasized by the children were that in the problem-solving lessons, they worked together with their classmates, thinking a lot, not using textbooks, but rather using manipulatives. Further, they emphasized that they had been surprised by the outcomes many times:

Ja, det har det nog faktiskt varit. När vi jobbar brukar vi inte använda saker så mycket. Vi jobbar mer i bok.
Yes, I actually think it has been. When we work, we usually don't use things that much. We mostly do bookwork.

Annars brukar vi mest få rita. Skitmycket!
Otherwise we usually draw. A great lot!

Vi brukar jobba med Trulle [namnet på deras lärobok]. Vi brukar göra matte, men inte så som du lärt oss.
We usually work with Trulle [the name of a textbook]. We do the math, but not the way you have shown us.

Det var väldig skillnad tycker jag. Är man hos dig får man lära sig mycket mer matte.
I think it was very different. When we were with you, we learned a lot more mathematics.

Conclusion and Discussion

The interviews reported in this chapter completed the first cycle in a design research study where young children were taught mathematics through problem solving. The starting point for the study was how much and in which way mathematics can be taught in Swedish preschool classes. The researcher wanted to initiate, explore, and investigate possibilities for and barriers to teaching mathematics through problem solving. The focus in this chapter and in this final section is on how the information from the interviews informs the aims of the first cycle in the study, that is, to develop knowledge of how to implement mathematics problem solving with young students who do not always know how to read or write and to develop knowledge of how young students learn and perceive problem solving. Answers on those questions are important information regarding if it is possible to teach mathematics through problem solving in preschool class.

The interviews show that young children are competent, both in terms of problem solving and in terms of reflecting on their own learning and the mathematics lessons they encounter. At the same time as the interviews said something about the problem-solving lessons, they also said something about the ordinary mathematics teaching in these preschool classes. When talking about the ordinary mathematics lessons, the children emphasized drawing and working in the textbook and learning mathematics individually. When talking about the problem-solving lessons, the children expressed that they worked together with classmates using manipulatives and that this was something different from what they usually did when working with mathematics. The ordinary way of teaching mathematics seemed to be in line with what was found in the Swedish national inspection of mathematics teaching made in 2009, which showed that the teaching of mathematics was dominated by individual calculating, with limited possibilities for students to develop their ability to solve problems (The Swedish Schools Inspectorate 2009). This influences not only which mathematics these children learn but also their experiences of how mathematics should be learned.

Based on the interviews, there are both similarities and differences between the ordinary teaching of mathematics in these preschool classes and the lessons in the intervention. These lessons were also perceived as work by the children. This is interesting given that preschool class is supposed to be a combination of school (with a tradition of learning) and preschool (with a tradition of play), and it might have been expected that building towers with blocks and on tablet computers would have been classified as play. Nevertheless, based on the interviews, the teaching of mathematics through problem solving was perceived as work.

In a study of children in the same age as in this study, Wing (1995) found that children are very clear about what is work and what is play. Markers that the children used to distinguish between work and play were the obligatory nature of the activity (work is considered as an externally controlled, obligatory activity), the expectations and involvement of the teacher (in work teachers are close to the children, setting the rules for the activity, giving directions and evaluations), the possibilities to quit (when working you have to finish), and the cognitive efforts required (work is hard and involves cognitive effort). Based on this, it seems logical for the children in this study to categorize the problem-solving lessons as work. In the study by Wing (1995), both work and play could be fun from the perspective of the children, which also seemed to be the case in this study. However, in the study by Wing (1995), the children expressed a continuum between work and play, which was not seen in this study. Because the distinction between work and play was not the focus of the interviews, it is possible that the categorization between work and play made by the children is not really a question of either/or but a continuum between the two. In the study by Wing (1995), one category was play-like work. Play-like work could be considered as the working method and pedagogy to be implemented in preschool class as a combination of preschool and primary school (The Swedish National Agency for Education 2014). If problem solving could be perceived as play-like, work by children is something that needs to be further investigated, however.

As mentioned, the aim for the Swedish preschool class is to support a smooth transition between preschool and primary school and to prepare children for further schooling. However, teaching young children mathematics should not only be about preparing for future schooling but also for providing "young children with rich and engaging intellectual stimulation" (Ginsburg 2009, p. 405). The children's answers in the interviews indicate that the problem-solving lessons were different because they were difficult and, based on that, the children "had to think." Further, the children said that they were surprised by the results many times. This indicates that the children were not used to working with demanding mathematics tasks. The children's talk about "thinking" is probably connected to the request of the researcher during the problem-solving lessons that the children were to explain how they were thinking when solving the problem-solving tasks, both to her and to each other. Based on how the children described their ordinary mathematics lessons, working in pairs was not their usual way of doing mathematics.

The interviews indicate that it is appropriate to give young children demanding problem-solving tasks. The information in the interviews shows that the children did learn from the problem-solving lessons. Their reflections on specific tasks indicate that they had learned about the mathematics imbedded in the problem-solving tasks. In the interviews, the children in the study evaluated the problem-solving lessons as "difficult but fun." As mentioned, only one of the 44 children initially said that he would need ten blocks to build the tower. Thus, the task was difficult for the children, and several of them expressed that the task was difficult. Still, they evaluated the lesson as fun. As such, difficult can be fun from the perspective of the children.

Before the intervention, the preschool class teachers of the two preschool classes said that they had limited knowledge of problem solving. Such limited knowledge is not unique to this study (Ginsburg 2009; Perry and Dockett 2008; Tsamir et al. 2014). As well, in Swedish preschool classes, there are no formal expectations of what mathematics to teach. However, it would seem that the ordinary teaching of mathematics in these two preschool classes was likely to have been affected by both the preschool teachers' limited knowledge of problem solving and by the absence of regulations or goals. According to the children's answers in the interviews, the ordinary mathematics teaching consisted of lessons in the textbook, with mostly nondemanding tasks. As mentioned, design research was used in the study to increase the impact, transfer, and translation of the research into practice. It is, however, too early to say whether these intentions will support changes in these preschool classes. It is possible, though, to say that the intervention was successful based on how it was perceived by the children.

Even if preschool class is a national phenomenon, the issues of content, design, continuity, progression, and transition in mathematics teaching are not. Thus, the content of this chapter may also be of interest in other educational contexts. The interviews indicate that the intervention in the design research study is in line with the national and international suggestions that children should learn mathematical knowledge by working with problem solving. This emphasizes the need for discussing how much and, especially, in which way mathematics is to be taught with young children. When discussing these topics, the voices of the children are important. As shown in this chapter, young children are competent, both in terms of problem solving and in terms of reflecting on their own learning and the mathematics teaching they receive. We need to discuss not only the how and what of mathematics we teach young children but also how the children themselves perceive the teaching they encounter and how that influences their ideas of what mathematics is, how teaching of mathematics is performed, and how one learns mathematics.

References

Agency for School Improvement. (2004). *Rapportering av regeringsuppdrag 'Integration förskola, förskoleklass, grundskola och fritidshem—Att bygga broar'* (Report of government mandate *'Integration preschool, elementary school and recreation centers—Building bridges').* Dnr. 2003:171.

Alderson, P., & Morrow, V. (2011). *The ethics of research with children and young people: A practical handbook.* London: Sage.

Anderson, T., & Shattuck, J. (2012). Design-based research: A decade of progress in education research? *Educational Researcher, 41*(1), 16–25.

Ball, D., Thames, M., & Phelps, G. (2008). Content knowledge for teaching. *Journal of Teacher Education, 59*(5), 389–407.

Björklund, C. (2013). *What counts in preschool?* Lund: Studentlitteratur AB.

Cai, J. (2010). Commentary on problem solving heuristics, affect, and discrete mathematics. A representational discussion. In B. Sriraman & L. English (Eds.), *Theories of mathematics education. Seeking new frontiers* (pp. 251–258). London: Springer.

Cross, C. T., Woods, T. A., & Schweingruber, H. (2009). *Mathematics learning in early childhood. Paths toward excellence and equity.* Washington, DC: National Research Council of the National Academics.

English, L. (2004a). Mathematical and analogical reasoning in early childhood. In L. English (Ed.), *Mathematical and analogical reasoning of young learners* (pp. 1–22). Mahwah, NJ: Lawrence Erlbaum.

English, L. (2004b). Promoting the development of young children's mathematical and analogical reasoning. In L. English (Ed.), *Mathematical and analogical reasoning of young learners* (pp. 201–213). Mahwah, NJ: Lawrence Erlbaum.

Ginsburg, H. P. (2009). Early mathematics education and how to do it. In O. A. Barbarin & B. H. Wasik (Eds.), *Handbook of child development & early education. Research to practice* (pp. 403–428). New York: Guilford Press.

Harris, K., Marcus, R., McLaren, K., & Fey, J. (2001). Curriculum materials supporting problem-based teaching. *School Science and Mathematics, 101*(6), 310–318.

Lesh, R., & Zawojewski, J. (2007). Problem solving and modeling. In F. K. Lester (Ed.), *Second handbook of research on mathematics teaching and learning* (pp. 763–799). Charlotte, NC: National Council of Teachers of Mathematics & Information Age.

Lester, F. K., & Lambdin, D. (2007). Undervisa genom problemlösning (Teach through problem solving). In J. Boesen (Ed.), *Lära och undervisa i matematik—internationella perspektiv (Learn and teach mathematics—international perspectives)* (pp. 95–108). Göteborg: NCM.

McKenney, S., & Reeves, T. C. (2012). *Conducting educational design research.* London: Routledge.

NCTM. (2000). *Principles and standards for school mathematics.* Charlotte, NC: National Council of Teachers of Mathematics & Information Age.

Perry, B., & Dockett, S. (2008). Young children's access to powerful mathematical ideas. In L. D. English (Ed.), *Handbook of international research in mathematics education* (pp. 75–108). New York: Routledge.

SOU. (2004:97). *Att lyfta matematiken: intresse, kunskap, kompetens. Matematikdeligationens betänkande (Lifting mathematics: interest, knowledge, competence. The report of the mathematics delegation).* Stockholm: Fritzes.

Stein, M. K., & Smith, M. S. (1998). Mathematical tasks as a framework for reflection: From research to practice. *Mathematics Teaching in the Middle School, 3*(4), 268–275.

The Design-Based Research Collective. (2003). Design-based research: An emerging paradigm for educational inquiry. *Educational Researcher, 32*(1), 5–8.

The Swedish National Agency for Education. (2010). *Curriculum for the preschool Lpfö98. Revised 2010.* Stockholm: National Agency for Education.

The Swedish National Agency for Education. (2011). *Curriculum for the compulsory school, preschool class and the recreation centre 2011.* Stockholm: Skolverket.

The Swedish National Agency for Education. (2014). *Preschool class—Assignment, content and quality.* Stockholm: Swedish National Agency for Education.

The Swedish Schools Inspectorate. (2009). *Undervisningen i matematik—utbildningens innehåll och ändamålsenlighet (The teaching of mathematics—The content and the effectiveness of the education).* Kvalitetsgranskning rapport 2009:5. Skolinspektionen: Stockholm.

Tsamir, P., Tirosh, D., Levenson, E., Tabach, M., & Barkai, R. (2014). Developing preschool teachers, knowledge of students' number conceptions. *Journal of Mathematics Teacher Education, 17*, 61–83.

Wing, L. A. (1995). Play is not the work of the child: Young children's perceptions of work and play. *Early Childhood Research Quarterly, 10*, 223–247.

Wyndhamn, J., Riesbeck, E., & Schoultz, J. (2000). *Problemlösning som metafor och praktik (Problem solving as a metaphor and practice).* Lindköping: Linköpings Universitet.

Part IV
Mathematical Content

Playing with Patterns: Conclusions from a Learning Study with Toddlers

Camilla Björklund

Abstract Play and learning is said to be intertwined in young children's mathematical development. Play is, however, a multifaceted practice—and so is mathematics. In this report from an empirical study of toddlers' mathematical play and learning in a goal-oriented preschool practice, the learning content is "patterns". The framework for the pedagogical work is play, however, with a theory-driven approach to how children's learning of the idea of patterns is facilitated. Observations and analysis of occurrences, responses and learning within designed activities give insights and broaden our understanding of "patterns", how toddlers experience the idea of patterns and what is made possible to learn in play-based but goal-oriented activities in preschool.

Learning Mathematics: The Exploration of Relationships

This chapter discusses patterns, play and mathematics teaching for the youngest children in preschool. The purpose is to present an analysis of an investigation of toddlers experiencing patterns in play activities, aiming at contributing to the field of knowledge concerning teaching that is highly sensitive to children's perspectives. A central question is, in addition, what play may contribute to the learning of patterns and children's concept development.

Development of basic mathematical competence is complex in character, but one requirement is arguably to experience and explore the surrounding environment and how phenomena are related to each other, in space, in shape and in number. The relationships between objects in space and quantities are closely connected to the everyday life of children's play, yet there is no guarantee that children explore these relationships on their own initiatives (Hannula 2005), nor that they develop an understanding of these on a conceptual level, by merely being in the environment (Ginsburg 2006; van Oers 2013a). Education, defined as "making the invisible visible to the child" (Pramling Samuelsson and Pramling 2013), plays therefore a crucial role for concept development. This way of seeing learning and teaching is the basis for the study and discussion that will follow.

C. Björklund (✉)
University of Gothenburg, Gothenburg, Sweden
e-mail: camilla.bjorklund@ped.gu.se

T. Meaney et al. (eds.), *Mathematics Education in the Early Years*,
DOI 10.1007/978-3-319-23935-4_15

A conceptual base of mathematical phenomena is considered important for the development of mathematical skills. There are many studies confirming that early mathematical knowledge has effects on later mathematical achievements, but little is known about *how* this is facilitated with the youngest children in preschool. How can a teacher make the invisible visible to the youngest children who do not yet express themselves verbally and for whom most mathematical notions are novel?

A research project in a preschool environment addresses this question of "how" and seeks to find out how children may be introduced to mathematical concepts while being responsive to the strong emphasis on children's perspectives, their own initiatives, creativity and play as promoted by the Swedish National Agency for Education (2011). The project integrates a theoretical conjecture how learning occurs with empirical work in preschools and authentic learning objects. One of the learning objects was chosen to be "pattern". A pattern is considered to be a repetitive phenomenon following a predictable order, which is an important aspect of pre-algebraic thinking. Thus, questions arose about how to design activities that are goal oriented and play centred and whether the framework constituted by play and a theory of learning added to or limited the learning process. The main interest in this study is therefore to understand what happens when children play with patterns in a goal-oriented activity and what play contributes to the learning process. In the following sections, the concept of play is discussed followed by a short introduction to the idea of mathematical patterns. This frames the ideas elaborated upon in the analysis of the empirical project and the three conclusions drawn by the participating teachers and the researcher.

What's So Special About Play?

There is hardly any early childhood practitioner or researcher who would not argue for the value of children's play in early childhood. At the same time, there seems to be different perspectives on how and why play is considered so important. Certainly there are different definitions of play which affect the way one perceives the meaning and usefulness of play. Sutton-Smith (1997) describes a variety of acts, objects and forms of play that includes both children's and adults' activities. He presents different perspectives and ideologies that dominate the discussions and assumptions about play, for example, that children develop cognitively through play, in contrast to children playing only for pleasure or that play is defined as an exercise of free choice. As a balance to the idea of play as a form of cognitive progress, Sutton-Smith advocates for children's play as an arena for power and identity, also connected to traditional celebrations and festivals in which both children and adults participate and where play works as a social glue. Sutton-Smith's argument is that play in itself is a matter of ambiguity, an argument that is confirmed when analysing theoretically the function and form of play, which reveals that the observer's view may be very different from the player's own point of view. This is important to remember when discussing children's learning and development in relation to play, since what is interpreted is not necessarily what is experienced, including within this inquiry of toddlers' play with patterns.

As stated by Sutton-Smith, play is multifaceted and works for social, cognitive and pleasure purposes. How then is mathematical development and reasoning related to play? Bishop (1988) argues for attention to the cultural similarities that can be found worldwide in play. He categorizes play in mathematical activities and ideas, since his understanding of mathematics is that it should not be seen as different topics to be learnt, but as products of various processes. His search for activities that are common in most cultures leads to a discussion on how mathematics is developed. Mathematics is expressed in different ways in different cultures, but the ideas are similar. Perhaps it is then possible to consider the mathematical structures that appear in play, which children participate in and make meaning of.

Mathematical activities that are addressed by Bishop (1988) are counting and measuring, perhaps the most obvious mathematical practices that human engage in all cultures. Locating and designing are two other activities that are connected to architecture and spatial structuring of various kinds. However, Bishop argues for a need to consider the relational aspect of mathematics, which appears in human interaction and social activities such as explanations and not least *play*. Bishop's understanding of play concerns social procedures with rules for performance and acts. This also includes activities where imagination and hypothetical thinking are significant, expressed in ideas characterized by "as if" or "what if". Rule-governed games are likely to have a mathematical structure which gives the participant both satisfaction and cognitive challenges to engage in. This way of connecting play and mathematical development is based on an anthropological and cultural perspective where play adds the opportunity to explore abstraction of reality, imaginative games and aesthetic appreciation. Play is for Bishop the investigation of mathematical structures, and as it is difficult to imagine mathematics learning without relation to basic human activities, play is then perhaps the most essential feature for cognitive development.

However, one should not disregard the aspect of pleasure for mathematical learning. Ginsburg (2006) argues that children use informal and intuitive mathematical knowledge embedded in play, which provides a good foundation for cognitive development, but children may also enjoy playing with patterns, number, shape and space, in other words engage in explicit mathematical objects.

According to Vygotsky (2004), children's play is the authentic creative process that constitutes human development. In play children recreate much of what they have experienced and what they already know, but these are never reproduced exactly nor precisely imitated. Creativity and fantasy in play adds a combinatory aspect where earlier experiences are integrated in new ways, creating a coherent whole with logical relationships. Vygotsky claims though that the boundaries of creativity in play are related to the richness of the earlier experiences, on which fantasies and content within play rely. For example, a child can fantasize about a house made of gingerbread even though the child is very well aware that houses in reality are not made of bread, but the child has experiences of gingerbread and of houses, which are combined into an imagination of a gingerbread house.

The pedagogical implication that follows Vygotsky's (2004) line of reasoning is that it is necessary to offer the child widening experiences as a base for creative and developing thoughts and skills. When their experiences and the fantasy products of their own creative minds are guided by another's experiences, the children may borrow experiences from the other to extend their own experience, enabling the fantasy product to get closer to the real event. Vygotsky further explains that fantasy goes through a circular process, where elements from reality are combined and brought together in order to return to reality as a new invention, which changes the person's view on reality. Fantasy and creativity brings to the learning process coherence and construction of new meaning. The inner logic frames what is possible to experience within a play, story or game. Coherence and differentiation are two central features; the experiences need to be differentiated from other experiences and related to the coherent whole, to be possible to merge with other experiences and create new meaning. According to Vygotsky, this ability is the foundation for abstract thinking and concept development. What direction the development takes is though relying on the context and available opportunities for experiences. All innovations are thereby social constructs, harnessed by the individual to take a gestalt.

Sutton-Smith (1997) also presents perspectives on play which include flexibility, imagination and creativity that can be recognized in several forms of play among both adults and children. According to Sutton-Smith, play should be understood as an attitude or a frame that may be directed towards anything. Play thereby has a framing function that includes intersubjectivity or shared attention. This intersubjectivity must, however, be established for a play to be functional and cannot be taken for granted.

From the previous discussion, it can be seen that playful activities enable children to interact in different ways and with different purposes, where mathematics may become a central feature. Mathematics education should according to van Oers (2013a) be elaborated in relation to the context of play that goes beyond adult-defined rule games. Van Oers (2013b) also takes a social and cultural perspective on mathematics learning, suggesting that mathematics education in the early years should focus on developing skills to communicate about number, quantity, space and relations. This is said to be accomplished when children are confronted with demands from the situation that require translation of experiences into mathematical language and objects. This skill helps the child to interpret situations from a mathematical point of view, in that he or she will know appropriate strategies to use, in other words knowing the mathematical orientation in a situation. Educational practices are suggested in which children's play is used as a frame for coherence, within which children are encouraged to communicate with and about specific mathematical concepts.

Pramling Samuelsson and Asplund Carlsson (2008) argue against separating play and learning. They argue that when children act, they play and learn simultaneously as they strive to make meaning, always directed at *something*. Play and learning both have content, something to act about and make meaning of, where the teacher's role may be to challenge children's understanding through engaging them in metacognitive dialogues. In high-quality pedagogical practice, one can see in children's play what they have been working with in their daily activities and also how content and themes that are appearing in children's play are picked up by the teachers in their planned pedagogical work. The teacher in this way offers a relevance structure, which is crucial for goal-oriented activities, as it frames the learning process as a coherent whole. The learning object then conveys meaning, and it is possible to direct children's attention to an intended object of learning.

In a preschool study by Björklund (2014), different contextual frameworks, even though taking a similar theoretical approach, produced different outcomes for learning and, most of all, for the goals that children experience as being of central interest in an activity. Traditional tasks and problem solving do not necessarily encourage the meaning making intended by the teachers. In contrast, learning objects embedded in narrative play and stories bring several features to the learning process which are essential for goal-oriented learning with young children. The narrative feature gives relevance to the acts and tasks within an activity (Burton 2002). Further, the narratives may be used to limit the possible alternatives and so reduce complexity. There are usually many alternatives to solve a problem or play a game, but not all alternatives are helping the child to discern a certain learning object. Within a narrative framework the teacher may limit the possibilities and exclude alternatives that he or she considers as not developing the chosen learning object and instead bring forth critical aspects of the learning object in meaningful situations (Björklund and Pramling Samuelsson 2013). This further facilitates shared sustained thinking where teacher and child direct their awareness towards the same phenomenon and both of their experiences develop and are extended (Sylva et al. 2010).

Based on the discussion above, it is possible to conclude that play is significant for young children's learning and mathematical development, due to some of the specific features that play contributes with. *Amusement and joy* is usually related to playful activities and brings forth a social aspect (see Sutton-Smith 1997), and *attractive settings and props* are often present in pedagogical settings to invite and motivate children to explore certain phenomena. Play may offer children a *relevance structure* within which mathematical phenomena are integrated and the play itself frames the mathematics that is possible to explore and make meaning of (Burton 2002; van Oers 2013a, b; Pramling Samuelsson and Asplund Carlsson 2008). Within this structure *fantasy and creativity is encouraged* (Vygotsky 2004), as play provides participants with equal rights to progress. The *communication* and *interactivity* aspects of both play and mathematics are also features worth considering, as these features build upon shared sustained thinking and joint perspectives (Sylva et al. 2010) which are essential in goal-directed activities.

Patterns, Algebra and Emerging Mathematical Skills

Mathematics is considered as structuring the surrounding world, where relationships are measured, compared and described in mathematical terms, both symbolic and verbal. Devlin (1994) argues that mathematics is the science of patterns, and Mulligan and Mitchelmore (2013) show empirical evidence of young children being able to generalize mathematical patterns and structures that involve a deeper understanding of internal relationships. Therefore, patterns are considered an important aspect of mathematics education as it is the basis for not only spatial and geometrical knowledge areas of mathematics but also for numerical structuring and development of numerical understanding. Young children's pattern making and understanding has only recently been the object for inquiry, even though there is considerable evidence of young children's abilities to generalize and abstract already in their early years (Cross et al. 2009; Garrick et al. 1999; Sarama and Clements 2009).

Patterns are commonly connected to algebraic thinking because of the abstraction and relational aspect that patterns build upon. Heeffer (2010) defined algebra as an analytical tool to solve a problem where the unknown unit is replaced with an abstract unit. To understand algebra the child needs to be able to discern the structure in order to handle the unknown, making the connection between patterns and algebra clear, as a pattern also is determined by a rule that explains the structure. Radford (2012) argues that the ability to discern and generalize patterns and mathematical structure in general does not develop spontaneously; rather it depends on cultural influence or some kind of education. Patterns and structure can be explored with the support of concrete manipulatives and artefacts which will facilitate the discernment of general ideas and relationships (Rivera 2013; Threlfall 1999). Many preschool children rely on the physical relationship between objects, which is expressed as children describe patterns as "blue, red, blue, red, blue and red" instead of a more general description "every other blue and red" (Björklund and Pramling 2014; Papic et al. 2011; Warren and Cooper 2008). This development towards a more advanced conceptual understanding is supposedly generated in interaction with a teacher who challenges children's reasoning and the underlying rule of the pattern.

Björklund and Pramling (2014) investigated the aspects that seem important for discerning the idea of patterns among 6-year-olds and found that the generalizing aspect is central for children's opportunities to create patterns on their own. Further, units need to be differentiated as parts of a whole, such as xo xo xo, where xo constitutes the unit of repeat. Generalization is facilitated when children are offered opportunities to discern rules for structuring a pattern using different objects and differently composed units. The composition of the units and the features of the objects, used as manipulatives, need to be paid attention to since the features may distract the children's ability to reason about abstract relationships. Focusing on features, both irrelevant and relevant, should be, in line with this reasoning, important for the building of further concept development. Empirical work with even younger children may thereby be a way to explore the nature of the concept "pattern".

Learning About Space: An Empirical Preschool Project

The empirical study used as the basis for this discussion derives from the research project *Learning about space—an educational study of teachers and children learning mathematics in and of space*. The wider project aims at developing mathematics education in preschool through close collaboration with preschool teachers in their preschool practice. This particular study is one part of the larger project. As noted previously, the interest in this current study was to observe how a specific mathematical topic was experienced by children in designed learning activities. Preliminary analyses revealed that the context and framework (play activities) had considerable effect on the learning outcome. This chapter thereby focuses on the issue of play in mathematics learning.

The model chosen for the empirical inquiry was Learning Study, as developed by Marton and Tsui (2004), Runesson (2008) and Holmqvist (2011). The characteristics of this kind of study is the theory-driven teaching activity where the special interest is how the learner perceives the learning object and what aspects of the learning object are possible to discern due to the activity's design. The theory underlying this design and analysis is Variation Theory of Learning that will be briefly described in the following section.

Variation Theory of Learning

Variation Theory of Learning derives from a methodological interest in how people experience the surrounding world and phenomena in this world. Multiple studies of people's varying experiences have given credence to the idea that phenomena are perceived in a specific way due to the aspects that are made possible to discern at a specific time (Marton 2014; Marton and Booth 1997). What aspects of the phenomenon that a person perceives depends both on the earlier experiences of the same phenomena and occurrences and what aspects are offered for a person to discern at a particular moment. Considering the phenomenon "pattern", regularity and the unit of repeat are aspects constituting the phenomenon. These are critical aspects of pattern, in addition to possible others as well, and have to be differentiated and "seen" by the child before he or she is able to imitate or create novel patterns. In accordance with this conjecture of how the world is perceived, learning means to experience the world in ways that a person has not previously been able to experience it. Teaching becomes an act of offering the learner opportunities to explore different aspects of a learning object that enables the learner to experience the learning object in new ways. In other words, the teacher offers a child such experiences that enable the child to discern critical aspects of patterns that the child has not previously been aware of. This approach to teaching and learning is closely related, not least ontologically, to developmental pedagogy (Pramling Samuelsson and Pramling 2013) where emphasis is set on making the invisible visible to the learner through shared attention and engaging children in meta-communication.

Experiencing variation is a key concept within the theoretical framework, meaning that content and meaning is only possible to discern if certain patterns of variation are brought to the foreground (Marton and Booth 1997; Marton 2014). In goal-oriented pedagogical practices, this implies that the child should be offered such experiences through carefully chosen activities, materials and communication that enables aspects of a learning object to emerge from the background into the foreground. The critical question for the current study is thereby what aspects of the phenomenon "pattern" that can be discerned by the children in designed play activities.

Empirical Inquiry

The empirical design used in this study is a model for exploring patterns of variation and how learning objects are perceived by the learners. This is accomplished through carefully planned activities that account for the learners' earlier experiences, current knowledge and understanding and how the learning object is handled by teacher and child. The design includes observations, cooperative planning and discussion among a group of teachers, and active engagement in authentic teaching activities that are documented and analysed before another discussion and further planning are commenced (Marton and Tsui 2004). The model is iterative and explorative, yet the analysis is done in strict accordance with the theoretical framework.

In this particular study, three teachers and a researcher focused on the idea of patterns and children's creative play. The intention was to learn about how children experience the phenomenon of patterns and how a theoretically founded playful teaching act may facilitate concept development. Each participating teacher conducted two learning sessions with three of the children, 1–3 years old, from their own child groups (nine children participating). The teachers discussed the chosen learning object together but planned and conducted individual activities. The children's acts and responses when encountering the learning object were studied and analysed in relation to the teachers' intended learning object. These observations generated the basis for further planning of activities emphasizing aspects of the learning object that were not yet discerned by the children. The empirical data collected during the learning sessions with the children consists of a total of 140 min video observations, six episodes between 4 and 60 min long. These episodes are analysed with focus on what aspects are made possible to discern. The teachers implemented the Variation Theory approach in their acts and interaction with the children, striving to make critical aspects discernable to them. Activities characterized by goal-oriented play were orchestrated by the teachers, giving the children opportunities to explore phenomena individually and in interaction, but guided by the materials and instructions of the teacher.

The model used for developing new knowledge in this project is iterative and theory driven but very close to the authentic practice. Important results are how children experience the object of learning as it is offered to them in play context and how this can be understood in relation to the theoretically designed teaching act. In the following section, there is a discussion of the learning study with special attention to the impact of the concept of play on the learning opportunities.

What Did We Learn?

In this study the teachers choose to work with patterns as a topic in designed play-based activities. The children are engaged in activities that to an observer most likely would be labelled as "play" or "games" due to the flexible nature of the interaction between adult and children. This approach allows the children to take initiatives within the frames of the designed activities; communication and interaction are characterized by interest in the child's intentions and collaborative problem solving with focus on patterns and aspects of the idea of patterns. The model for inquiry enables scrutiny of both intentions and outcomes of the process, whereas the main outcomes are presented in the following as three conclusions.

First Conclusion: Differentiating the Basics of Patterns

Understanding the idea of patterns means that you can discern features of objects and how they may constitute categories and subcategories in different ways. More experienced children may organize different patterns with the same objects, for example, first by colour, blue-red-blue-red, and then by size, big-small-big-small (Björklund and Pramling 2014). This ability relies on the child discerning several features of the object and then deciding which feature will be in the foreground for the rule of every second that is repeated, leaving the other features in the background. In order to do this, children need to first discern the features of the objects.

The teachers in this study quickly became aware of the necessity of breaking down the learning object to more basic competences For example, they focused on supporting children to discern an individual feature and the values within this feature. A visual feature commonly discussed in daily life is colour, of which there are many different colours, which are necessary to distinguish in order to use these features in a deliberate pattern.

In one episode, the teacher invited the children to help her collect red buttons from several piles of buttons that the children had sorted previously. The invitation includes exploration of features such as colours. By offering a pattern of variation where colours are differentiated from other features such as shape or size, the children are enabled to "see" this feature as possible to categorize.

Lotta (teacher): Do you have any red ones? [turns towards Noel, 2 years, 2 months old]
Noel: Yes, there!

Noel lifts the lid of a jar and several buttons pop out, spreading over the table. He points at a large blue button.

Lotta: A large blue one, do you have any red ones?

Noel looks into Lotta's jar and picks up a button.

Lotta: That's my button, do you have any red ones?
Noel: There.

Noel gives a red button from his jar to Lotta.

Lotta: I get a red button from you, thank you, do you have any more?

Noel takes two buttons, one red and one yellow at the same grab from his jar and put them into Lotta's jar. Lotta points at the yellow button in her jar. Noel laughs.

Lotta: Look there! I know, we can have yellow there.

Lotta moves the yellow button to an empty jar. Noel holds out his arm with another yellow button to Lotta, letting it drop down into the jar with yellow buttons.

Noel: One, one!

The teacher and child explored the feature "colour" and the different values within the feature. This was done by contrasting buttons. In accordance with Variation Theory, the meaning of "red" is not apparent if only experiencing red buttons that differ in size and shape; however, when contrasted with a button that is *not* red, the meaning may emerge. At the beginning, most of the children mix colours and sizes and shapes presumably without meaning (although we cannot be completely sure of this lack of purpose). The teacher began a collecting game with the children, encouraging them to collect similar buttons and emphasizing contrasts in colour when they appeared. She encouraged the children to participate by saying "I collect red buttons, what do you collect?" The children helped the teacher, giving her mixed buttons, and the teacher described every button she was given. When a button that was not red was given, she focused on this button, describing the colour explicitly, in relation to the red ones and not comparing sizes or shapes.

The children's behaviour in the activity changed as the children become more aware of the selected feature, colour. The child Mia is very active in sorting and helping other children collect specific coloured buttons. First she searched for the same colour until most of the same coloured buttons were found and then her attention moved towards another colour, which then became the centre of her attention. After a while, though, she began to work with two colours simultaneously.

Lotta (teacher): Do we need more buttons to sort?

Lotta grabbed a couple of handfuls from a box of buttons. Mia [3 years, 2 months old] picks out red buttons and gives them to Vilhelm [2 years, 11 months old] who puts them in a jar with mostly red buttons.

Mia: There is a red and there is a red.
Lotta: Here are yellow buttons.

Lotta points at a jar with yellow buttons. Mia then continues picking out yellow buttons from the pile on the table, leaving them close to Lotta's jar.

Mia: And here is a yellow one and here is a yellow. And there is a red one!

The rule of the game or play is clear to the child Mia; she knows the sorting game is about colours. The challenge though is to keep focus on the specific feature colour that may vary but still has to follow the rule.

The teachers made use of another pattern of variation as well, when they explored other dimensions or features, for example, when describing buttons as large or small, in the shapes of flowers, hearts or squares. Some children adopted this idea and started sorting new groups of similar features (such as large buttons of different colours).

Fig. 1 Sorting buttons in different shapes

Bringing in several features contributed another complexity, as the child discerned several features simultaneously but needed to decide which feature was in the foreground and which in the background (Fig. 1).

Carol (teacher) points at a row with buttons shaped as different animal figures that Maja (3 years, 3 months old) has sorted.

Carol: You had a whole lot of figures in a row too!

Maja took two large round buttons and placed them next to each other between the row of animal figures and a row of heart-shaped buttons. She then put another large round button in the same row and another one on each side of the two large buttons. Maja selected five round buttons in different colors and sizes, placing them right above the row with large buttons.

In this episode, the child has made her own categorization and decided that animals, large-sized round buttons and heart-shaped buttons constitute different features. A more challenging task is when two features are mixed, as in the following excerpt:

Lotta (teacher): Here is a square. And here is a flower.

Mia [3 years, 2 months old] picks up the yellow square.

Mia: Where put it?
Lotta: It can be yellow, it can be put there. Square-yellow!

Mia lets the square-shaped button go into the jar with yellow buttons.

Vilhelm [2 years, 11 months old]: What was it?
Lotta: The square?

Lotta picks up the square and gives it to Vilhelm. Vilhelm looks closely at the button and puts it into his jar with mixed buttons. Noel [2 years, 2 months old] reaches his hand towards the others showing a button in his hand.

Noel: Too, too.
Lotta: Look at this, Noel also found a yellow square! Maybe we should have a jar with yellow squares instead. Is there someone else who has more yellow squares?
Vilhelm: I have!

Vilhelm gives his yellow square-shaped button to Lotta.

Lotta: Then we make a jar with yellow squares. Look, now we have two.
Mia: Is this one?

Mia picks up a round yellow button.

Lotta: It is yellow but not a square.
Mia: Is it not a square?

Mia leaves the button on the table. Lotta shows her the jar with the squares to Mia.

Lotta It is a circle, look.

These children are not exploring patterns, in that no units were discovered or repeated. However, when scrutinizing the concept of pattern and finding the aspects of repetition, generalization and part-whole relationship crucial, the experiences that these children are offered do play a significant role in the development of the concept. In order to generalize an abstract idea, the idea has to be discovered and recognized in several expressions. In order to generalize such an idea, one has to recognize features of units that differentiate them from each other. Discerning similarities and differences as well as systematically sorting in one-dimensional rows thereby seems to be crucial for developing emerging pattern presentations. Systematic comparison through carefully offered patterns of variation as seen in the episodes lets the children experience the differentiated aspects that are the basic understandings about patterns. This is made possible in these play activities where children are encouraged to extend their understanding as in trying out category belonging and when "borrowing experiences", as Vygotsky (2004) says is necessary for new inventions of ideas.

Second Conclusion: Young Children Do Not Imitate Patterns – They Imitate Ideas

The teachers in this study offered children systematically organized objects as inspiration and challenges. Children who are given models and opportunities to "borrow" experiences (Vygotsky 2004) from a more experienced teacher could be expected to show these as imitations in their own activities. None of the participating children picked up on the teachers' idea at the beginning and none imitated exactly the offered models. This brings focus to the relevance structure (Pramling Samuelsson and Asplund Carlsson 2008), how the children perceive the meaning and goal of the activity or offered model and also what teaching means when working with very young children. However, when studying children's actions and initiatives, it is possible to see them imitating the idea, rather than the actual pattern that they are offered as model (see Vygotsky 1978).

What is most interesting is that some children, who do not imitate a pattern on the teacher's request, are nonetheless made aware of ideas that they seemingly have not previously been aware of or made use of. The framework of play seems to encourage this approach, as the child can make his or her own decisions and is free to pick up ideas that others are expressing. This is seen, for example, in the episode below, where the teacher introduced the children to exploring similarities that may be generalized to new situations.

Maja (3 years, 3 months old) is showing a heart-shaped button before she adds it to a row of three other heart-shaped buttons that she lined up before.

Maja: Another heart
Carol (teacher): That many. Four hearts in a row!

Carol lined up long rows of similar buttons on the table. Jenny (2 years, 11 months old) hands over a handful of mixed buttons to Carol.

Carol: Can you help me find some that look alike? Look, it makes a long, long row.

Maja gives more buttons to Carol, adding to the row of similar buttons. Jenny shows a large button that they previously discussed and compared with a steering wheel.

Carol: Look, that was the big one. Where is the small one?

Jenny and Maja searched the boxes on the table. Maja found the smaller steering wheel button. Carol put the buttons side by side.

Carol: Look, here is big and small, the same!

Maja looked at her own rows of buttons; she took a small button and said small heart and added it to her row of larger heart-shaped buttons.

Carol: A small heart, can it be together with the large hearts?

Rather than imitating an exact pattern or be inspired to copy a model of a peer or adult, children can discern an underlying idea and generalize this idea to another situation, which is a far more complex idea than imitating a model. This would seem to be one of the major aspects that supports pattern recognition as well as later pattern construction. Play seems to be central to this kind of learning in that the play or game does not require a correct imitation. The children can be creative and extend their experiences and develop their ideas. This is likely to be more powerful for the child's learning than pure imitation since the acts build upon the discerned abstract relationships rather than feature similarities.

Third Conclusion: Coherence Is Key

After the first learning session, the teachers felt confident that the children had experienced that features of objects may vary but could be organized in categories of similar features. They then experienced that the children lacked an understanding of how to discern the systematic organization, such as how objects are sorted according to a rule. This brings in a spatial dimension to the object of learning and calls for a framework where the features are not in focus, but the structure of the ordering of objects. However, this produced some unexpected challenges.

One of the teachers planned an activity in which the children were encouraged to print patterns on paper with cut-up fruit: bananas and apples that when dipped in paint and pressed on a piece of paper would leave circle-shaped imprints, smaller circles from the banana imprint and larger circles from the apple imprint. The pedagogical intention was to direct attention to systematic presentation of similar shapes that differ in size. The children, accustomed to painting with hands and brushes, did watch the teacher intensively, but did not attempt to make any patterns of their own (in accordance with the second conclusion described above). Instead, the children painted with the pieces of fruit, as if they were brushes. This is interesting in its own sense, since the children have experiences of cutting and eating fruit, of painting with brushes, but not painting nor printing with fruit. They nevertheless created a connection and made meaning of the situation by starting to paint with the fruit in ways they were familiar with. Children make their own meaning when they are not offered "borrowed" experiences (to use Vygotsky's terms) that may guide them towards the teacher's intended meaning. A thorough analysis of the teaching act reveals that in accordance with Variation Theory, there might have been a too big a leap from discerning features towards exploring aspects of patterns with different objects in two directions (one row of equal banana prints, one row of equal apple prints followed by another row of banana prints and so on). The children were not given the opportunity to experience what systematic sorting means, where the critical aspect probably is the regularity but relies on the child discerning the unit of repeat. The aspect of regularity is difficult to discern in the activity, since the teacher provides a row of large imprints and then another row of small imprints underneath. To discern this systematic structure as a pattern, the child would have to discern the rule for changing features related to the part-whole structure.

One teacher did, however, find a way to explore the systematic aspect of patterns. There seems to be a relationship between the idea of repetition and features of the provided items. A single row of units is not bringing fore the idea of patterns as a row of identical things is repeating indirectly (the gaps between the objects in the row is as important as the units of the physical objects). If objects within the row are varying, there may be a better chance of discerning the systematic rule, when attention is brought towards what is similar and what is different. Once again, variation is necessary for discerning the phenomenon. Nevertheless, if too many dimensions are brought in, it may be confusing for determining what to take account of and what to keep in the background. The teacher Carol figures out a way to bring attention to the coherent whole and at the same time enable exploring how this whole can be structured in different ways.

Fig. 2 One of the models used for emphasizing systematic ordering of shells

The teacher Carol sits with Maja (3 years, 3 months old), Wendy (2 years, 5 months old) and Gloria (3 years old) at a table. She offers to the children a tray with shells, all different but some belonging to the same kind of shells. She also offers cards where the same shells are printed in natural sizes. The activity begins with the children exploring the shells and cooperating in finding shells of the same kind as the printed ones.

This activity offers the children opportunities to discern similarities and differences. It directs attention to the shapes of the shells, which is critical in order to work more systematically with the features of the shells. The cards show single shells or up to five shells placed in different but systematic order. The activity is framed as a sorting game, where the children's choices are used as a basis for talk and problematization. Together with their peers and teacher, the children compare features of different shells and choose which ones fit their desired pattern.

Carol offers a card with a pattern of shells (see Fig. 2): Wendy (2 years, 5 months) quickly finds the two twisted shells and puts them on the picture, remaining three round white shells uncovered.

Carol: What are we going to do with the others? Do you have an idea Maja?

Gloria (3 years old) takes three blue clam shells, one at a time, and fills the three remaining units on the picture.

Wendy gets another picture with a pattern of four shells: white-blue-white-blue. Wendy points at the blue clam shells and looks at the teacher.

Carol: Can Wendy borrow two blue shells which she has on her card?

Wendy gets the two blue shells, puts them on the card on corresponding picture and then takes all different shells for the remaining shells on the card.

Wendy seems to discern the twisted and then the blue shells as standing out from the row of shells. This is probably a very important aspect for her discerning the systematic idea of units constituted by similar features and repeated in a specific order. The other children work in similar ways when they get the same pattern card, for example, first picking out the blue clam shells and then the other ones. They are very accurate in making sure that they get the same shape, size and orientation as the shells on the card. The type of shell, however, may be different (blue clam shells instead of white seashells). Nevertheless, the systematic order of "every other of the same kind" is apparent.

The conclusions drawn from these activities are that coherence that frames the idea of the systematic and repetitive aspect of patterns is not easily understood by young children. There seems to be a necessity to offer very concrete patterns that make it possible to discern specific similarities and differences. Play adds to this process through the children's expressed joy in sorting and comparing shapes, sizes and colour in different ways. According to Ginsburg (2006) this would be an example of playing with mathematics, which is orchestrated in the coherence offered by the activity's structure.

Conclusion

Bruner (1996) said that any child, at any age, may learn anything. If that is so, then no content or concept is too difficult or unintelligible to work with in preschool. At the same time, there are of course limits to what can be expected in terms of skills and knowledge. A reasonable interpretation of Bruner's claim and in accordance with the theoretical approach adopted in this study is that it should be possible to work with topics like patterns and pre-algebra in early childhood education, even with toddlers of 2 or 3 years old. This brings attention to the meaning of patterns and pre-algebra, the foundations and emerging skills that are necessary as a basis for further conceptual understanding.

The design of this study includes a theoretical inquiry of the chosen concept. Teachers may investigate the concept and find several learning objects to work with: pre-algebra includes the meaning of the equal sign, abstraction and not least the following of a rule. These are potential learning objects, but the teachers need to choose a specific learning object, such as the following of a rule, and then investigate what it takes for the children to understand that learning object. It is crucial that the teacher identify the child's current understanding which may be expressed in different ways and guide the child through offering activities and experiences that promote exploration of the aspects that are necessary to discern. Offering such experiences is, however, a delicate practice, and the relevance structure is crucial to establish.

While studying the children's exploring activities and expressions of developing understanding, play emerges as a vehicle for developing conceptual understanding towards the idea of patterns. This is shown, for example, early in the study when children create and take initiatives to a larger extent than first assumed by the teachers. The freedom within play is powerful, but in order to use play for pedagogical purposes, this presumes that the teacher is aware of the complexity of understanding mathematical patterns.

So what are the children doing when they are playing with patterns? They are playing with patterns, in the sense of trying to make meaning in the games, play and structures of activities they are offered. The children who were engaged in a painting activity made use of their earlier experiences of painting and used pieces of fruit as they previously had used brushes. The idea of making patterns with imprints of the fruit remained invisible to the children. The goal of the activity was different for the children than the intended goal of the teacher, which limited the possibilities to explore the learning object (see also Björklund (2014) for similar results).

The activity with the shells and printed cards with the shells sorted in different patterns could be seen as a game with rules in which the children had to find shells of the same kind and make pairs. The children figured out the need for one-to-one matching quickly, which seemed to help them direct their attention to the features of the shells. This play or game brought in both earlier experiences and abilities to discern similarities, but required a new approach, the structuring of the shells. This structuring is explored by the children, and the mathematical content is actually the core of their play. The children approached the activity in different ways; some children very precisely covered the printed pictures, while others made use of the idea, first picking out shells of one kind and then of another kind. They were thus making meaning based on their previous experiences in a new setting.

Variation Theory helps in interpreting the process of conceptualizing patterns, as there seems to be some crucial aspects that children have not yet discerned but are exploring in their play with patterns. The ability to categorize and discern different dimensions of features within objects is presumably crucial. It facilitates the generalizing of ideas, which is seen in the activities where the children paid attention to peer's and adults' ideas and implement the abstracted idea in their own project.

The children were observed playing with patterns. Play means in this sense that children were making meaning, striving for coherence and doing this on their own terms eagerly supported by the teachers who followed their acts and expressions. According to Sutton-Smith (1997), play has a framing function that includes intersubjectivity, meaning that attention of two interactive subjects is directed at one and same object of interest, which is of utmost importance for goal-oriented teaching. The teacher creates a relevance structure in which the child is invited to explore certain phenomena and play contributes to sustained shared thinking, as the teacher is genuinely involved in the play together with the children.

Results from the analysis show that reproduction should not be primarily the goal to strive for; instead, there are more basic relationships and abilities that seem to be prerequisite for the development of understanding the concept of pattern. By making relationships visible to the children, based on the children's own initiatives in their explorative play, understanding seems to be facilitated. The results thereby present a broader understanding of patterns and the learning of the concept of patterns. Taking the Vygotskian perspective on pattern concept development and play-based activities, there is a legitimate pedagogical aim: young children may not have the ability to create mathematical patterns as such, but in the act of play, the teacher may lend his or her knowledge and experiences of how to combine features and follow a rule. These borrowed experiences are interpreted by the children; logical relationships are discerned (logical from the child's point of view) and integrated with the child's own experiences.

This chapter will end with a quote from van Oers (2013b, 271) that summarizes the conclusions drawn also from the current study:

> The future of mathematical thinking in young children strongly depends on the quality of early years teachers to recognise mathematical actions in children, to see the mathematical potential of play activities and play objects, and to guide children into the future where they can still participate autonomously and creatively in mathematical communications.

Acknowledgement This study was conducted with financial support from the Swedish National Research Council (grant no. 724-2011-751).

References

Bishop, A. (1988). *Mathematical enculturation. A cultural perspective on mathematics education.* London: Kluwer.

Björklund, C. (2014). Powerful teaching in preschool—A study of goal-oriented activities for conceptual learning. *International Journal of Early Years Education, 22*(4), 380–394. http://dx.doi.org/10.1080/09669760.2014.988603.

Björklund, C., & Pramling, N. (2014). Pattern discernment and pseudo-conceptual development in early childhood mathematics education. *International Journal of Early Years Education, 22*(1), 89–104. doi:10.1080/09669760.2013.809657.

Björklund, C., & Pramling Samuelsson, I. (2013). Challenges of teaching mathematics within the frame of a story—A case study. *Early Child Development and Care, 183*(9), 1339–1354. http://dx.doi.org/10.1080/03004430.2012.728593.

Bruner, J. (1996). *The culture of education.* (3rd ed.). Cambridge, Mass.: Harvard University Press.

Burton, L. (2002). Children's mathematical narratives as learning stories. *European Early Childhood Education Research Journal, 10*(2), 5–18.

Cross, C., Woods, T., & Schweingruber, H. (Eds.). (2009). *Mathematics learning in early childhood. Committee on early childhood mathematics: National Research Council.* Washington, DC: National Academic Press.

Devlin, K. (1994). *Mathematics: The science of patterns. The search for order in life, mind and the universe.* New York: Scientific American Library.

Garrick, R., Threlfall, J., & Orton, A. (1999). Pattern in the nursery. In A. Orton (Ed.), *Pattern in the teaching and learning of mathematics* (pp. 1–17). London: Cassell.

Ginsburg, H. (2006). Mathematical play and playful mathematics: A guide for early education. In D. G. Singer, R. Michnick Golinkoff, & K. Hirsh-Pasek (Eds.), *How play motivates and enhances children's cognitive and socio-emotional growth.* New York: Oxford University Press.

Hannula, M. (2005). *Spontaneous focusing on numerosity in the development of early mathematical skills.* Turku: Turku University.

Heeffer, A. (2010). Learning concepts through the history of mathematics: The case of symbolic algebra. In K. François & J. P. Van Bendegem (Eds.), *Philosophical dimensions in mathematics education* (pp. 83–103). Dordrecht: Springer.

Holmqvist, M. (2011). Teachers' learning in a learning study. *Instructional Science, 39*(4), 497–511.

Marton, F. (2014). *Necessary conditions of learning*. London: Routledge.

Marton, F., & Booth, S. (1997). *Learning and awareness*. Mahwah, NJ: Lawrence Erlbaum.

Marton, F., & Tsui, A. (Eds.). (2004). *Classroom discourse and the space of learning*. Mahwah, NJ: Lawrence Erlbaum.

Mulligan, J., & Mitchelmore, M. (2013). Early awareness of mathematical pattern and structure. In L. D. English & J. T. Mulligan (Eds.), *Reconceptualizing early mathematics learning, advances in mathematics education* (pp. 29–45). Dordrecht: Springer. doi:10.1007/978-94-007-6440-8_3.

Papic, M., Mulligan, J., & Mitchelmore, M. (2011). Assessing the development of preschoolers' mathematical patterning. *Journal for Research in Mathematics Education, 42*(3), 237–268.

Pramling Samuelsson, I., & Asplund Carlsson, M. (2008). The playing learning child: Towards a pedagogy of early childhood. *Scandinavian Journal of Educational Research, 52*(6), 623–641.

Pramling Samuelsson, I., & Pramling, N. (2013). Orchestrating and studying children's and teachers' learning: Reflections on developmental research approaches. *Education Inquiry, 4*(3), 519–536.

Radford, L. (2012). On the development of early algebraic thinking. *PNA, 6*(4), 117–133.

Rivera, F. (2013). *Teaching and learning patterns in school mathematics: Psychological and pedagogical considerations*. Dordrecht: Springer.

Runesson, U. (2008). Learning to design for learning: The potential of learning study to enhance learning on two levels: Teachers' and students' learning. In T. Wood & P. Sullivan (Eds.), *The international handbook of mathematics teacher education* (Knowledge and beliefs in mathematics teaching and teaching development, Vol. 1, pp. 153–172). Rotterdam: Sense.

Sarama, J., & Clements, D. (2009). *Early childhood mathematics education research. Learning trajectories for young children*. New York: Routledge.

Sutton-Smith, B. (1997). *The ambiguity of play*. Cambridge, MA: Harvard University Press.

Swedish National Agency for Education. (2011). *Curriculum for the preschool Lpfö98. Revised 2010*. www.skolverket.se.

Sylva, K., Melhuish, E., Sammons, P., Siraj-Blatchford, I., & Taggart, B. (Eds.). (2010). *Early childhood matters: Evidence from the effective pre-school and primary education project*. New York: Routledge.

Threlfall, J. (1999). Repeating patterns in the primary years. In A. Orton (Ed.), *Pattern in the teaching and learning of mathematics* (pp. 18–30). London: Cassell.

van Oers, B. (2013a). Challenges in the innovation of mathematics education for young children. *Educational Studies in Mathematics, 84*(2), 267–272.

van Oers, B. (2013b). Communicating about number: Fostering young children's mathematical orientation in the world. In L. D. English & J. T. Mulligan (Eds.), *Reconceptualizing early mathematics learning, advances in mathematics education* (pp. 183–203). Dordrecht: Springer. doi:10.1007/978-94-007-6440-8_3.

Vygotsky, L. S. (1978). *Mind in society*. Cambridge, MA: Harvard University Press.

Vygotsky, L. S. (2004). Imagination and creativity in childhood. *Journal of Russian and East European Psychology, 42*(1), 7–97.

Warren, E., & Cooper, T. (2008). Generalising the pattern rule for visual growth patterns: Actions that support 8 year olds' thinking. *Educational Studies in Mathematics, 67*, 171–185.

Development of a Flexible Understanding of Place Value

Silke Ladel and Ulrich Kortenkamp

Abstract In this chapter, we highlight the importance not only of an understanding of place value but the importance of a flexible understanding. We describe the principles of our decimal place value system and the development processes of children. Embedded in artefact-centric activity theory, we present an education-oriented design of a virtual place value chart and its potential to support this development and understanding. We also present results of a qualitative study with second graders as well as results of a quantitative study with third graders that can guide further research in that area.

Introduction

This article reflects on the notion of *flexible understanding* of place value and presents some evidence that it might help children in early number learning and arithmetic. After introducing the key elements of a flexible understanding of place value and the associated development processes, we consider the artefact "place value chart" both as an existing, touchable thing and as a virtual realisation in an app. Within the framework of artefact-centric activity theory, we analyse the underlying actions.

A textbook analysis, briefly presented here, served as a basis for a qualitative study with second graders and a quantitative study with third graders. The qualitative study leads to a categorisation of mistakes when working with place value, while the second study gave rise to a categorisation of students' answers in nine

S. Ladel (✉)
Universität des Saarlandes, Saarbrücken, Germany
e-mail: silkeladel@gmail.com

U. Kortenkamp (✉)
Universität Potsdam, Potsdam, Germany
e-mail: kortenkamp@cinderella.de

T. Meaney et al. (eds.), *Mathematics Education in the Early Years*,
DOI 10.1007/978-3-319-23935-4_16

categories. Using statistical implicative analysis, we see that while a flexible understanding is not a necessary condition for children to succeed in other tasks, it still does help them to solve them.

Flexible Understanding of Place Value

The understanding of place value and in particular the *flexible* understanding of place value plays an important role in learning and understanding mathematics. We define the flexible understanding of place value as the ability to switch between different possibilities to split a whole in parts, whereupon the parts are multiples of different powers of ten, e.g. "19 hundreds 77 ones" is the same as "1 thousand 9 hundreds 7 tens 7 ones" is the same as…. A flexible understanding of place value is not only the basis of understanding the written calculation methods of addition, subtraction, multiplication and division, but it is also important for applying advanced calculation strategies or even to master everyday's life, as outlined below.

For written arithmetic, it is necessary to know, for example, that 3 hundreds 14 tens 7 ones is the same as (3 + 1) hundreds 4 tens 7 ones (see Fig. 1). This helps to understand what *the little 1's* (the carry-overs) mean.

With a flexible understanding of place value, it is also possible to easily divide a number like 361 218 by 6, avoiding the formal algorithm: "36 ten thousands 12 hundreds 18 ones" divided by 6 is the same as "6 ten thousands 2 hundreds 3 ones," which in turn is 60 203. A flexible understanding of place value is even the basis to understand polynomial division in later years, which remains a justification for teaching written division algorithms.

Apart from being an aid in written and mental arithmetic, nonstandard partitions are part of everyday life. We have to travel "fourteen hundred" kilometres (14 hundreds = 1400), the trip costs "twelve hundred" dollars (12 hundreds = 1200), in the year "nineteen hundred seventy two" (19 hundreds 7 tens 2 ones = 1972) and so on. There are numerous situations where we use a flexible representation instead of the standard form. So it is important for children to understand these and to be able to switch between different forms of representation of one and the same number fluently.

Fig. 1 Addition in written arithmetic

	H	*T*	*O*
	1	*7*	*2*
+	*2*	*7*	*5*
	3	*14*	*7*
	4	*4*	*7*

In the following section, we describe the development of the decimal part-whole concept that is the basis for our decimal number system. We then explicate the principles of our decimal number system.

Development and Principles of the Decimal Number System

Children develop a general part-whole concept before they acquire the decimal part-whole concept (Resnick et al. 1991; Ladel and Kortenkamp 2011, 2014), by experiencing the conceptual base of elementary arithmetic that all numbers are additive compositions of other numbers. "This compositional character of numbers provides an intuitive basis for understanding fundamental properties of the number system" (Resnick et al. 1991, p. 375). The part-whole concept manifests itself in the *additive property* (AP) principle, which states that the quantity represented by the whole numeral W is the sum of the values P_i that are represented by the individual digits (Ross 1989, p. 47):

$$P_1 + P_2 + \cdots + P_k = W. \quad (1)$$

Here, the P_i are whole numbers, as is W. On this general level, it is only important that each digit will contribute additively to the whole.

In order to use the additive property for the notation of numbers, special parts have to be built, namely, multiples of powers of ten. This is because of the *base-ten property* (B10) of our number system that states that the values of the positions increase in powers of ten from right to left. To get the value of an individual digit, we have to multiply the face value of the digit with the value assigned to its position. That is the *multiplicative property* (MP) principle. All the parts are in the special form $n_i \cdot 10^i$ for each of the k places.

$$n_{k-1} \cdot 10^{k-1} + n_{k-2} \cdot 10^{k-2} + \cdots + n_o \cdot 10^0 = W. \quad (2)$$

This representation so far does not impose any restriction on the n_i apart from the fact that they are integers. In a positional number system, it is necessary not only to bundle but to bundle maximally, that is, to continue bundling until it is not possible any more (*principle of continued bundling*, PCB). When doing so, all face values n_i will become single digit ($0 < n_i < 10$).

$$n_{k-1} \cdot 10^{k-1} + n_{k-2} \cdot 10^{k-2} + \cdots + n_o \cdot 10^0 = W, \; and \; n_i < 10 \; for \; all \; i. \quad (3)$$

The understanding of continued bundling is a necessary condition for understanding place value. Nevertheless, the understanding of place value and its acquisition has to be distinguished! We can demonstrate this in the enactive form of representation. The action of *bundling* (see Fig. 2, left) has the meaning of "changing ten ones into one ten" or the meaning of "sticking ten ones together to become one ten". The *place* of the ones and tens does not matter, as we can see in Fig. 2 (lower left).

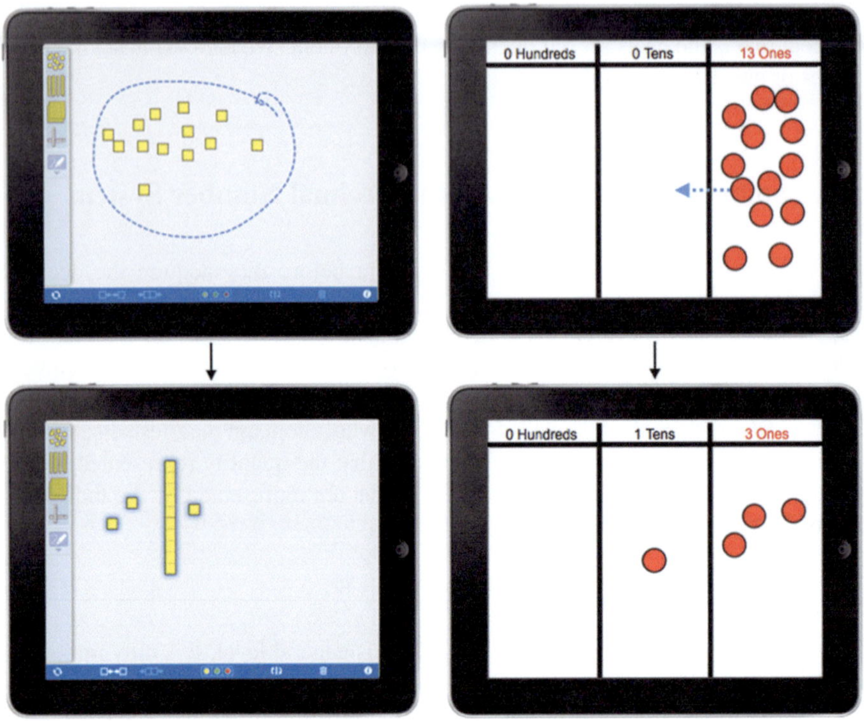

Fig. 2 Bundling and place value

However, the action of *changing the place*, e.g. move one token from the ones column to the tens column (see Fig. 2, right), means that either the value of this token changes from a one into a ten and hence the value of the represented number changes (by $-1+10=9$) or that we need nine additional ones to make a ten so that the value of the number stays the same. In this case, the *place* of a token decides about its value.

When learning numbers and place value, the children come into contact with different forms of symbolic and iconic number representations. For example, we have:

- Numbers indicated by bundle units, e.g. 4 ones 23 tens
- Numerals, e.g. nineteen hundred seventy-seven
- Tokens in a place value chart
- Numbers in a place value chart
- Numbers in standard notation

The constraint that the parts in Equation (3), above, have to be single digit is a specification that is only needed if numbers are represented in the last mentioned form, e.g. 1971. In all the other forms of representation, the face values n_i can also be larger than 9, as in Equation (2).

Only numbers in standard notation need the *positional property* (PP) where the quantities represented by the individual digits are determined by the position they hold in the whole numeral and not by the indication of the bundle units, not by the indication words like "-teen" and not by the column and its designation.

As we stated earlier, the flexible understanding of place value involves switching between different possibilities to split a whole into multiples of different powers of ten [Equations (2) and (3)], in several forms of number representation. Concerning the different kinds of partitioning, we distinguish standard partitions [Equation (3)] that result from continued bundling and nonstandard partitions [Equation (2)]. The nonstandard partitions we distinguish again in strong and not strong nonstandard partitions (see Fig. 3). A strong partitioning fulfils the additional rule that the parts do not "overlap": For two summands $n_i \cdot 10^i$ and $n_j \cdot 10^j$ with $i<j$, also $n_i \cdot 10^i < 10^j$ holds if $n_j > 0$. As a consequence, a strong partitioning only uses the digits of the standard partitioning (and maybe some additional zeros).

The Place Value Chart and the Meanings of Actions

There are different possibilities of the meaning that an action with an artefact may have. Bartolini (2011) calls it the *semiotic potential* of the artefact and refers to the triangle of artefact, task and mathematics knowledge. There is:

> a double semiotic link between the artefact and a task on the one hand and the artefact and mathematical meanings on the other hand. The former is within the reach of students whilst the latter emerges from the epistemological analysis made by teachers and experts. (Bartolini 2011, p. 96)

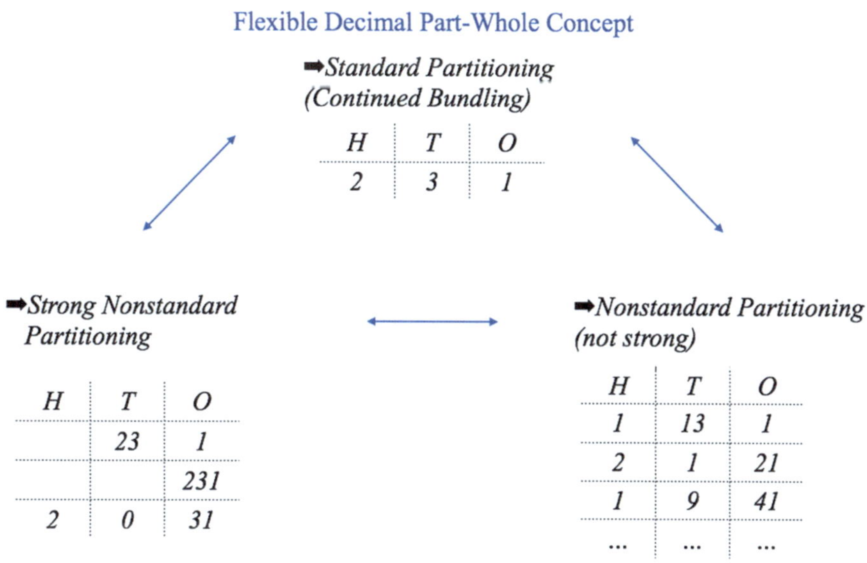

Fig. 3 Flexible decimal part-whole concept

It is the teachers and experts who put their knowledge about mathematics in an artefact. According to that an artefact can "behave" or externalise in different ways when actions are done with it.

Askew (2012, p. 13) discusses the co-construction of meaning when introducing place value. We would like to contrast this with a stronger focus on the artefact, given the extended affordances of digital tools.

In a place value chart, the action of moving a token can have different mathematical meanings. With real material, moving a token in the place value chart often means a change of value, e.g. one token in the tens becomes one token in the ones, so the change of value is $-10+1=-9$ because there is one "ten" less (-10) but one "one" more ($+1$). Another meaning of moving a token is a change of representation, e.g. one token in the tens becomes ten tokens in the ones that means an unbundling with constant value (see Kortenkamp and Ladel 2014; Ladel and Kortenkamp 2014).

Because of the different meanings of moving a token in the place value chart, the children internalise different mathematical knowledge depending on the externalisation of the artefact. If one token remains one token—whether it is in the tens column or in the ones column—the value of the number changes. So this will be the mathematics that a child learns while using this artefact. But if place value is connected to bundling and unbundling and hence there are, e.g. ten tokens in the ones column instead of one token in the tens column, the child will internalise another aspect of the mathematics behind place value (see Fig. 4).

The internalisation/externalisation pairs along the main axis need some explanation: The student (subject) externalises his or her understanding of the object through actions using the mediating artefact. The artefact includes an internalisation of the object, and it is able to externalise this through visualisation and behaviour. The interpretation of the action depends on the context and on the material the children

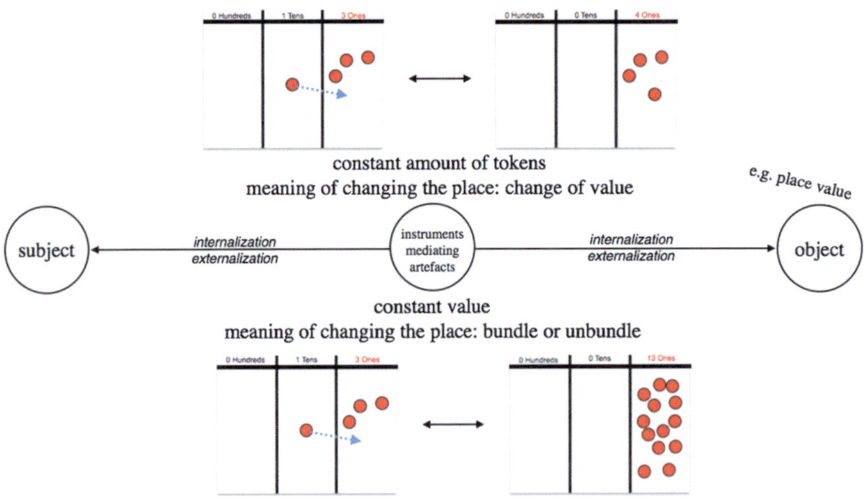

Fig. 4 Main axis of artefact-centric activity theory

are working with. The different behaviours of the token(s) and externalisations of the artefact focus on different principles of our number system. While an artefact that behaves in the way of changing the value suggests to the child that if the position of a number changes its value changes, too, an artefact that keeps the value constant suggests to the child that if the position of a token changes, then its value has to be multiplied or divided by multiples of the power of ten so that the value is constant.

Textbook Analysis

Before working with children, we analysed eight German mathematics textbook series for grades 1–4 in regard to number representations and kinds of partitioning. All tasks that apply to bundling or/and place value were categorised according to the kind of number representation and the kind of partitioning. Although all kinds of partitioning were found, about 80 % of the tasks only dealt with standard partitionings. However, the fact that all kinds of number representation and all kinds of partitioning, standard and nonstandard, strong and not strong, were found in textbooks (see some examples in Table 1) emphasises the importance of a proper understanding and the ability to change between various representations and partitions.

Table 1 Examples of tasks in textbooks for standard and nonstandard partitionings

Number Representation	Standard Partitioning	Example of Non-Standard Partitioning in Textbooks (strong and not strong)
numbers indicated by bundle units	3T 6O	*Enter it in the place value chart.* *a)* 7 O, 14 T *b)* 12 T, 6 H T H T O a) 1 4 7 b)
numerals	thirty six	*Enter these numbers in the place value chart and write them down.* *a) From Munich to Barcelona there are approximately forteen hundred kilometers.*
points in a place value chart	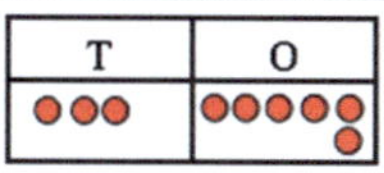	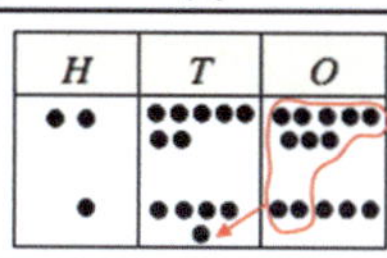
numbers in a place value chart	T O 3 6	H T O 3 5 9 + 1 6 4 4 11 13 4 12 3 5 2 3
numbers	36	

Qualitative Study

The empirical data comes from two studies, a quantitative one and a qualitative one. In the qualitative study, we examined 52 children from Germany at the end of second class. The framework for the qualitative study was based on a guided interview with four main tasks (Ladel and Kortenkamp 2014).

At first the children should circle the number that is larger in each pair[1]:

(1a)	**5T 3O**	**4T 15O**
(1b)	**1T 14O**	**2T 8O**
(1c)	**73O**	**7O 3T**
(1d)	**4T 9O**	**1T 29O**

The intention of this task was to find out whether the children paid attention to the bundle units or not. Therefore, we provided two levels of difficulty. On the one hand, we purposely mixed up the order of the ones and tens (see 1c) to see if the children just wrote the numbers from left to right or if they took care of the bundle unit. We also included standard partitions as well as nonstandard partitions. To compare the numbers, the children first of all have to bundle and add (see 1d right). After circling, we asked the children to describe the reasons for their decisions. Analysis of the answers showed four types of errors:

Type 1: No Bundling

The numerals are simply written from left to right without paying attention to the bundle units, e.g. (1b) 1T 14E as 114.

Type 2: Largest Bundle Unit

Only the largest bundle unit (here tens) is noted, e.g. (1a) 5T 3O is more than 4T 14O, because 5T is more than 4T.

This strategy works if the bundles are single digit (continued bundling as in Equation (3), above) and is the correct and used strategy to compare two numbers (without the indication by bundle units, having the same number of digits).

---

[1] In the German version, Z stands for "Zehner" (instead of T for tens) and E for "Einer" (instead of O for ones). This reduces the risk of confusing "O" (the letter capital o) with a "0" (the digit zero).

Type 3: Bundles as Separate Numbers

The children do not see the partition as a number but each part of it as a separate number. As a consequence, the first item shows four numbers, 5, 3, 4 and 15. The children resorted to comparing 5 and 3 first and then 4 and 15, instead of adding the bundles and comparing the composed wholes.

Type 4: No Bundle Unit

The bundle units are ignored, and the decision is only made on the basis of the largest number, e.g. (1a) 15 O(nes), because 15 is the largest.

Secondly, we asked the children how many tokens they would need to represent the number 35 in the place value chart, 35, 17, 8 or 26 tokens and if there are different ways to represent 35. After answering, they were given manipulatives so that they could show it. Besides incorrect representations that did not care about the place of the tokens at all (e.g. showing 35 single tokens in the tens column), there were two types of representations which occurred frequently: the *base-ten-block representation* and the *colour representation*. These two categories are interesting because they result from the work with didactical material.

In connecting the principle of bundling to the principle of place value, base-ten blocks are often placed on a place value chart (Fig. 6, left). This representation is misleading, as it becomes false when the students understand place value. In Fig. 5, there are two ten blocks in the tens column, so there are altogether 20 tens—hence 200. But what is the meaning of a ten block in the hundreds' place? Or a hundreds block in the one's place? Mathematically correct and better from a didactic perspective is the representation in Fig. 5, right. Here the base-ten blocks represent the ones, tens, etc., in the header row and can be replaced by them later on. The amount of ones, tens, etc., is represented by counters (or tokens).

A lot of the children interviewed laid 30 single tokens in the tens column, saying that there are 30 tens and 5 ones (which in fact would be the number 305). However, some of them laid 3 tens in the tens column (see Fig. 6, left). At this point, the interviewer asked if it is also possible to take just three tokens in the tens column (see transcript 1). With this question, we could investigate the understanding of the child with respect to place value:

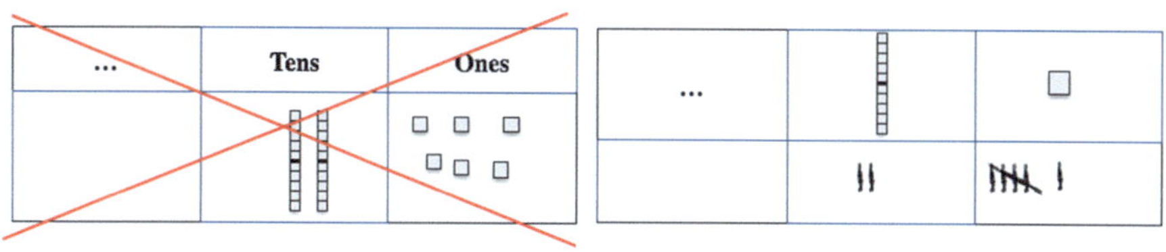

Fig. 5 Connecting bundling with place value

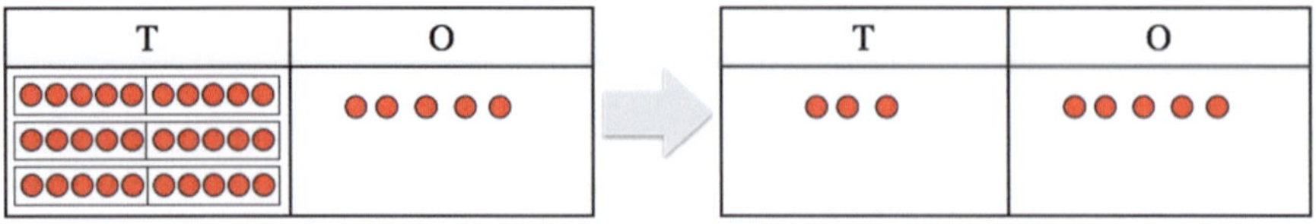

Fig. 6 Base-ten block representation

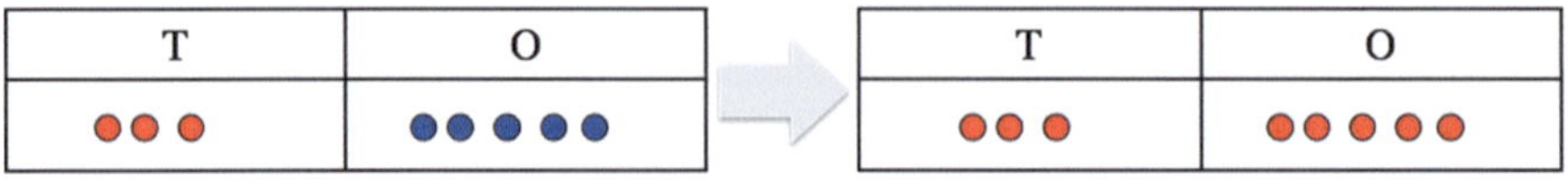

Fig. 7 Colour representation

Transcript 1

I "Can we also show 35 like that?" (taking away the 30 and placing 3 (Fig. 6, right))
P "Then we have 8 tokens." (with emphasis on 8)
I "But those are in the tens-column and the others are in the ones-column. Doesn't that make a difference?"
P "Hm, then we also have 35, because those are 3 tens and those are 5."

First of all, student P did not note the place of the tokens except that he laid the tens as 30 in the tens column and the 5 in the ones column. After replacing 30 by 3, he summed up the token numbers to end up with a total of 8 counters. P did not apply the multiplication principle. Only after the interviewer (I) mentioned the column he thought and agreed that 3 tens in the tens column represent 30.

Another representation that results from the work with didactical material is that ones and tens can be represented with different colours (see Fig. 7). Again the question is what happens if a blue token is in the tens column or a red token in the ones column? If a child already is able to abstract, he may answer like this boy in the interview: "The colour doesn't matter, it is just to see it better".

The aim of the third question in the interview was to cause a cognitive conflict. We showed the children 23 in a place value chart (two tokens in the tens column and three tokens in the ones column) and asked them "What happens if you move one token from the tens to the ones?" Most of the children answered that this will represent 14. But this is only one possible meaning of changing the place (see Fig. 4). The children then worked with a digital place value chart (Ladel and Kortenkamp 2013)[2] and were asked the same question. Only a few pupils were able to explain and connect the place value chart with bundling. An example can be seen in the following extract (transcript 2, video access available on request from the authors):

Transcript 2

I "And if I move one token from the tens to the ones, what will happen?"
P "Then this is not a ten anymore but a one… or will it be a ten?"
I "I don't know."
P "I think it is a one. Then it would be ten, fourteen."

[2] The digital tool is available at http://kortenkamps.net/index.php/Place_Value_Chart.

I "Please write it down here."
I "So let's try it with this. We do have twenty three. Move one token from the tens to the ones."
P "Oh, ay caramba!"
I "Ay caramba. What happens there?"
P "There are many."
I "There are numbers written on top. Maybe you can have a look at them."
I "Move again. What number is this?"
P "Thirteen ones."
I "And there is one ten left."
P "Weird."
I "What happened there?"
P "These are ten singles!"
I "Why that?"
P "It wouldn't work. Look, if you take one token and if you do not have ten singles, it would only be one. And if you have ten singles, it would be tens."
- P moves one token from the ones to the tens. -P "Oops, what happened there?"
I "What happened there?"
P "The others moved there. Because if only one moved to the tens it would be only eleven and then always nine follow in addition and nine plus one is ten and then it would be again two tens, twenty and three, twenty three."

It is important to connect the knowledge about bundling to the knowledge about place value. This girl was able to do the connection but she was an exception. However, to develop a flexible understanding of place value, it is important to care about this connection.

Quantitative Study

The quantitative study took place with 255 third graders resident in Halle and Saarbrücken (Germany) and Luxembourg. We created a 30 min test of three parts. When administering the test, the children were allowed to work for 10 min on each part.

The first part consists of two tasks of different types. The first task is to compare two numbers each given as certain numbers of hundreds (H), tens (T) and ones (O). The children should circle the number that is larger. For completeness, we list all eight subtasks:

(1a)	**2H 5T 3O**	**2H 4T 17O**	**(1c)**	**735O**	**7O 3T 5H**
(1b)	**3H 13T 5O**	**4H 6T 5O**	**(1d)**	**1H 43T 9O**	**63T 9O**
(1c)	**735O**	**7O 3T 5H**	**(1g)**	**7H 3O 6T**	**7H 6T 1O**
(1d)	**1H 43T 9O**	**63T 9O**	**(1h)**	**8H**	**91T 3O**

For example, in task (1a) students should circle 2H 4T 17O, as this represents 257, which is larger than 2H 5T 3O which represents 253. Note that both strong and non-strong partitions are used and the order of hundreds, tens and ones differs between subtasks.

The second task is to write a number given in the notation shown above into standard notation. This again involves both reordering and simple calculations when non-strong partitions are used:

(2a)	**3H 6T 1O**	(2c)	**1H 32T 4O**	(2e)	**7H 3T**	(2g)	**7O 31T**
(2b)	**3T 23O**	(2d)	**3O 2H 5T**	(2f)	**279O**	(2h)	**3H 3T 14O**

The second part again consists of two different tasks. The first task is to write down a number that is shown in a place value chart (Fig. 8). We only show the image of the first number (3H 8T 2O) of task 1a:

Three other representations are shown with the numbers 1H 2T 13O (task 1b), 2H 12T 13O (task 1c) and 5H 15T 3O (task 1d).

The second task of the second part is the inverse task to the first one: Children should draw tokens into an empty place value chart for the numbers 314, 163, 542 and 304 (tasks 2a–2d of part 2). In addition, we ask them whether they can think of a different way to represent the number in the place value chart (German: "Kannst du die Zahl auch anders in der Stellenwerttafel darstellen?"). No further explanation of what "different" means was given.

The third part of the test repeats the questions of the first part with different numbers:

(1a)	**1H 7T 2O**	**1H 8T 15O**	(1e)	**51T 23O**	**4H 3T 8O**
(1b)	**3H 25T 7O**	**2H 9T 4O**	(1f)	**3H 4T 27O**	**45T 6O**
(1c)	**385O**	**3O 8T 5H**	(1g)	**8H 1O 3T**	**8H 3T 1O**
(1d)	**2H 27T 8O**	**53T 7O**	(1h)	**7H**	**84T 2O**

(2a)	**2H 8T 3O**	(2c)	**2H 41T 5O**	(2e)	**8H 1T**	(2g)	**5O 53T**
(2b)	**4T 51O**	(2d)	**2O 3H 7T**	(2f)	**329O**	(2h)	**6H 2T 35O**

About half of the children had access to an iPad with an interactive place value chart (Ladel and Kortenkamp 2013) while working on part 2. No further instruction besides basic usage of the place value chart app (adding, deleting and moving tokens) was given, and it was the first time the children had contact with that app. In our data, we recorded whether the children had access to the iPad, but we did not monitor how the children used it or whether they used it at all.

For the analysis, we marked the tasks in parts 1 and 3 as well as in part 2, task 1, for correctness. The inverse operation of representing numbers in the place value chart was analysed by categorising the answers for the first subtask ("represent 314 in the place value chart" and "can you find a different representation"). The answers for the three other numbers usually did not fall into another category, but it

Fig. 8 Place value chart of task 2

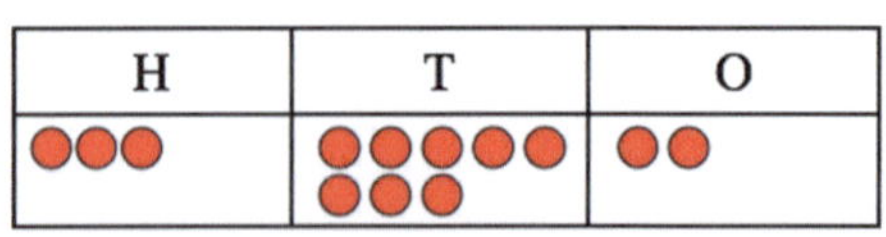

happened that they were often left unanswered due to time or other constraints. We found nine different categories:

1. *Flexible answer (N=56)*: Students who gave three representations of 314 that were correct and used different (nonstandard) partitions. Note that only a small percentage of students gave only two representations (see Fig. 9).
2. *Flexible answer with errors (N=31)*: Students who gave three representations of 314 using different (nonstandard) partitions but made minor errors that can be attributed to wrong counting or miscalculations.
3. *Base-ten-block errors (German: "Mehrsystemfehler", N=19)*: Students who made mistakes that can be explained by mixing base-ten blocks and the place value concept, for example, putting 10 tokens into the "tens" cell to represent 10 (see Fig. 10).
4. *Other symbols (N=24)*: Students who used other symbols like flowers instead of circles when giving a "different" representation.
5. *Permutation (N=69)*: Students who just permutated the 3, 1 and 4, thus producing wrong representations showing, for example, 413 or 341.
6. *Only one representation (N=16)*: Students who just gave the standard representation.
7. *Other arrangement (N=5)*: Students who used the same symbol (a circle) for other representations but only changed the arrangement within the cell, not the number.
8. *Value changing (N=4)*: Students who used the same number of tokens (3+1+4=8) but moved them into other cells, thus creating representations of other numbers.
9. *Non-categorisable (N=26)*: Students who gave other representations that could not be categorised at all. Representations were showing other numbers with other token counts.

In addition, 5 students did not answer at all, giving a total of 255 answers.

Analysis of the Quantitative Data

We used statistical implicative analysis (SIA, Gras et al. 2008) to analyse the data. With SIA, we can find implications between (binary) variables. The algorithm calculates an implication intensity between two variables that measures the "surprisingness to observe a small number of counter-examples" (Gras et al. 2008, p. 16). The intensity is a number between 0 and 1, where 1 means that it is no surprise at all to see a small number of counter examples, corresponding to the fact that we suspect an implication between these two variables. Note that we are indeed looking at implications, as opposed to mere correlations.

In Fig. 11, you see the dependency graph created by our own analysis tool (Ruby code available on request from the authors). Inputs were the binary variables for correct and incorrect solutions of subtasks (TxAyz is true when task yz in part x was solved correctly; NTxAyz is true when task yz in part x was solved incorrectly or

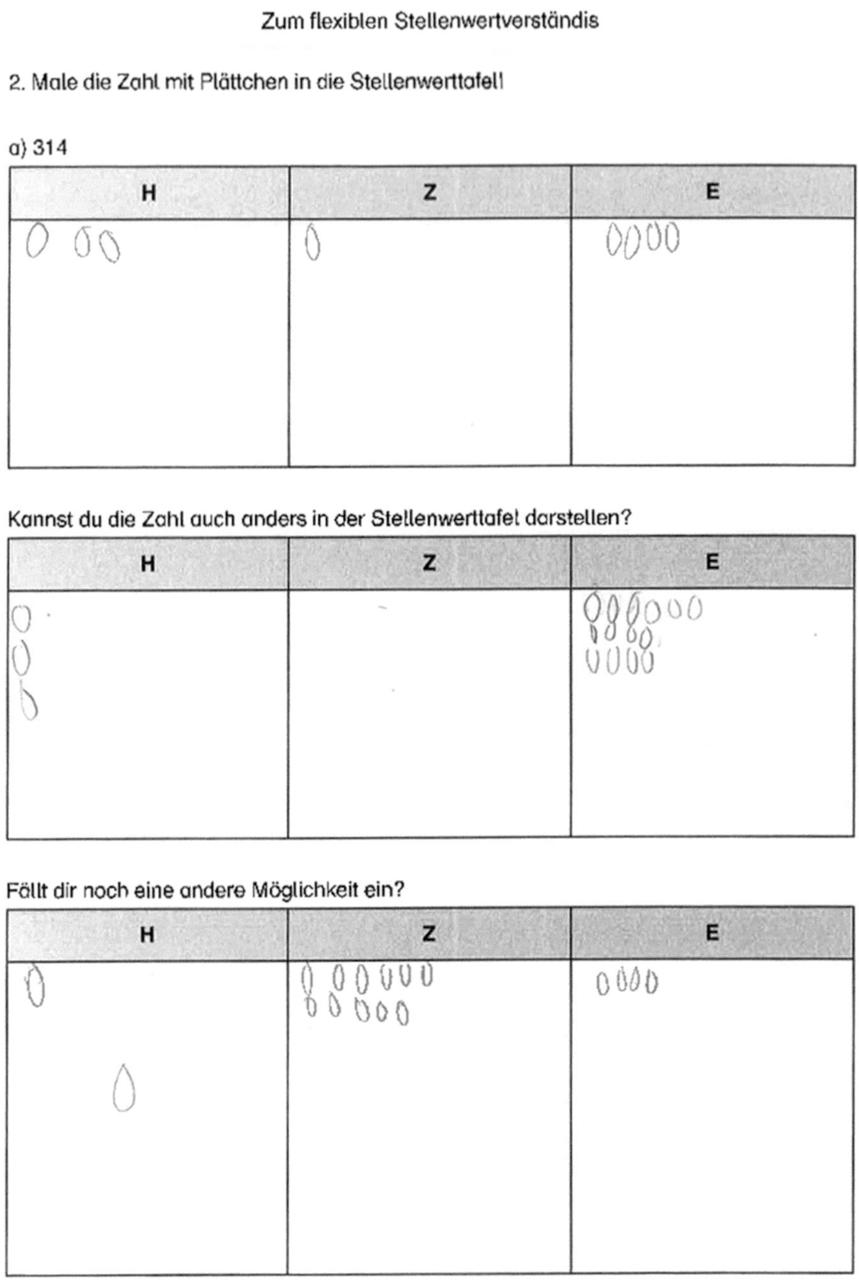

Fig. 9 A flexible answer in part 2

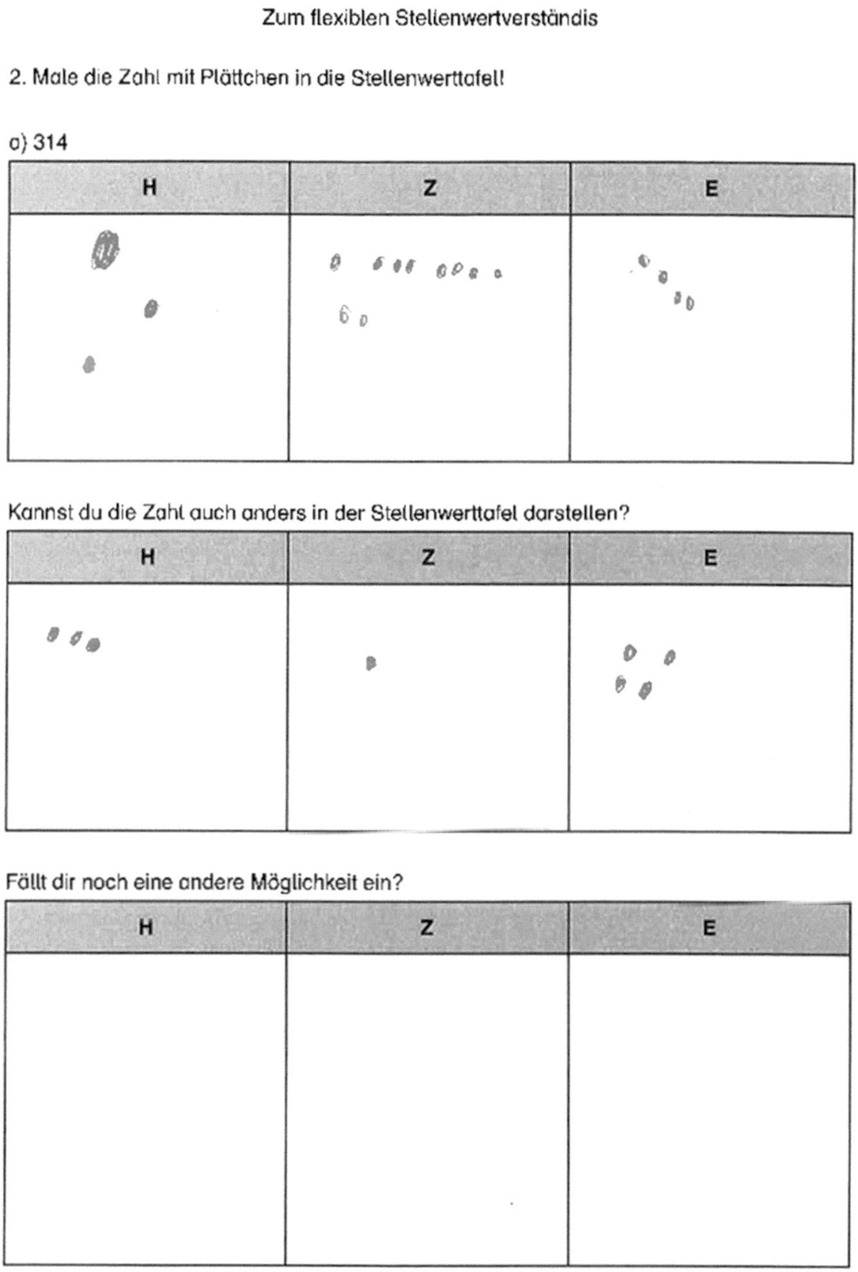

Fig. 10 A base-ten-block error of a student

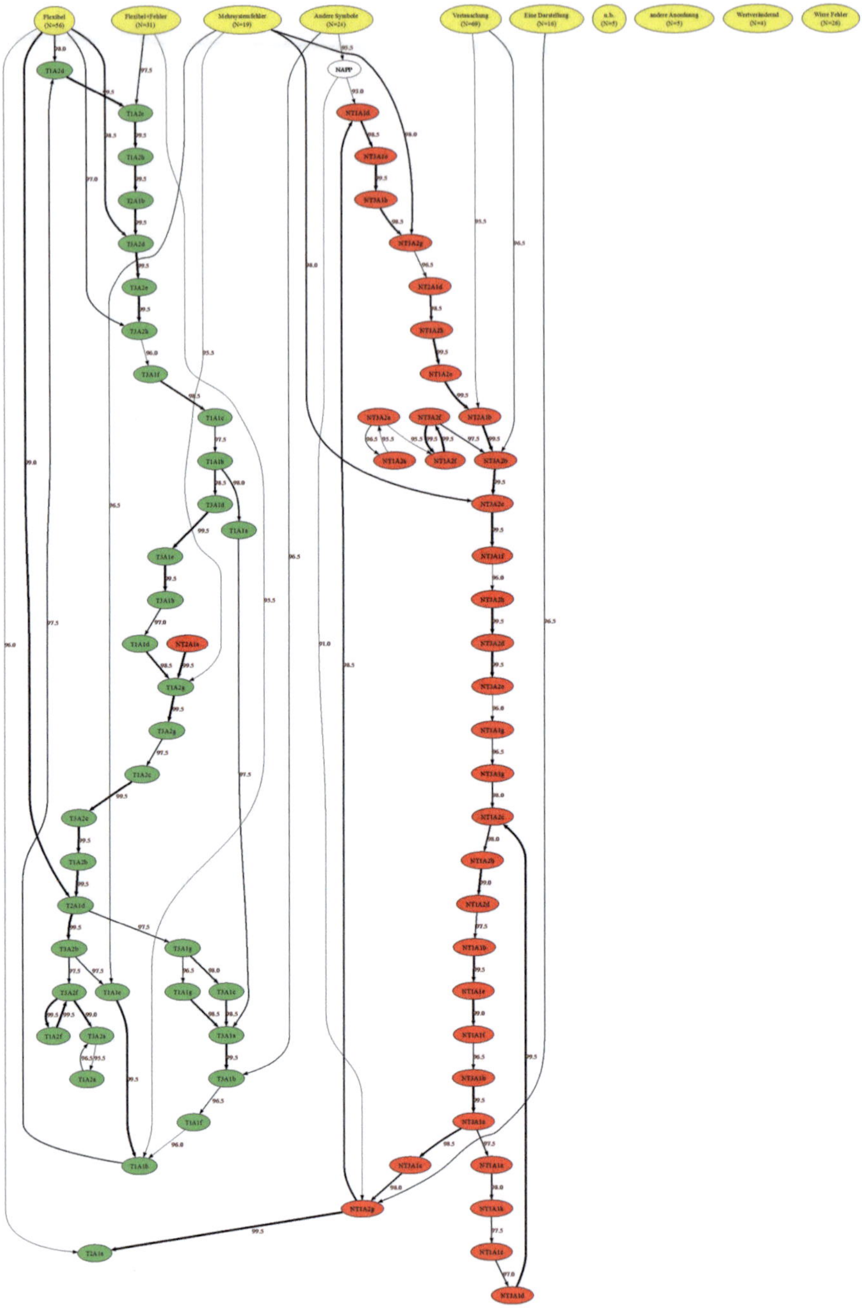

Fig. 11 Dependency graph of the quantitative data

not at all), the binary variables APP and NAPP that are true when the students had access to the iPad app in part 2 resp. did not have access to it and binary variables for each of the ten categorisations (including "no answer") of the answer for task 2a in part 2. Category variables are marked in yellow, correct answer variables in green and wrong answer variables in red. The NAPP variable is shown in white. Implication intensities between variables are shown by arrows labelled with the values rounded to .5 %. We removed transitive implications for better readability of the graph.

We only show implication intensities that reached a level of 95 %, with the exception of those that are connected to APP or NAPP, where we show all three implications that have an intensity larger than 90 %. APP is not represented in the graph, as it was not connected to any other variable with an intensity of more than 90 %.

A first glance immediately shows that—unsurprisingly—the "correct answer" items and the "incorrect answer" items are highly connected, showing a high internal consistency of our test. The only two outliers are NT2A1a → T1A2g and NT1A2g → T2A1a, both with a very high intensity of 99.5 %. In T2A1a, the 382 in standard partitioning should be read off a place value chart and so can be considered a very easy and basic task. In T1A2g, children were asked to translate 7E 31Z into 317, which includes both nonstandard partitioning and the permutation of places as additional hurdles. The data show that students who fail at the easier (actually, the easiest in this test) task are likely to master the harder one, and vice versa, students who fail at the harder task master the easy one. That students who fail at the harder task master the easier one can be explained by the fact that the easier task can be solved by all students (but the gifted ones).

Apart from this internal consistency, it is impossible to conclude anything from the students' answers in parts 1 and 3 and the first task of part 2, as the implication intensity of implications towards the category variables is below the threshold. Just looking at the correctness of answers in the easy-to-check tasks is not enough to diagnose the student. The categories "other arrangement", "value changing" and "non-categorisable", as well as "no answer", are not connected by presumed implications at all, which can easily be attributed to the low numbers of cases and the non-categorisability.

On the other hand, there do seem to be implications originating at the categories towards the correct answers or incorrect answers cluster. With high implication intensities of 96 %, 99 %, 98 %, 97 % and 98.5 %, the "flexible thinkers" solved T2A1a, T2A1d, T1A2d, T3A2h and T3A2d correctly. Recall that we removed implications that are implied due to transitivity from the graph, so these are just the "entry points" into the cluster of correct answers. Even students in the "flexible with errors" category end up solving T1A2e and T1A1b correctly, albeit with lower implication intensities of 97.5 % and 95.5 %. Still, there is no above-threshold implication into the cluster of wrong answers for any of the flexible categories.

This is different for the "base-ten-block error" category: With implication intensities of 98 %, those students give wrong answers in T3A2g and T3A2c. There is also a presumed implication into the correct answer block but at lower intensities of 96.5 % (T1A1e) and 95.5 % (T1A2g). It is safe to say that students who put 10 tokens into the tens column to represent 10 are more likely to solve the tasks incorrectly.

Students in the "other symbol" category show implications both to the "correct answers" cluster and, very weakly, to the "incorrect answers" cluster. The latter implication is only via the NAPP variable, which is shown as implication of "other symbol" with 95.5 % intensity. As we chose the children who were working with the app and those who did not randomly, we cannot draw conclusions here. In fact, the wording of the question ("Can you show a different way to represent that number?") is very open and might be interpreted in various ways by students. In those cases, only a qualitative design could give more insight.

Finally, it is likely that an item belonging to the categories "Permutation" (95.5 % to NT2A1b, 96.5 % to T3A2b) and "Only One Representation" (96.5 % to NT1A2g) again implies that there will be wrong answers in the other tasks.

Conclusions

The qualitative data allows us to set up types of error and to gain further insight in the thinking and actions of the children. We could see that the connection between different kinds of number representation, in particular from numbers indicated by bundle units to numbers, is not stable. That might be a reason why many children had difficulties comparing two numbers.

Another very important fact we can conclude from the qualitative data is that we have to take care when using didactic material, particularly when connecting bundling to place value. For example, the base-ten-block error is an error that can result from the misusage of base-ten block in the place value chart (Fig. 10).

The quantitative data shows that there is a connection between flexible representations in the place value chart and correct solutions of tasks that involve interpretation of numbers given in nonstandard representations. Even students who make mistakes when giving several different representations (with "different" being "different partitioning") were likely to solve the other tasks correctly. The implication analysis shows that it seems to be sufficient to be aware of the fact that numbers have different representations in the place value chart.

Students who had difficulties with representations in the place value chart were shown to be more likely to fail at tasks that do not involve the chart itself but need competences in place value and nonstandard partitions. Note again that we analysed implications—nothing is being said about students who solve the tasks successfully; they could lack the place value chart skills or could be flexible thinkers. But our data shows that working with place value charts *and nonstandard partitions* could be a way to improve students' performance.

We can also see that the transition from sorting base-ten blocks into place value charts and abstracting from the blocks into tokens for counting is crucial for a proper understanding. Students who mix up "ten" as a quantity and "ten" as a place value showed problems with solving other tasks. We suggest not to use base-ten blocks with place value charts at all. Instead, we suggest using a place value chart for counting marks, each mark representing one block of the proper size. This way there is no confusion of the block representing the number 10 (or 100) and the counting token for "a ten" (or "a hundred").

Outlook

We could show that students who show a *flexible* understanding of place value are more likely to solve tasks that involve nonstandard partitions of numbers into hundreds, tens and ones. Such tasks appear in textbooks and in everyday life and are as well a basis for addition and subtraction algorithms.

As a next step, we will gather more qualitative data related to the quantitative results. For this, a follow-up study with third graders in Halle is planned, in which students are interviewed based on their results in a new quantitative study using the same test as the one presented here.

The textbook analysis has shown that there is a lack of emphasis on the flexible interpretation of place value charts, while there are still tasks in the book that require this understanding. Based on these considerations and the results of this study, we will design activities involving the interactive place value app that could improve students' understanding, and we will test these activities as treatments in a pre-post design.

References

Askew, M. (2012). *Transforming primary mathematics*. Abingdon: Routledge.

Bartolini Bussi, Maria G. (2011). Artefacts and utilization schemes in mathematics teacher education: place value in early childhood education. J Math Teacher Educ (2011), 14:93–112, DOI 10.1007/s10857-011-9171-2.

Gras, R., Suzuki, F., Guillet, F., & Spagnolo, F. (2008). *Statistical implicative analysis*. New York: Springer.

Kortenkamp, U., & Ladel, S. (2014). Flexible use and understanding of place value via traditional and digital tools. Research report. In P. Liljedahl, C. Nicol, S. Oesterle, & D. Allan (Eds.), *Proceedings of PME 38* (Vol. 4, pp. 33–40). Vancouver, Canada: PME.

Ladel, S., & Kortenkamp, U. (2011). Finger-symbol-sets and multi-touch for a better understanding of numbers and operations. In *Proceedings of CERME 7*. Rzeszów.

Ladel, S., & Kortenkamp, U. (2013). Designing a technology based learning environment for place value using artifact-centric Activity Theory. In A. M. Lindmaier & A. Heinze (Eds.), *Proceedings of PME 37* (Vol. 1, pp. 188–192). Kiel, Germany: PME.

Ladel, S., & Kortenkamp, U. (2014). Handlungsorientiert zu einem flexiblen Verständnis von Stellenwerten—ein Ansatz aus Sicht der Artifact-Centric Activity Theory. In: S. Ladel & Chr. Schreiber (Hrsg.), *Von Audiopodcast bis Zahlensinn. Band 2 der Reihe Lernen, Lehren und Forschen mit digitalen Medien in der Primarstufe*. Münster: WTM.

Resnick, L. B., Bill, V., Lesgold, S., & Leer, M. (1991). Thinking in arithmetic class. In B. Means, C. Chelemer, & M. S. Knapp (Eds.), *Teaching advanced skills to at-risk students: Views from research and practice* (pp. 27–53). San Francisco: Jossey-Bass.

Ross, S. H. (1989). Parts. Wholes, and place value: A developmental view. *The Arithmetic Teacher, 36*(6), 47–51.

The Relationship Between Equivalence and Equality in a Nonsymbolic Context with Regard to Algebraic Thinking in Young Children

Nathalie Silvia Anwandter Cuellar, Manon Boily, Geneviève Lessard, and Danielle Mailhot

Abstract In this chapter, we analyze the process of developing algebraic thinking in children, as it relates to the necessary conceptualization giving meaning to the ideas underpinning the basic rules of algebra. In recent years, the National Council of Teachers of Mathematics and the Ontario Ministry of Education continue to provide resources for the development of algebraic reasoning, starting in early childhood. In early childhood, the relationship between equivalence and equality is a key element that integrates different facets of the development of numerical. We analyze algebraic thinking by examining mathematics-related tasks completed by twenty-one 5-year-old children. Our purpose is to highlight the use of landmark strategies, big ideas, and models, in regard to equivalence and equality and their role in the development of early algebraic reasoning.

Introduction

The difficulty surrounding the transition from arithmetic to algebra is well documented (Alibali et al. 2006; Falkner et al. 1999; Jacobs et al. 2007; Kaput 1998, 1999; Knuth et al. 2006). For instance:

> During the last decade, more and more mathematics educators suggest initiating the study of algebra at the primary level. They argue that this is not early teaching of algebra at the secondary level nor is it "pre-algebra" […], but rather help[ing] students develop algebraic thinking without necessarily using the textual high-level algebra language. (Squalli 2002, p. 4)

N.S. Anwandter Cuellar (✉) • G. Lessard • D. Mailhot
Université du Québec à Outaouais, Gatineau, QC, Canada
e-mail: Nathalie.Anwandter@uqo.ca; genevieve.lessard@uqo.ca; maid03@uqo.ca

M. Boily
Université du Québec à Montréal, Gatineau, QC, Canada
e-mail: boily.manon@uqam.ca

T. Meaney et al. (eds.), *Mathematics Education in the Early Years*,
DOI 10.1007/978-3-319-23935-4_17

While several research articles are aimed at the development of early algebraic reasoning (Carpenter et al. 2003; Sáenz-Ludlow and Walgamuth 1998; Squalli 2002), most studies consider algebraic learning at the secondary and primary levels. However, as promoted by the Ontario Ministry of Education (OME):

> Algebraic reasoning should be promoted and cultivated in kindergarten. We all have the ability to think algebraically, for algebraic reasoning is essentially how humans interact with the world. (OME 2013, p. 3)

In accordance with the research community, "[u]nderstanding and using algebra is dependent on understanding a number of fundamental concepts, one of which is the concept of equality" (Knuth et al. 2006, p. 297). Thus, one of our team's areas of focus is to study the development of algebraic reasoning in early childhood, with a view to examining equivalence and equality.

Analytical Tools

The Equal Sign as an Indicator of Mathematical Equivalence and Equality

Several approaches can be used to develop early algebraic reasoning in students; the OME (2013) highlights two approaches for preschool education:

a) Functional thinking which consists in analyzing regularities and patterns (numerical and geometrical) to identify a change and recognize the relationship between two sets of numbers (Beatty and Bruce 2012)
b) Generalization of mathematics, which is based on the reasoning behind operations and properties associated with numbers (Carpenter et al. 2003)

According to Kieran (1996):

> Algebraic thinking can be interpreted as an approach to quantitative situations that emphasizes the general relational aspects, with tools that are not necessarily letter-symbolic, but which can ultimately be used as cognitive support for introducing and for sustaining the more traditional discourse of school algebra. (Kieran 1996, p. 275)

Thus, algebraic thinking requires a refocusing of the meaning of the equal sign (Kieran 2004). At preschool, it can be introduced to compare the cardinality of sets, through either equivalence or equality. However, Carpenter et al. (2003) show that research has documented evidence of elementary grade children's misconceptions of the equal sign as "mean[ing] that they should carry out the calculation that precedes it and that the number after the equal sign is the answer to the calculation" (Falkner et al. 1999, p. 233). Indeed, at school, the equal sign is often shown in canonical problems (e.g., $a+b=c$); this can reinforce the operator notion of the equal sign as meaning "the answer comes next" or "do something," rather than the relational notion that both sides are equivalent (Alibali 1999; Carpenter et al. 2003; Kieran 1981; McNeil and Alibali

2005; Sáenz-Ludlow and Walgamuth 1998; Sherman and Bisanz 2009). As Kieran (1981) points out:

> the symbol, which is used to show equivalence, the equal sign, is not always interpreted in terms of equivalence by the learner. In fact, as will be seen, an equivalence interpretation of the equal sign does not seem to come easily or quickly to many students. (p. 317)

In mathematics, equivalence is defined as any relationship that is reflexive, transitive, and symmetrical. Equivalently, for all a, b, and c in X:

- $a \sim a$ (reflexivity).
- If $a \sim b$, then $b \sim a$ (symmetry).
- If $a \sim b$ and $b \sim c$, then $a \sim c$ (transitivity).

And equality is the most elementary relationship of equivalence.

In preschool, a relationship of equivalence is often found in the form of "the same number of elements," therefore a quantitative equivalence. In that particular relationship, two sets can have the same number of elements without necessarily consisting of identical objects. On the other hand, equality is a quantitative equivalence in a specific case where two sets have the same number of elements and these elements are exactly the same. According to Theis (2005), the expression "2+3=5" may correspond to a situation of equality where a child assembles sets of elements, for example, a child may start with two marbles, win three during a game, and end up with five. However, where two children compare the number of marbles, for example, one has two in his right hand and three in his left, and the other child has five marbles, it is a matter of quantitative equivalence because we do not compare the same elements physically. Nevertheless, the formal expression "2+3=5" describes two sides of the "=" sign as representing exactly the same number, which means a numerical equality (Theis 2005). Thus, the equal sign alone does not distinguish the type of situation under consideration.

Sherman and Bisanz (2009) undertook research on equivalence (using numbers and the equal sign) in symbolic and nonsymbolic contexts (using objects). Using data from two studies, these authors demonstrated that "Grade 2 children are quite capable of using reasoning relationally for complex, algebra-like problems when the problems are presented in a non-symbolic context" (p. 98). Indeed, in this research, the results revealed that children solve problems differently, depending on whether the context is symbolic or nonsymbolic. Success rates for children who solved equivalence problems in a nonsymbolic context were much higher than for children who solved the same problems in a symbolic context. Similarly, children in nonsymbolic groups were much more prone to using equivalence with a relational perspective than the children in the symbolic group. In a second study, they observed:

> children who solved non-symbolic equivalence problems in the first session went on to have impression accuracy with the same problems presented symbolically, a week later. This result supports the idea that early success, specifically success in solving non-symbolic equivalence problems, can be useful for improving performance in subsequent symbolic problems. (p. 99)

Along the same line of thinking, research by Mix et al. (1996) has found that children of preschool age recognize numerical equivalence between sets of similar objects (e.g., black records and black dots) but not between sets of very different objects (e.g., sounds and dots). Mix's research (1999) also suggests that the students' success depends on the heterogeneity of objects, in the context of comparing sets of identical or different objects, for example.

Given the results of research, it seems reasonable to hypothesize that the conceptualization and the construction of meaning of concepts are important to develop before their symbolization. The goal of this article is to present evidence regarding the relationship between students' understanding of equivalence and equality in a nonsymbolic context and in the development of algebraic thinking.

Conceptual Model of Equality and Equivalence

In his thesis, Theis (2005) conducted a conceptual analysis of the relationship between equivalence and equality, drawing on the model of understanding by Bergeron and Herscovics (1988). These authors underscore two logico-physical levels which imply an understanding of equivalence, in which the child uses concrete objects for his reasoning, and logico-mathematical intelligence that requires numerical equality, in which the child reasons in terms of numbers. Furthermore, four types of understanding are differentiated for these two levels:

- Intuitive understanding: the child draws on visual perception and does not use mathematical procedures.
- Procedural understanding: the child uses early mathematical procedures.
- Abstract understanding: the child understands the construction of invariants, reversibility, the composition of transformation, and generalization.
- Formal understanding: the child is able to use mathematical symbolism.

Based on this initial research model, we have combined the levels and models of understanding, with strategies, models, and key concepts from the perspective of learning milestones by Twomey Fosnot and Dolk (2010). According to these authors, "[K]ey concepts are central ideas and frameworks for mathematics—principles define mathematical order" (p. 13), for example, equivalence, equality, quantity, or number.

> Models are representations of relationships that mathematicians have built over time, by reflecting on how something can be transformed into another and generalizing ideas, strategies and representations from various contexts. [...] When viewed from a certain angle, the models create conceptual maps used by mathematicians to organize activities, solve problems or explore relationships. (Twomey Fosnot and Dolk 2010, p. 84)

To analyze skills, which is to "restore equality or equivalence," we synthesized different elements in the conceptual grid shown in Table 1.

Table 1 Conceptual grid for "restore equality or equivalence" skills analysis

Levels	Types of understanding	Strategies	Key concepts	Models
Logico-physical (equivalence in concrete objects)	Intuitive	Visual perception	Quantity Equivalence	For example, connecting cubes to 10
	Procedural	One-to-one correspondence	Equivalence	
	Abstract		Quantity Invariance of quantity Reversibility Operations Transformations	
Logico-mathematical (numerical equality)	Procedural	Counting Double counting	Number Equality	
	Abstract		Number Equality Number Invariance Reversibility Operations Transformations	
	Formal		Number Equality	

Method

This study was carried out in early 2014, following a request by the Ministry of Ontario to support a collaborative enquiry into the development of algebraic thinking at the junior and senior kindergarten levels. Twenty-one 5-year-old children participated in this study. For this article, we chose to show the diversity of student procedures, with the help of a few examples. Types of understanding among students about equality and equivalence were defined, and Table 1 was developed based on this knowledge. Two tasks were developed that required students to "restore equivalence" in a nonsymbolic context. One task was contextualized as a problem-type situation—The Story of Fafounet and the Easter Egg Hunt (D'Aoust 2011)—and the other was a decontextualized task (with cubes, inspired by Squalli 2007). This study was conducted during regular school hours. Students had to complete tasks in one class period. Both tasks were done the same day. First, all students were asked to do task 1 in pairs. Then, once the first task was completed, students were asked to do the second task, this time individually. They took about 10 min to complete the tasks, about 5 min for each task. The researchers conducted research with the regular class teachers being observers who did not participate in the study.

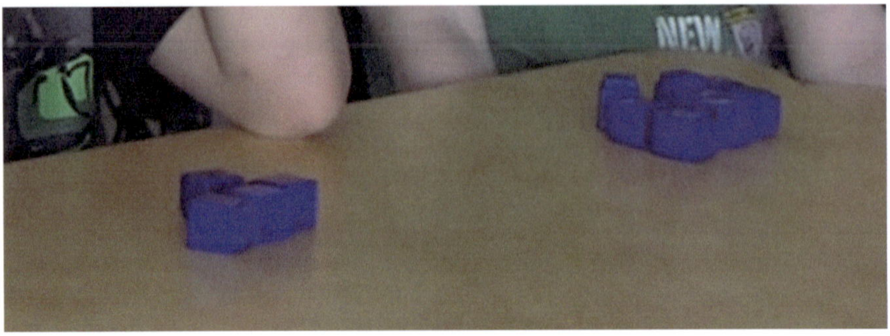

Fig. 1 Cubes given to each student to represent the chocolate eggs in task 1

Task 1: The Story of Fafounet and the Easter Egg Hunt

In this story, the main character, Fafounet, invites his neighbor Fafoundé to take part in a chocolate Easter egg hunt his mother has organized for him. The idea of the game is to respect his mother's golden rule: "you must find the Easter eggs together" and "at the end of the game, I will divide the eggs in equal parts between the two of you." At the end of the story, Fafounet finds eight chocolate eggs and Fafoundé finds four.

The researcher, Geneviève Lessard, reads a part of the story to the students. Then, in pairs, where each student portrays a character, the researcher hands out eight cubes and four cubes to symbolize chocolate eggs found by the characters (Fig. 1). Finally, the researcher asks, "What should you do to respect the golden rule?"

Task 2: Restoring Equality with Cubes

The researcher gives this task to students. He shows them two groups of cubes, one composed of ten cubes and the other of eight cubes (Fig. 2). The idea is that the students cannot easily calculate the number of cubes through visual recognition.

Then, he asks if there are as many cubes in each group, how many there are, and what should be done to have the same quantity in both clusters.

Results

Like Theis (2005), we undertake a conceptual analysis of the relationships of equivalence and equality, based on the criteria established by Theis. On that point, we will identify the types of understanding of equivalence and equality put forward by some of the students, to solve the problem situation.

Fig. 2 Two groups of cubes given to students for task 2

Part I: Contextualized Problem Situation

In the contextualized problem situation, two collections of cubes are presented to two children. The first collection contains eight cubes, while the second consists of four. The challenge for the students is to share the cubes among themselves "equally" so that everyone has the "same amount."

First pair of students Procedural understanding (counting) at the logico-mathematical level

Initially, the first girl (the one who takes the lead in solving the problem) looks at the two sets and decides to give away two cubes to her teammate. Then, she counts her cubes (six) and her teammate's (six) and determines that there is equivalence, that is to say, that each has the same amount of cubes.

When the researcher asks her how she achieved that outcome, the girl says she immediately noticed that she had eight and her teammate had four. Accordingly, she knew she had to give some to her teammate. When the researcher asks whether she knew she had to give two, she said that she did not know the exact number but that she knew she had to give some away.

To accomplish this, the girl decides to give her teammate two cubes and use counting to verify if they have the same amount. These actions are connected to thinking at the logico-mathematical level and to reasoning in terms of the number. Furthermore, when a child is encouraged to verify a task or to verify the effectiveness of the operation they have just carried out, they perform "a procedure." In this regard, some procedures associated with the logico-mathematical level, such as counting, may be used by the child, to determine if the same number is present in both groups. In the problem situation, the girl counted the cubes in each group, after the action was taken, the sharing of two cubes, to determine that there was the same number of items. In her view, there is thus equivalence, since there is the same number (six) of objects in both collections and both girls; therefore, they had six cubes each. The girl's recognition of the relationship of equivalence between the two sets of objects may be related to the development of algebraic thinking that could give a relational meaning to the equal sign. However, her procedure rests on a trial-and-error strategy with a numerical verification (by counting). From this perspective, her procedure seems to be situated in the construction of more general arithmetic strategies typically used in elementary schools for the solving equations and so may not support the more general properties associated with algebra.

Second pair of students Intuitive understanding (visual perception) and procedural understanding (one-to-one correspondence, grouping of two cubes at a time) at the logico-physical level

A girl and a boy are paired in this activity. When she first gets her cubes, the girl counted them and said, "I have four." She then proposed to cut the cubes into pieces. Symbolically, this can be represented as "$8+4=12/2+12/2$." This strategy would be applied generally to any divisible and physically countable number of objects by using the following property "$a+b=(a+b)/2+(a+b)/2$)" and, in this way, could be used for the development of algebraic thinking in later classes. However, the concrete material does not allow for the division of the unity into two parts. As a result, the girl changed her mind and instead suggested to "put an equal number." The researcher asks them how they could achieve "an equal number." The boy offered to count "how many do we have." The girl disagreed saying, "No, let's put two and two" and started to separate cubes into pairs. The little boy agreed to this suggestion and also started dividing his cubes into groups of two, and then he counted two by two (touching his cubes and those of his teammate) saying, "I have two, you have two, I have two more, you have two more, now two more for you (he gives her two cubes) and two more for me (he takes two more cubes)" and shouted, "it's equal." No counting was performed to verify the outcome.

In the mind of the girl, it was important to arrange their respective cubes into "an equal number." The method she decided on was to put them into pairs of two. Her thought process pointed toward grouping the cubes "two by two." Then, the boy took over the lead. He used matching pairs to compare his cubes to those of his teammate. This procedure corresponds to logico-physical thinking. Theis (2005) pointed out: "At the logico-physical procedural understanding level, it is mathematical procedures such as one-to-one correspondence that enable the establishment of groupings" (p. 47). At the end of the activity, when the researcher asked how they managed to achieve an "equal" amount, the girl explained that the boy gave her two because he had "a lot" and she had "less." She indicated that he "alone" had two. Her explanation of the situation showed that her understanding is more at the logico-physical level and that she is "intuitive" because she used the terms "more and less" to compare the quantities in the two collections and not numbers. However, intuitively (for the little girl) and by making a one-to-one correspondence (2×2), based on a logico-physical understanding, the children are able to establish a relationship of equivalence between two sets of objects and thus demonstrate an early development of algebraic thinking. This represents what Squalli (2007) describes as "pre-letter algebra, thereby putting the emphasis on thinking and not mathematical content." Here, the children recognized that the two quantities were equivalent without having to use numbers and without calculating. All they did was base their reasoning on a logico-physical procedure like the one-to-one correspondence (2×2), which required the little boy to use a mathematical procedure of giving one pair of two cubes to the little girl, to achieve two equivalent quantities, more specifically, two collections of equivalent cubes. The recognition of the relationship of equivalence of two sets of objects, regardless of the chosen procedure, depends on the development of algebraic thinking. Functional thinking consists in analyzing regularities and patterns (numerical and geometrical) to identify a change and recognize the relationship between two sets of numbers (Beatty and Bruce 2012).

Part II: Decontextualized Problem Situation

The second task was a decontextualized problem situation in which the researcher places ten cubes to the right of a board and eight cubes to the left. First, the researcher asks each student individually if there are "as many" cubes on the right side as on the left side. Then, in a second phase, Lessard asks them to ensure that there are "as many" cubes in one set as the other.

In the examples described, several types of thinking and procedures are used by the children to judge a situation of equivalence and to restore the equivalence between two sets of cubes. Some of the children used intuitive thinking and relied on visual perception to judge equivalence, while others used either counting, double counting, or a measurement device made on-site, to establish equivalence between two collections of objects. In this regard, they used a thought process more at the logico-mathematical level. Among the examples, one child, a girl, appears to access more abstract thinking associated with an understanding of the conservation of equivalence. In fact, she knows that, even though a transformation has occurred in the number of objects in the sets, they are still equal, since the same number of objects has been added on either side.

First student Entrenched intuitive thinking, coupled with an auto-add strategy, instead of sharing (adding and subtracting), to establish equivalence

The first student looks at both sets on the board and, pointing to the right, he says: "there." When the researcher reviews the meaning of "as many" with him and asks if they are "equal" on both sides, the child looks at both sets again and says: "yes." When the researcher asks him how he is able to know this, he says: "I thought in my head."

In this situation, the child's gestures underlay an understanding associated with intuitive thinking based on visual perception. It is at the heart of the logico-physical level and warrants no verification such as those associated with the one-to-one correspondence or counting found at the logico-mathematical level. The child therefore relied on what he sees. Theis (2005) points out that "intuitive understanding which, by nature, appears only at the logico-physical level, is the first and also the most rudimentary form of understanding. Most often, it relies on visual perceptions and does not yet imply the use of procedures" (p. 49).

Moving on to the next phase of the activity, the researcher decided to ask the child if he knew a way to check and make sure that there were as many cubes on each side. He said: "We could count." To that end, the child counted ten cubes on the right and eight cubes on the left, and, smiling, he said: "It's not equal." The researcher asks: "What can we do to make it 'equal'?" He answers: "We could add one." He then stood up to look for other cubes. He does not take those on the board. He takes three other cubes and places them on the board with the eight cubes on the left. When the researcher asks him if it is now "equal," he counted the cubes on both sides of the board. He counted 10 on the right and 11 on the left and then removed one cube. When the researcher asked if it was now "equal," he replied, "Yes." Initially, the child did not seem to need to use procedures to verify if the quantities of cubes were equivalent in both sets. Rather, it was the prompting of the researcher which led him to count the cubes. We can sense a benchmark here for a strategy informing the child's understanding, with regard to the concepts of equivalence and equality. Moreover, it is an "auto-add" strategy that the child used, as opposed to a "remove and add" strategy to achieve equivalence in both sets of objects. In fact, this auto-add strategy led the child to do a single "calculation" (that of adding to the set with fewer objects) instead of two-step calculation (removing from a set and adding to the other).

Second student Using a measurement model to verify equivalence

In this activity, the second student looked at both sets of cubes on the board and pointed to the right saying: "as many." When the researcher reviewed the meaning of the term "as many" with the child and asked her if both sides were "equal," looking at both sets, the child said: "No." The researcher then asked: "What could we do in order to have as many on the right as on the left?" The child replied: "We could add some." Then, she began to assemble all the cubes on the left side of the board and to build a tower. She repeats the exercise with the cubes on the right and assembles them into a tower. She counts the cubes in the right tower and comes up with eight. At that moment, she decided to remove one cube from the left tower and put it on a tray. She measured the two towers and realized that the left tower is still higher. She seemed surprised, and to check, she turned the towers upside down and measured them again. Then, she observed that she must remove another cube from the left tower. Again, she measured both towers, smiling. The researcher asked her if they were now "equal" to which she nodded "yes." The researcher asked the child: "How did you know they were equal?" The child took the two towers and measured them saying to the researcher: "Because I did that." With her strategy, the little girl brought the problem back to a measuring context. It was no longer about comparing the cardinality of the two sets of objects for her, nor was she interested in comparing the lengths of the two objects. From this perspective, we could associate certain properties with functional thinking, as the measure is defined by the connection of a set of objects to the set of numbers.

Nonetheless, her strategy is limited by the choice of objects. Indeed, we can compare the number of objects on each side by measuring them because they were the same stackable cubes, but the procedure cannot be generalized in all cases in order to conceptualize algebraic properties. To relate the girl's thinking to equivalence between sets, the researcher asked if she knew how many cubes were in each tower. The child replied nine, adding that she did not count. There are eight cubes in each tower at this point in time. This child's thinking seemed to borrow from both the logico-physical and the logico-mathematical levels. The child had to rely on her visual perception since she used the measurement of the towers to verify equivalence. Nevertheless, she started using procedures associated with the logico-mathematical level as they related to counting. It appeared, however, that there is a predominance of thinking at the logico-physical level since the child seemed satisfied when measuring the two towers and because she did not count at the end to ensure that both towers have an equal number of cubes.

Third student Understanding the conservation of equivalence

The third student looked at the two sets of cubes arranged on the board. Pointing to the left she said: "There are more cubes here." Pointing to the right she said: "Here, there are less." The researcher asked: "What can we do to have as many?" The child seemed reflective and unresponsive. The researcher reviewed the meaning of the term "as many." Pointing to the left the child said: "You take away four here and then they are equal." The child then proceeded to remove four cubes from the left. The researcher asked: "Why four?" The child replied: "It would be almost equal, and we also remove four from there," pointing to the right. Then, she changed her mind and said: "No, maybe three, perhaps now it will be equal." She said this looking once again at the number of cubes on the left and the number of cubes on the right. By examining both sides of the board, she decided to add one cube on the right saying: "Maybe one here and it would be equal." The researcher asked her how she can be sure that both sides are "equal." The child counted six cubes on the right and six cubes on the left. She said, "They are equal." Then the researcher asked the child what she could do with the six cubes left in her hands. She shared them one by one, into each of the sets. At this point, the researcher asked if she was certain they were still "equal." The child replied "yes." The researcher asked how she knew, to which the child replied, "Because I added one cube to each set." Then, she counted nine cubes in the set on the left and said, pointing to the right, "I believe there were nine here." To this end, she counted nine cubes. In this situation, it appeared that the child was already accessing abstract thinking connected to an understanding of the conservation of equivalence. Since she had actually counted six cubes to the right and six to the left and she had added cubes onto each side, one by one, the child knew that there was always a situation of equivalence. She knew that, in spite of the change in the number of objects in each set, they were always equal since the same number of objects has been added to both sides. This type of thinking could support the more general algebraic properties, particularly those associated with the solving of equations. For example, if "$a = b$," then "$a + x = b + x$."

Fourth student Use of two types of thinking—logico-mathematical thinking related to counting followed by the logico-physical thinking associated with one-to-one correspondence to verify equivalence

The researcher asked the child (a little girl) if there were "as many" cubes here (pointing to the left) as here (pointing to the right). The child counted the cubes in both sets in her head. She says nine on the left side and seven on the right side. The researcher asked if there are "as many" on this board and that board. The child pointed to the left. When the researcher verified the meaning of the term "as many" and asked if there were "as many," she replies "no." The researcher asked what she could do to make them "equal." The child counted both sets again (several times) and said ten and eight. Then, she makes a one-to-one correspondence saying "1-1, 2-2, 3-3…" up to 8 and puts one on the right as she says: "8 and 8." She counted again (several times because of miscounting) and then reached 9 and 9 on both sides. The researcher asked if both sides were "equal" now. She replied "yes." We note here that the child used several procedures at her disposal to confirm a situation of equivalence. We assumed that the fact that she often miscounted may have prompted her to use a procedure whereby she could refer to something more tangible, such as visual perception or one-to-one correspondence. Even if the student situated herself in a kind of logico-physical thinking, by using one-to-one correspondence, this gave a tangible meaning to the definition of equivalence between two sets: two sets have the same number of elements if we can find a bijective application between these sets. The relational aspect of equivalence, which serves the construction of the meaning of the equal sign, supports algebraic thinking.

General Discussion

From a child's first years at school, the OME (2013) suggests that the development of algebraic thinking implies the understanding of equality and equivalence. Children need experiences of using different models, concepts, and strategies. Most of the time, however, equality and equivalence are taught from a numerical perspective, and the equal sign is introduced early on. We have therefore proposed tasks which focus on the meaning of the equal sign prior to its formalization. Two non-symbolic tasks dealing with the concept of equivalence were proposed to 21 preschool students.

Our first results display the diversity of strategies, which could be found in the same class. Indeed, our model (Table 1) helped us characterize the students' procedures at two levels, logico-physical and logico-mathematical, and into four types of understanding: intuitive, procedural, abstract, and formal. In our class, the students' strategies could be situated on two levels, and their understanding could be interpreted as intuitive or procedural. Our analyses also revealed to us that the logico-physical level seems more elementary than the logico-mathematical level from the Bergeron and Herscovics (1988) model and, in relation to equivalence, diverse strategies which we designate as logico-physical show us a relational conceptualization of the notion of equivalence, which could favor the development of algebraic thinking. What comes to mind here is the third girl in part II who made the connection between the changes found in the sets and the conservation of equivalence in each set. Her thinking seemed to presuppose an ability to build relationships and to generalize. In that sense, and in that situation, the girl knew that regardless of the number of cubes she added, if that number is the same in both sets of objects, the quantities will always remain equivalent.

The reciting of numbers was very powerful for children throughout the experiment, and even the child who has difficulty touching each object, without skipping or double touching, will have no difficulty respecting the established order in the numbers of the counting rhyme. As Bergeron et al. (1987) stated: "Children may often know how to recite the sequence without necessarily coordinating it correctly in their enumeration process" (p. 348). Reciting numbers underpinned the counting used throughout the activities, in our research project. It could be said that the reciting of numbers is an oral model children use to support the strategy of synchrony: one word, one object, essential to the procedural logico-mathematical understanding in children. According to Mix (1999), this implies that the development of numerical competence has certain advantages. However, the results of this study indicated that numerical competency might create obstacles to the generalization of properties on the path to the development of algebraic thinking (e.g., the first pair of students). Indeed, the importance given to numerical strategies of counting at the preschool level may have constrained students so that they were enticed to use a trial- and-error procedure and verification through counting. This may have taken longer and be less efficient than certain intuitive procedures, which could be applied generally to other contexts.

As Theis (2005) showed, the meaning of the equal sign depended on the situation, whether it be a comparison (equivalence) situation or an addition (equality) situation. In addition, Sherman and Bisanz's (2009) study has revealed that students' success in a situation depends on whether the context is symbolic or nonsymbolic. This is similar to the example of the second student in part II, as she transferred the context of the set to measurement. However, her strategy is valid only because it is a case of identical objects. In the context of sets of different objects, it would not have been possible to compare measurements. Therefore, the context of the proposed situations and the used material are important tools in evaluating the relational reasoning of preschool children but also in developing the meaning of the notion of equivalence and relational thinking (Taylor-Cox 2003).

This study marks the first step in the identification of factors favoring the development of algebraic thinking, with the help of equivalence and equality situations. Firstly, it appears that children can reason in a relational manner on equivalence in a nonsymbolic context. This could help later in the conceptualization of relational aspects associated with equal signs. Secondly, we find that tasks related to notions of equivalence and equality can be treated at two levels, the logico-physical level and the logico-mathematical one. In any case, using counting procedures would not be more favorable to a relational understanding and to the generalization of properties than logico-physical procedures. Thirdly, it seems that the context of the situation and the material used condition the students' procedures and consequently the meanings related to equivalence and equality.

According to the Ministry of Ontario (2008), in whose jurisdiction this study was carried out, it is important that the concepts of equality and equivalence be concretely used in everyday life examples before being used in a symbolic manner in a mathematical statement. We believe that the importance of giving 4- and 5-year-old students concrete examples, containing an equality and equivalency relationship, specifically before they use this concept in a mathematical statement, is related to their way of learning. As Piaget (1994) highlighted it, being at the preoperational stage, preschool-aged children use symbolic thinking in the moment, a kind of thinking that is a representational thinking of the elements, people, etc., which make up reality. Each concept must be expressed through a concrete representation, before being represented symbolically. Once the concept has been explored through the manipulation of concrete objects, it can be represented in the child's thinking and subsequently used, even if it is still at the preconceptual stage, as a symbol. Thus, when a child is introduced to mathematical statements in elementary school, they will be able to use the concepts of equality and equivalence, assimilated through concrete experiences acquired in preschool. Furthermore, allowing students to deal with equality and equivalence situations through concrete examples and material will lead them to eventually better understand the meaning of the equal sign and to establish existing relationships between numbers, variables, and unknowns in equations and, subsequently, in algebraic formulas. As Sherman and Bisanz's research has shown, our study suggests that children can reason about the meanings of the equal sign, through the notion of equivalence in a relational sense, and this, as early as preschool, in this way, well before formal algebra, is introduced into the classroom. Activities can be used without using symbolic letters, and they can be specified at any time to include symbolic letters to conceptualize a nonsymbolic or pre-symbolic approach to algebraic thinking in elementary classes (Kieran 2004). It would therefore be very valuable to present nonsymbolic situations of equivalence and equality in contextualized and non-contextualized ways to students, before they start elementary school.

References

Alibali, M. W. (1999). How children change their minds: Strategy change can be gradual or abrupt. *Developmental Psychology, 35*, 127–145.

Alibali, M. W., Knuth, E. J., McNeil, N. M., & Stephens, A. C. (2006). Does understanding the equal sign matter? Evidence from solving equations. *Journal for Research in Mathematics Education, 37*, 297–312. Retrieved from http://www.region11mathandscience.org/archives/files/equality/Trainers/EqualSignJRMEKnuth.pdf.

Beatty, R., & Bruce, C. (2012). *From patterns to algebra: Lessons for exploring linear relationships*. Toronto, ON: Nelson.

Bergeron, J. C., & Herscovics, N. (1988). The kindergartners understanding of discrete quantity. In A. Borbas (Ed.), *Proceedings of the twelfth annual conference of the International Group for the Psychology of Mathematics Education, Veszprem, Hungary* (pp. 162–169). Retrieved from http://files.eric.ed.gov/fulltext/ED411128.pdf.

Bergeron, A., Herscovics, N., & Bergeron, J. C. (1987). Kindergartners' knowledge of numbers: A longitudinal case study, Part I: Intuitive and procedural understanding. In J. C. Bergeron, N. Herscovics, & C. Kieran (Eds.), *Proceedings of the 11th psychology of mathematics education conference* (Vol. II, pp. 88–97). Montreal: University de Montreal.

Carpenter, T. P., Franke, M. L., & Levi, L. (2003). *Thinking mathematically: Integrating arithmetic and algebra in the elementary school*. Portsmouth, NH: Heinemann.

D'Aoust, L. (2011). *Fafounet et la chasse aux cocos de Pâques [The story of Fafounet and the Easter Egg Hunt]*. Montréal, QC: Les Malins.

Falkner, K. P., Levi, L., & Carpenter, T. P. (1999). Children's understanding of equality: A foundation for algebra. *Teaching Children Mathematics, 6*, 232–236. Retrieved from http://ncisla.wceruw.org/publications/articles/AlgebraNCTM.pdf.

Jacobs, V. R., Franke, M. L., Carpenter, T. P., Levi, L., & Battey, D. (2007). Professional development focused on children's algebraic reasoning in elementary school. *Journal for Research in Mathematics Education, 38*, 258–288. Retrieved from http://homepages.math.uic.edu/~martinez/PD-EarlyAlgebra.pdf.

Kaput, J. (1998). *Transforming algebra from an engine of inequity to an engine of mathematical power by "algebrafying" the K-12 curriculum*. Paper presented at the Algebra Symposium, Washington, DC.

Kaput, J. J. (1999). Teaching and learning a new algebra. In E. Fennema & T. A. Romberg (Eds.), *Mathematics classrooms that promote understanding* (pp. 133–155). Mahwah, NJ: Lawrence Erlbaum.

Kieran, C. (1981). Concepts associated with the equality symbol. *Educational Studies in Mathematics, 12*, 317–326.

Kieran, C. (1996). The changing face of school algebra. In C. Alsina, J. Alvarez, B. Hodgson, C. Laborde, & A. Pérez (Eds.), *8th The international congress on mathematical education: Selected lectures* (pp. 271–290). Seville: S.A.E.M. Thales.

Kieran, C. (2004). The core of algebra: Reflections on its main activities. In K. Stacey, H. Chick, & M. Kendal (Eds.), *The future of the teaching and learning of algebra. The 12th ICMI Study* (pp. 21–34). New York: Kluwer.

Knuth, E., Stephens, A., McNeil, N., & Alibali, M. (2006). Does understanding the equal sign matter? Evidence from solving equations. *Journal for Research in Mathematics Education, 37*(4), 297–312.

McNeil, M. M., & Alibali, W. M. (2005). Knowledge change as a function of mathematics experience: All contexts are not created equal. *Journal of Cognition and Development, 6*, 285–306. Retrieved from http://www3.nd.edu/~nmcneil/McNeilAlibali05a.pdf.

Mix, K. S. (1999). Similarity and numerical equivalence: Appearances count. *Cognitive Development, 14*, 269–297.

Mix, K. S., Huttenlocher, J., & Levine, S. C. (1996). Do preschool children recognize auditory–visual numerical correspondences? *Child Development, 67*, 1592–1608.

Ontario Ministry of Education. (2008). *Guide d'enseignement efficace des mathématiques de la maternelle à la 3e année Modélisation et algèbre. Fascicule 2, Situations d'égalité [A Guide to Effective Instruction in Mathematics Kindergarten to Grade 3, Patterning and Algebra]*. Toronto, ON: Queen's Printer for Ontario.

Ontario Ministry of Education. (2013). *Paying attention to algebraic reasoning: K-12*. Toronto, ON: Queen's Printer for Ontario.

Piaget, J. (1994). *La formation du symbole chez l'enfant, imitation, jeu et rêve, image et représentation [The formation of symbols in infancy. Imitation, play, and dream. Image and representation]* (8th ed.). Neuchatel: Delachaux et Niestlé.

Sáenz-Ludlow, A., & Walgamuth, C. (1998). Third graders' interpretations of equality and the equal symbol. *Educational Studies in Mathematics, 35*, 153–187.

Sherman, J., & Bisanz, J. (2009). Equivalence in symbolic and non-symbolic contexts: Benefits of solving problems with manipulatives. *Journal of Educational Psychology, 101*, 88–100.

Squalli, H. (2002). Le développement de la pensée algébrique à l'école primaire : un exemple de raisonnements à l'aide de concepts mathématiques [The development of algebraic thinking in elementary school : An example of reasoning using mathematical concepts]. *Instantanés mathématiques, 39*, 4–13.

Squalli, H. (2007). Le développement de la pensée algébrique à l'école primaire : un exemple de raisonnements à l'aide de concepts mathématiques [The development of algebraic thinking in elementary school : An example of reasoning using mathematical concepts]. *Site de Math Vip, Mathématique virtuelle à l'intention du primaire*. Retrieved from http://spip.cslaval.qc.ca/mathvip/article.php3?id_article=56.

Taylor-Cox, J. (2003). Algebra in the early years? Yes! *Young Children, 58*(1), 14–21. Retrieved from www.journal.naeyc.org/btj/200301/Algebra.pdf.

Theis, L. (2005). *Les tribulations du signe "=" dans la moulinette de la bonne réponse [The trials and tribulations of the "=" sign in the mill of the correct answer]*. Coll. "Mathèse". Montréal: Éditions Bande didactique.

Twomey Fosnot, C., & Dolk, M. (2010). *Jeunes mathématiciens en action [Young mathematicians at work]*. Montréal, QC: Chenelière.

Developing a Mathematically Rich Environment for 3-Year-Old Children: The Case of Geometry

Pessia Tsamir, Dina Tirosh, Esther Levenson, Ruthi Barkai, and Michal Tabach

Abstract This chapter describes an integrated program in Israel for 3-year-old children and their caregivers. For the caregivers, the aim of the program was to increase their mathematical and pedagogical knowledge for teaching geometric concepts. For the children, the aim of the program was to introduce geometry into the different spaces of the classroom, at different times in the daily schedule, and with different activities. Care was taken to introduce mathematical language and encourage communication skills. In addition, caregivers were encouraged to share their experiences and try out activities with the children. Questions and dilemmas are discussed.

Introduction

Many children attend day care centers from a very young age. While a prime aim of these centers is to provide a caring environment for young children, many also seek to actively provide literacy and numeracy experiences. On the one hand, children engage with mathematical ideas during everyday play (Baroody 1987). On the other hand, play may not be enough (Clements and Sarama 2013; Ginsburg et al. 2008). One of the challenges of early childhood mathematics education (in this chapter considered to be for children aged between 3 and 5 years (Ginsburg et al. 2008)) is finding a balance between spontaneous play and adult guidance. In this chapter, we describe and examine ways in which the natural play of children may merge with guided activities to afford early foundations for geometry. More precisely, the cases we present are semi-directed; the child is led to play in an instructional context and through playing (which is natural for the child), but under guidance and instructions from adults, the first foundations of geometry emerge. Taking into consideration that many caregivers of 3-year-old children do not receive professional development related to mathematics, and even less are prepared to engage children with geometrical

P. Tsamir • D. Tirosh • E. Levenson (✉) • R. Barkai • M. Tabach
Tel Aviv University, P.O. Box 39040, Tel Aviv 6997801, Israel
e-mail: pessia@tau.ac.il; dina@post.tau.ac.il; levensone@gmail.com; TabachM@post.tau.ac.il

T. Meaney et al. (eds.), *Mathematics Education in the Early Years*,
DOI 10.1007/978-3-319-23935-4_18

activities (Clements and Sarama 2011), an additional challenge is guiding caregivers in their endeavor to create a mathematically rich environment. For these reasons, in this chapter, we also describe some of the ways in which our program addressed this additional challenge.

Background

Early Childhood Mathematics Education: Importance and Diversity of Approaches

Early childhood care matters. In a recent longitudinal study which followed young children under the age of 4½ years till they reached the age of 15, results indicated that the quality of non relative child care is linked to adolescent functioning including cognitive-academic achievement (Vandell et al. 2010). Vandell et al.'s study investigated children from diverse economic backgrounds and followed previous studies which showed how quality child care interventions can impact on children from disadvantaged backgrounds (Barnett 2011; Havnes and Mogstad 2011). Regarding mathematics, studies have found that young children from disadvantaged homes exhibited lowered levels of both number and geometrical knowledge than children from advantaged homes, even before starting school (Starkey et al. 2004). Some of the reasons cited for these differences included the home mathematics practices reported by parents such as providing games, toys, and computer software that promote mathematical activities as well as the frequency of engaging children with these activities (LeFevre et al. 2009; Starkey et al. 1999). In addition, it was found that levels of early quantity–number competencies can be seen as mathematical precursor abilities on mathematical achievement in elementary school (Krajewski and Schneider 2009). Mathematical achievement is affected by spatial skills (e.g., Ansari et al. 2003), and thus it is important to promote spatial skills from an early age. One way to promote spatial skills is through the exploration of geometrical concepts. Geometric knowledge is also highly related to proportional reasoning, judgmental application of knowledge, concepts and properties, and managing data. Thus, geometry may be considered a gateway to higher-order thinking skills (Clements and Sarama 2011). Taking into consideration all of these studies, the importance of providing early childhood mathematics education, even before entering school, becomes evident.

Several countries have come out with guidelines for preschool education. In the United States, the National Council of Teachers of Mathematics (NCTM) outlined specific content emphases for children ages 4–5 years (NCTM 2006). In geometry, the focus includes developing spatial reasoning by examining and identifying shapes and describing them. In Australia, *Belonging, Being and Becoming: The Early Years Learning Framework for Australia* (EYLF) (Department of Education, Employment and Workforce Relations [DEEWR] 2009) provides guidelines for promoting learning among children in all early childhood education and care settings from birth to five years. Within these guidelines, there is emphasis on promoting numeracy which includes "understandings about numbers, patterns, measurement, spatial awareness and data as well as mathematical thinking, reasoning and counting" (p. 43). In addition, the framework recognizes the importance of language and encourages educators to use rich mathematical vocabulary to describe children's mathematical thinking. While the above frameworks relate to general guidelines, others offer specific learning goals for children of different ages. For example, the Israel National Mathematics Preschool Curriculum (INMPC 2008) states that children between 3 and 4 years should be able to identify, name, and categorize two-dimensional shapes, as well as identify, name, and draw lines that are straight and lines that are not straight. The curriculum advises to begin with simple shapes such as triangles, circles, and squares and then to move on to other shapes such as ellipses and other quadrilaterals.

The Program Framework

In the past few years, we have provided professional development programs for early childhood teachers based on the Cognitive Affective Mathematics Teacher Education (CAMTE) framework (Tsamir et al. 2014b). The eight-cell framework (see Table 1) is based on theories of teachers' knowledge (Ball et al. 2008; Shulman 1986) and Bandura's (1986) social cognitive theory of self-efficacy, taking into consideration both cognitive and affective issues related to professional development. For example, in one study of preschool teachers' knowledge and self-efficacy for teaching triangles and pentagons, we requested through a questionnaire that the teachers identify a series of figures as examples or non-examples of each figure and give their reasons (Tsamir et al. 2014a). This type of knowledge is related to Cell 1 (producing solutions) of the framework. After this part of the questionnaire was answered and handed in, the teachers were presented with the responses of a fictitious 5-year-old boy named Yossi to the same questions. For example, "Yossi was shown the following figure (a hexagon) and claimed that it was not a pentagon because it has too many sides." Teachers were requested to assess Yossi's evaluation of the figures as being correct or incorrect. This would be knowledge related to Cell 2 (evaluating solutions) of the framework. In this chapter, we focus mainly on

Table 1 The Cognitive Affective Mathematics Teacher Education (CAMTE) Framework

	Subject matter		Pedagogical content	
	Solving	Evaluating	Students	Tasks
Knowledge	Cell 1: Producing solutions	Cell 2: Evaluating solutions	Cell 3: Knowledge of students' conceptions	Cell 4: Designing and evaluating tasks
Self-efficacy	Cell 5: Mathematics self-efficacy related to producing solutions	Cell 6: Mathematics self-efficacy related to evaluating solutions	Cell 7: Pedagogical-mathematics self-efficacy related to children's conceptions	Cell 8: Pedagogical-mathematics self-efficacy related to designing and evaluating tasks

Cells 1 and 2 of the framework but touch briefly on Cells 3 and 4. In other studies (e.g., Tsamir et al. 2014b), we focused on additional cells.

Relating the framework to triangles, Cell 1 concerns solving geometrical tasks such as identifying intuitive and non intuitive examples and non-examples of triangles (Tsamir et al. 2008), defining a triangle, and explaining why some figure is or is not a triangle. Cell 2 concerns evaluating solutions such as comparing different definitions for a triangle and evaluating explanations for why some figure is or is not a triangle.

Although the framework in Table 1 was developed as an organizational tool for planning and researching professional development, Cells 1 and 2 also depict knowledge which may be promoted at the children's level. As noted in the earlier section, several early childhood curricula suggest having children engage in spatial reasoning activities such as identifying and describing two- and three-dimensional figures. In addition, several curricula advocate encouraging communication skills which can include evaluating others' descriptions and explanations. Young children may be expected to reach the first and second van Hiele levels of geometric reasoning (van Hiele and van Hiele 1958). At the first level, visualization, children often identify figures based on visual reasoning, taking in the whole shape without considering that the shape is made up of separate components. Children at this level can name shapes and distinguish between similar looking shapes (van Hiele and van Hiele 1958). At the next level, analysis, children begin to notice the different attributes of shapes, but the attributes are not perceived as being related. Attributes may be critical or non critical (Hershkowitz 1989). In mathematics, critical attributes stem from the concept definition. We also note that research has suggested that the van Hiele levels may not be discrete and that a child may display different levels of thinking for different contexts or different tasks (Burger and Shaughnessy 1986).

Setting

In Israel, most municipal-run preschools include children from the age of 3 till 6, learning in separate age groups or in heterogeneous groups. In some cities, children aged 3–4 years learn in day care centers together with younger children or in separate groups. The site of this study was a day care center in a middle to low socioeconomic neighborhood in a major city. The center cared for children between the ages of approximately 3 months to 3 years, separating children into different classrooms based on age. In September, there were 33 children between the ages of 24 and 34 months learning in the same classroom. We began interacting with the children in the beginning of May and then only with the children who were nearing or had already turned three. The spaces allocated to this group included their main classroom setup with tables and chairs and various corners (e.g., a doll corner, a music corner, etc.), a large indoor gym area with mats on the floor, and a large outside play area. Four caregivers tended this group of children. The caregivers had completed a course in early childhood care and were certified by the government to care for children from age 3 months to three years, but did not have an academic degree in early childhood education.

Like other programs we have conducted in preschools (e.g., Tirosh et al. 2011), this program integrated professional development for the caregivers with enrichment for the children. A major aim of the professional development part of the program was to increase the caregivers' content knowledge and pedagogical-content knowledge for teaching number and geometry concepts and skills in preschool. Toward that aim, the caregivers and two teacher-educators (TEs) met together for six 1-hour sessions at the day care center, during the children's nap time. The first two authors of this chapter were the TEs in these sessions. Ten caregivers attended these sessions, the four who worked with the 3-year-olds and six others who worked with younger children. Over the years, different caregivers were assigned to different age groups, and thus it was important to work with all the caregivers. Furthermore, those caring for younger children could perhaps adapt some of the activities to the children in their age group.

The TEs had several roles in the program. First, when meeting with the caregivers, the TEs demonstrated different mathematical activities that could be implemented with young children in various situations using common items found in the center. During these sessions, the TEs also promoted the caregivers' knowledge of mathematics, for example, by emphasizing differences between critical and non critical attributes of a triangle. They also promoted the caregivers' knowledge of children by play acting children engaged in mathematical activities. A second role of the TEs was to advise the caregivers on how to create a mathematically rich environment and to work with the caregivers in creating this environment. In addition, as the program progressed, caregivers began to describe their endeavors to engage children in mathematical activities as well as their observations of mathematical activities which arose spontaneously during play (such as a child naming the shapes she was using or drawing) during the group sessions. The role of the TEs was then to listen and to help the caregivers analyze the activities. In addition to the sessions with the caregivers, each time the TEs arrived at the center to engage the children, one or two caregivers would observe and then participate in the children's activities run by one of our team members, gaining valuable practical experience and guidance. At different times, different authors of this chapter took on this role with two or three TEs always present at the same time. Thus, an additional role for the TEs was to model possible caregivers' roles during mathematical activities. It is important to note that the five authors of this chapter have worked and researched together for several years on issues involving early childhood mathematics education (e.g., Tsamir et al. 2014b).

For the children and the classroom, a major aim of the program was to introduce mathematics into the different spaces of the classroom, at different times in the daily schedule, making it a routine part of the day. We (two or more TEs) came nine times to the center, once every week or so, each time meeting with approximately three groups of four to six children, for 15–20 min at a time. All meetings, with the caregivers and the children, were videotaped and transcribed. (Consent from the parents to participate in the program was attained prior to beginning the program.) The videos had several purposes. First, they allowed us to review the activities implemented with children and accordingly prepare further activities. Second, although time and technology constraints did not allow the caregivers to view the videos, the videos allowed the TEs to review several sessions at a time and to choose from the videos specific pertinent episodes to relate and discuss with the caregivers. The sessions with the caregivers and TEs were also videotaped, allowing us to review those sessions and prepare future ones. Finally, the videos allowed us to study ways in which young children may engage in mathematical activities and ways of promoting caregivers' knowledge for teaching mathematics to young children.

The mathematical content dealt with during the program included number concepts (e.g., counting, recognizing number symbols) and basic geometrical shapes (e.g., triangles, squares, and circles). The content was chosen based on previous research (e.g., Sarama and Clements 2009) which showed that young children are capable of learning this content, by reviewing guidelines from several countries for promoting mathematics to young children and by specifically adapting the Israeli preschool guidelines to younger children. This chapter focuses on episodes revolving around triangles.

To summarize, our program objectives were to promote caregivers' knowledge for teaching number and geometry concepts for 3-year-old children and to enrich the mathematical environment for the children through a program, which, although structured, called for much spontaneous behavior. The research aim of the study was to investigate possible ways of introducing caregivers to early childhood mathematics education possibilities and to investigate ways of enriching a day care center's mathematical environment. The aim of this chapter is to describe typical cases of these attempts within the context of learning geometry and share some insights and dilemmas. In the next sections, we discuss some episodes and dilemmas from different parts of the program related to teaching and learning about triangles. The first section focuses on our meetings with the caregivers. The second section demonstrates how different spaces can be used when engaging children with triangle activities.

Results

Working with Caregivers

"Is an upside down triangle still a triangle?" This question was asked by Cecile, one of the caregivers in our program, during our first session. It illustrates the power of the concept image over the concept definition (Tall and Vinner 1981) as well as the necessity to strengthen the mathematical knowledge of those working with children. In working with the caregivers, we took into consideration that they had received little or no preparation to guide children in their mathematics learning, but at the same time, we also recognized that most adults have some knowledge regarding triangles. Cecile's question prompted a discussion among the participants and TEs regarding the critical and non critical attributes of triangles. For example, it was emphasized that having straight and not curved lines is critical, but orientation, being upside down, is not critical. In addition, the importance of using precise mathematical language, such as calling the points of a triangle vertices, was introduced. This discussion was especially important in enhancing the caregivers' ability to explain why a shape was or was not a triangle. In the following example, we demonstrate the development of knowledge related to Cell 1 (producing solutions) of the CAMTE framework (see Table 1). The TE showed the participants several intuitive and non intuitive examples and non-examples of triangles. After being shown a pentagon stretched to look like a triangle , the following discussion ensued:

TE	Is it a triangle?
Several participants	No.
TE	Why?
Belinda	Because these corners are (pointing to the two vertices between the oblique lines and the vertical lines) are like this… the vertices here have to be straight. That is, straight and not have two vertices here. Do you agree with me? (Several participants murmur yes and nod their agreement.)
Liz	It has too many vertices. It has five vertices.

In the above excerpt, the participants are all able to identify the figure as a non-triangle. Belinda and Liz also explain why the shape is not a triangle, referring to the two extra vertices. Belinda's explanation, however, does not make use of precise mathematical language. She points to the parts of the figure which she knows are problematic, but her explanation reveals some confusion regarding vertices and being able to "straighten" vertices. In addition, her explanation is specific to the figure (e.g., "*these* corners," "two vertices *here*") and does not relate to the critical attribute of all triangles, having exactly three vertices. It is Liz who explains in a more precise manner why the figure cannot be a triangle, because it has five (and not three) vertices.

At the end of the session, Cecile reverted back to her question about the upside down triangle and made a point of letting everyone know that her question arose from an interaction with a child in the day care center that included drawing a Star of David[1] in which appeared an upside down triangle.

Cecile	Why did I ask you if a triangle standing on its head is a triangle? I answered them regular [Cecile means that she answered him that it was a triangle]. I was making with them a Star of David and then I drew the triangle (upside down), so the boy asked me, "is that also a triangle?"
TE	How nice.
Cecile	I said to him, yes, the first triangle has the point on top and the second has the point on the bottom. That's why I asked you. The children are very smart. I told them that it is also a triangle but that it's an upside down triangle.
TE	Great. Why did we decide to talk with you about triangles? Because it's familiar and yet there is still a lot to learn about it.

[1] The Star of David is a figure made up two triangles and is found on the Israeli flag.

The above dialogue illustrates how even young children are attentive to and curious about geometrical shapes. It also illustrates the role of the teacher in being able to respond to a child's query. But if Cecile was able to respond to the child, then why did she bring up the question to the other caregivers and to the TEs? Possibly, Cecile was unsure that she had responded correctly and thus sought out verification of her judgment. We also note that Cecile's acknowledgment of the boy's dilemma demonstrates that she is learning to recognize children's conceptions and misconceptions of triangles (Cell 3 of the framework—Knowledge of students' conceptions).

Another interesting geometrical discussion occurred during the second session when the participants were discussing materials they could use to help children learn about triangles.

Cecile	Someone had the idea of putting together two triangle halves to make a whole triangle.
TE	Is the triangle half also a triangle?
Cecile	How? It is one triangle that you cut in half.
Nancy (answering the TE)	No.
TE	If it's (the triangle) cut into two, is each half a triangle?
Nancy	No.
Helen	It is a triangle!
Gila	Why? Half half…
Cecile	(Cecile draws a prototypical isosceles triangle in the air and with her finger draws a line from the top vertex to the side opposite.) If you cut the triangle in two, this half and this half make a whole triangle.
TE	But the half is also a triangle.
Cecile	(thinks about it some more and laughs.) Yes, it is a triangle! Yes. (Other caregivers talk and laugh.)
TE	In this instance, the triangle half is also a triangle. But, see how even for us it can be difficult.

The above discussion relates to composing and decomposing shapes, an integral component of several early childhood mathematics curricula. In the above case, the caregivers' difficulties might have been related to visualizing the problem. It might also have been related to their conception of a half and that a half of something cannot be the same as that something. Although the triangle half is not congruent to the original triangle, in the specific case above, it is still a triangle.

Taken together, the excerpts above illustrate different aspects of knowing about triangles. The discussions related to triangle halves and upside down triangles are related to identifying various examples of triangles. The discussion revolving around the triangle-like elongated pentagon is related to explaining why some shape is or is not a triangle. These are aspects of teachers' knowledge related to Cell 1 of the CAMTE framework (producing solutions) (see Table 1). As Belinda explains why the pentagon is not a triangle, the participants learn about evaluating others' explanations. This is an example of knowledge related to Cell 2 (evaluating solutions). In the next section, we illustrate how similar aspects of knowing about triangles were promoted among the young children.

Working with the Children and the Environment

One of our aims was to incorporate mathematics into the physical environment. Toward this aim, we decorated the classroom with shape mobiles hanging from the ceiling and with cut out transparent shapes taped onto the glass window of the classroom (see Fig. 1a, b). These efforts were not merely decorative; they were also educational. Caregivers could use these decorations as reference points when discussing shapes. Children could look at the triangles from different angles, becoming familiar with the idea that a triangle remains a triangle even though it may twist and turn on the string or it may look different depending on which side of the window they are standing.

Evidence of the children's awareness of these decorations and how the children spontaneously incorporated them into their activities can be seen from the following episode. Four children were sitting on a carpet along with one TE in a quiet corner of the classroom. The TE laid out several cards on the carpet, each card having a drawing of a figure, either an example or non-example of a triangle. Taking turns, each child had to look for a card with a triangle on it, pick it up and show it to the rest of the children, and say why he or she thought it was a triangle. If it was indeed a triangle, the child could then stick the card on the board on the wall.

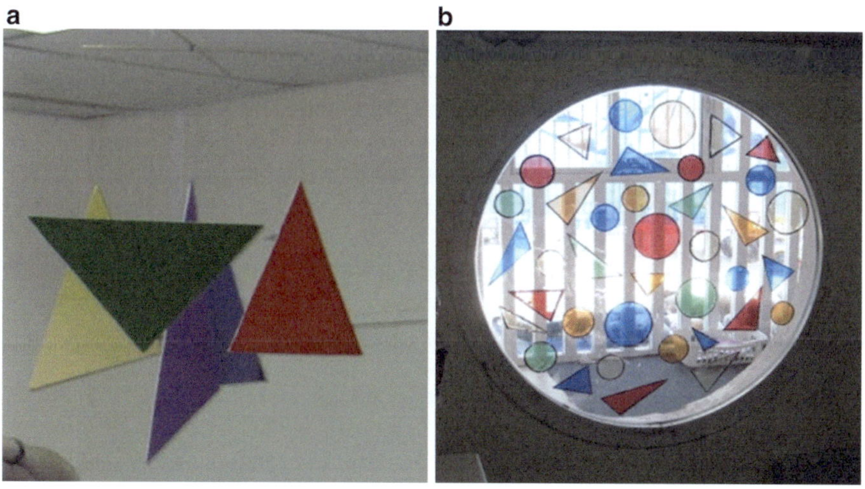

Fig. 1 (**a**) Triangle mobile and (**b**) window decorated with shapes

TE	Ok. Arie, it's your turn.
Arie	(picks up a card with a triangle.)
TE	So, why is this a triangle?
Shena	It has vertices.
TE	And straight lines. Right. So, Arie, you can stick it on the board. Now it's Olga's turn.
Rina	(Rina points to a triangle hanging from the mobile on the ceiling pictured in Fig. 1a). That's a big triangle.
TE	Yes, it is. Now…
Shena	Here is a big triangle, high high up.
TE	Yes, it is high.
Shena	And this (pointing to a circle on the circle mobile) doesn't have vertices.
TE	Right. They (meaning all of the circles on the mobile) don't have vertices. So, are they triangles?

In the above episode, the TE had not planned to refer to the shapes hanging from the ceiling. Indeed, it was Rina, speaking out of turn, who spontaneously pointed to the mobiles, identifying the mobile with triangles. Shena then looked up, looked at the triangle mobile, and then looked at the circle mobile, pointing out that the circle mobile had figures without any vertices. In other words, she noted the critical attribute of having vertices and that some shapes have them and some do not. Shena may be said to be operating at the second van Hiele level of geometric reasoning (analytic reasoning).

Children at this age also form concepts by sorting items into categories (Smith et al. 1974). The features of a new stimulus are judged against features of a known category in order to determine if it belongs to that category, meaning that it is an example of the concept. Thus, before children learn formal scientific definitions, they learn to sort food items into, for example, fruit and vegetables. Similar activities may be carried out with geometrical shapes. We created very large cardboard shapes of triangles, squares, and circles of different sizes and colors and attached to them Velcro strips. The bottom half of the walls of the indoor gym were carpeted and perfect for attaching these shapes. We then spread out the cardboard shapes on the floor and played a music disc. While the music was playing, the children danced around the shapes. When the music was stopped, the TE or caregiver called out the name of a shape, for example, circle, and the children had to find a circle on the floor and stick it on the first wall (see Fig. 2). This was repeated by calling out "square," having the children run to find squares, and having them stick the squares on the second wall. Eventually, all the shapes on the floor were sorted and the triangles, squares, and circles were hanging on different walls of the indoor gym.

In the following episode, we describe an activity that included instances where the children were requested to evaluate the identifications and explanations of other children. Four children sat around a table along with one TE. The TE used the same set of cards as described in the previous activity, laid them out on the table, and asked each child, one at a time, to find a triangle and explain why it was a triangle. The other children were asked to say if they agreed or disagreed with the choice and explanation of the first child.

Fig. 2 Children stick cardboard circles on the wall

TE	Now it's Gila's turn. Gila, please pick up a card with a triangle.
Gila	(picks up a card with a triangle-like figure on it  .)
TE	(turning to the other children at the table) What did Gila pick? Is it a triangle? What do you say?
Jordan	It's a clown hat.
TE	But was Gila right? Is it a triangle?
Shena	No. It's folded here (pointing to the curved line).
TE	The line is curved.

Jordan's comment above demonstrates that he is operating at the first van Hiele level, taking in the whole figure without looking at the separate attributes. In addition, his claim that the figure is a clown hat does not let the TE know if he agrees or disagrees with Gila's choice. Thus, the TE asks again for the children to comment on Gila's choice (promoting the children to solve problems and evaluate others' solutions as per Cells 1 and 2 in Table 1). Shena responds and claims that Gila did not choose a triangle and explains in her own language why she disagrees. At this point, she does not have the vocabulary to explain the problem, so the TE helps her.

Enriching both the caregivers' and children's mathematical language was a central part of the program. Our expectation was that both the caregivers and the children would integrate mathematical language into their activities. For example, in one activity, we incorporated children's fine motor skills and counting skills while promoting geometrical language. Four children sat around a table with the TE and caregiver and were each given a large card with a large triangle drawn on the card. A pile of bottle caps was placed in the middle of the table, and each child, in turn, was requested to place a bottle cap on a vertex of the triangle (see Fig. 3). The group activity encouraged children to watch each other and at times correct each other. Children practiced using the word vertex when answering questions from the TE

Fig. 3 Placing bottle caps on the vertices of triangles

such as: did you cover the vertex? Is the cap on the side of the triangle or on the vertex of the triangle? It also reinforced the triangle's critical attribute of "threeness" as children counted the number of bottle caps they had at the end, which matched the number of vertices they had just covered.

The caregivers also incorporated geometrical activities during play time, demonstrating the development of their knowledge related to tasks and teaching (Cell 4—Designing and evaluating tasks—of the CAMTE framework). During one of the sessions with caregivers, Cecile describes an activity she did with a group of children in the outside play area.

Cecile	I took a piece of chalk and drew on the ground a giant triangle and I said to the children, let's walk around the triangle and when I say the word vertex, you stand next to a vertex and when I say sides, you stand on the straight line. Who was with me? I think Lily (another caregiver).
Lisa	Like the game sea and land.
Cecile	So we walked around the triangle and then suddenly I said, vertices, and you should have seen them, a group here (pointing to an imaginary space), a group there, and group there and then we walked around the triangle again and again and then I said, sides, and then they stood in a line on the sides.
TE	Wonderful.

There are several important issues to note in the extract. First, Cecile has adopted the use of precise geometrical language, calling the points of the triangle by their geometrical term, vertices, even when working with the children. Second, Cecile also makes sure to say that the sides of the triangle are "straight lines," emphasizing this critical attribute. Third, the caregivers are making use of the outdoor play area and thinking of ways for children to use their bodies when playing geometrical games. Finally, the caregivers have adapted a known child's game (called "sea and land" by Lisa) in order to fit their geometrical teaching goals.

Summary and Discussion

Geometrical shapes are part of the world around us and are part of the children's world as well. This makes it natural, on the one hand, to enhance young children's knowledge of shapes and makes it a good starting point for introducing other mathematical ideas, such as differentiating between critical and non critical attributes of a shape. On the other hand, one may think that it is not necessary to specifically spend time teaching such young children about triangles because, after all, they can basically name shapes such as triangles, squares, and circles by the age of three or four, and knowing much more than that is not necessary. As can be seen from the episodes above and from previous studies (Gutiérrez and Jaime 1999; Hershkowitz 1989), intuitive notions of triangles (e.g., all triangles must have a vertex on top) may be difficult to uproot in later years. Fischbein (1993) considered the figural concepts an especially interesting and complex situation where intuitive and formal aspects interact. The image of the figure promotes an immediate intuitive response. Yet geometrical concepts are abstract ideas derived from formal definitions. While it may not be appropriate to introduce young children to formal minimal definitions, children can learn to distinguish between straight and not straight lines. Just as children learn that an apple remains an apple whether or not the stem is on the top or the stem is on the bottom, children can learn that a triangle remains a triangle whether or not the vertex is on the top or the vertex is on the bottom. Uprooting intuitive misconceptions at an early age, before they become rigid, is essential (Fischbein 1987).

There are many dilemmas working with such young children and their caregivers. What is appropriate content? What are appropriate materials for promoting knowledge of this content? What is the correct amount of guidance versus informal play? How can early childhood caregivers be encouraged to take part in professional development programs and how should these programs be organized and delivered?

Regarding content and materials, as noted in the background, several curricula now mention educational guidelines for early childhood education, including the preschool years. To be sure, promoting geometrical knowledge is but one of many educational aims. It is equally important to encourage language and communication skills (DEEWR 2009). However, such skills may also be promoted, as demonstrated in the episodes described in this chapter, through appropriate geometrical activities. Motor skills and listening skills may also be promoted through appropriate geometric activities, such as the activity carried out in the indoor gym and the outside play area. Balancing between adult guidance and informal play is an open question, usually left to the caregiver's discretion. Yet, as mentioned in the background sections, some prospective and practicing early childhood teachers might hold several misconceptions regarding teaching geometry, and thus, professional development for early childhood caregivers is critical. By integrating geometrical activities in several ways and in several places, we were able to overcome some of these obstacles. We also note that although this chapter focused on Cells 1 and 2 of the CAMTE framework (promoting teachers' mathematical knowledge for solving tasks and evaluating solutions of tasks), it is equally important to promote their knowledge of students and tasks (Cells 3 and 4). A hint of these aspects can be seen in the episodes provided earlier as the caregivers pay attention to their children's queries and as they adapt familiar children's games to promote geometrical learning.

Several dilemmas remain regarding professional development for early childhood caregivers. One is the timing of such a course. Taking into consideration that many childcare centers are open throughout the year, including the summer, it is difficult to find an optimum time of year when caregivers are free to take part in professional development. This was one reason why we decided to provide the professional development sessions at the center. Another reason to come to the center was to get a feeling for how the center operates, including the physical environment. On the other hand, the caregivers were sometimes distracted at the center by the dynamics of their schedule. The duration of a course and incentive for participating are two additional questions. In our program, duration was dictated by scheduling and funding, and the caregivers were mandated by the day care center management to participate in the program. That being said, the caregivers did give up their rest period to participate. In addition, it was not only the caregivers of the 3-year-olds who joined and observed the TEs engaging children with mathematical activities. Caregivers of the younger children requested the director of the center to allow them to leave their groups for short periods of time in order to observe both the number and geometry activities in the 3-year-old classroom. The director, in charge of both financial and educational planning, supported this. In fact, the director of the day care center participated in part of the sessions with the caregivers and the TEs. This type of support is essential in order to encourage change.

Finally, in analyzing the organization and delivery of the program, we conclude that each element of the program contributed to the overall success. Obviously, sessions with the caregivers were necessary for introducing them to mathematical concepts, processes, and language. However, our active involvement with the children and the environment did not only promote the children's knowledge but also contributed to teachers' knowledge of children's conceptions. In turn, these engagements affected caregivers' beliefs regarding the possibility of enriching children's mathematical environment. In the future, we would attempt to meet more frequently with the caregivers, as the time allotted to us was short indeed. If caregivers are to recognize and encourage children's spontaneous engagement with mathematics, they need additional knowledge and support. The program described here was initiated in order to examine possible ways of promoting more mathematical activities among young preschool children. We are encouraged by the small but sturdy strides made by the caregivers and children and call out to mathematical educators to continue researching ways to promote mathematical learning during the important early years.

Acknowledgment This research was supported by the Israel Science Foundation (grant No. 654/10).

References

Ansari, D., Donlan, C., Thomas, M. S. C., Ewing, S. A., Peen, T., & Karmiloff-Smith, A. (2003). What makes counting count? Verbal and visuo-spatial contributions to typical and atypical number development. *Journal of Experimental Child Psychology, 85*, 50–62.

Ball, D., Thames, M., & Phelps, G. (2008). Content knowledge for teaching. *Journal of Teacher Education, 59*(5), 389–407.

Bandura, A. (1986). *Social foundations of thought and action: A social cognitive*. Englewood Cliffs, NJ: Prentice Hall.

Barnett, W. S. (2011). Effectiveness of early educational intervention. *Science, 333*(6045), 975–978.

Baroody, A. J. (1987). *Children's mathematical thinking*. New York: Teachers College.

Burger, W., & Shaughnessy, J. (1986). Characterizing the van Hiele levels of development in geometry. *Journal for Research in Mathematics Education, 17*(1), 31–48.

Clements, D. H., & Sarama, J. (2011). Early childhood teacher education: The case of geometry. *Journal of Mathematics Teacher Education, 14*(2), 133–148.

Clements, D. H., & Sarama, J. (2013). Rethinking early mathematics: What is research-based Curriculum for young children? In L. D. English & J. T. Mulligan (Eds.), *Reconceptualizing early mathematics learning* (pp. 121–147). Dordrecht: Springer.

Department of Education, Employment and Workforce Relations (DEEWR). (2009). *Belonging, being and becoming: The early years learning framework for Australia*. Canberra: Commonwealth of Australia.

Fischbein, E. (Ed.). (1987). *Intuition in science and mathematics*. Dordrecht: Reidel.

Fischbein, E. (1993). The theory of figural concepts. *Educational Studies in Mathematics, 24*(2), 139–162.

Ginsburg, H. P., Lee, J. S., & Boyd, J. S. (2008). Mathematics education for young children: What it is and how to promote it. *Social Policy Report, XXII*(I), 1–22.

Gutiérrez, A., & Jaime, A. (1999). Preservice primary teachers' understanding of the concept of altitude of a triangle. *Journal of Mathematics Teacher Education, 2*(3), 253–275.

Havnes, T., & Mogstad, M. (2011). No child left behind: Subsidized child care and children's long-run outcomes. *American Economic Journal: Economic Policy, 3*(2), 97–129.

Hershkowitz, R. (1989). Visualization in geometry—Two sides of the coin. *Focus on Learning Problems in Mathematics, 11*(1), 61–76.

Israel National Mathematics Preschool Curriculum (INMPC). (2008). Retrieved April 7, 2009, from http://meyda.education.gov.il/files/Tochniyot_Limudim/KdamYesodi/Math1.pdf.

Krajewski, K., & Schneider, W. (2009). Early development of quantity to number-word linkage as a precursor of mathematical school achievement and mathematical difficulties: Findings from a four-year longitudinal study. *Learning and Instruction, 19*(6), 513–526.

LeFevre, J. A., Skwarchuk, S. L., Smith-Chant, B. L., Fast, L., Kamawar, D., & Bisanz, J. (2009). Home numeracy experiences and children's math performance in the early school years. *Canadian Journal of Behavioural Science/Revue canadienne des sciences du comportement, 41*(2), 55.

NCTM. (2006). *Curriculum focal points for prekindergarten through grade 8 mathematics: A quest for coherence*. Reston, VA: National Council of Teachers of Mathematics.

Sarama, J., & Clements, D. (2009). *Early childhood mathematics education research: Learning trajectories for young children*. New York: Routledge.

Shulman, L. S. (1986). Those who understand: Knowledge growth in teaching. *Educational Researcher, 15*(2), 4–14.

Smith, E. E., Shoben, E. J., & Rips, L. J. (1974). Structure and process in semantic memory: A featural model for semantic decisions. *Psychological Review, 81*(3), 214.

Starkey, P., Klein, A., Chang, I., Dong, Q., Pang, L., & Zhou, Y. (1999, April). *Environmental supports for young children's mathematical development in China and the United States*. Paper presented at the meeting of the Society for Research in Child Development, Albuquerque, NM.

Starkey, P., Klein, A., & Wakeley, A. (2004). Enhancing young children's mathematical knowledge through a pre-kindergarten mathematics intervention. *Early Childhood Research Quarterly, 19*(1), 99–120.

Tall, D., & Vinner, S. (1981). Concept image and concept definition in mathematics, with special reference to limits and continuity. *Educational Studies in Mathematics, 12*, 151–169.

Tirosh, D., Tsamir, P., & Levenson, E. (2011). Using theories to build kindergarten teachers' mathematical knowledge for teaching. In K. Ruthven & T. Rowland (Eds.), *Mathematical knowledge in teaching* (pp. 231–250). Dordrecht: Springer.

Tsamir, P., Tirosh, D., Barkai, R., Levenson, & Tabach, M. (2014a). Subject-matter and pedagogical content knowledge and self-efficacy for teachers: The case of kindergarten teachers and geometry. *Maof V'Maeseh, 16*, 19–42 [In Hebrew].

Tsamir, P., Tirosh, D., Levenson, E., Tabach, M., & Barkai, R. (2014b). Employing the CAMTE framework: Focusing on preschool teachers' knowledge and self-efficacy related to students' conceptions. In C. Benz, B. Brandt, U. Kortenkamp, G. Krummheuer, S. Ladel, & R. Vogel (Eds.), *Early mathematics learning—Selected papers from the POEM 2012 Conference* (pp. 291–306). New York: Springer.

Tsamir, P., Tirosh, D., & Levenson, E. (2008). Intuitive nonexamples: The case of triangles. *Educational Studies in Mathematics, 69*(2), 81–95.

van Hiele, P. M., & van Hiele, D. (1958). A method of initiation into geometry. In H. Freudenthal (Ed.), *Report on methods of initiation into geometry*. Walters: Groningen.

Vandell, D. L., Belsky, J., Burchinal, M., Steinberg, L., & Vandergrift, N. (2010). Do effects of early child care extend to age 15 years? Results from the NICHD study of early child care and youth development. *Child Development, 81*(3), 737–756.

MaiKe: A New App for Mathematics in Kindergarten

Anna Susanne Steinweg

Abstract Increasing availability of tablets at home—even for the youngest ones—is a fact. Therefore, research on mathematical play and learning apps is of growing importance. The project MaiKe (Mathematik im Kindergarten entdecken) develops its own and mathematically sound app. MaiKe regards and reflects in concept, development and design both developmental psychology and mathematics educational findings and research. MaiKe takes into account the broad range of competencies which are considered to promote a successful school beginning. The concept and design information about the app as well as concrete examples of technical realisation and use are given.

Introduction

When it comes to ICT and early mathematics for kindergarten children, many different research approaches can be noticed. On the one hand, research has focused on understanding the interaction of children, adults and the software or hardware (e.g. Hundeland et al. 2013). On the other hand, programmes and apps concerning the opportunities offered for mathematical learning have been analysed (e.g. Lange and Meaney 2013). Both kinds of research investigate apps for early mathematics which are already on the market. In contrast, our co-operative project between Thomas Weth (University of Erlangen-Nuremberg) and the author starts from the assumption that mathematics education research is a 'design science' (Wittmann 1995). Therefore, designing a mathematically sound learning environment or material—here an app—is the first and important step. The design of learning materials needs to be based on a theoretical framework.

Our design is theoretically embedded in research results on early mathematical learning and ICT. This is discussed before some design realisations are outlined, followed by a brief description of the insights gained from the design and case studies on the use of the app. First of all, the specific situation of kindergarten education in Germany is briefly outlined in the next section.

A.S. Steinweg (✉)
University of Bamberg, Bamberg, Germany
e-mail: anna.steinweg@uni-bamberg.de

T. Meaney et al. (eds.), *Mathematics Education in the Early Years*,
DOI 10.1007/978-3-319-23935-4_19

Kindergarten Education in Germany

Although the majority of 4- to 6-year-olds attend kindergarten in Germany, it is not compulsory. Parents, educators and the German society consider kindergarten to be a place to play and not to learn. Consequently, kindergarten is not led by teachers but by educators. Kindergarten educators, apart from some rare examples (see Benz 2016), have neither college nor university degrees, let alone mathematics or mathematics education degrees or courses. It is therefore not surprising that different research studies on attitudes of kindergarten educators show an uncertainty about what mathematical content should be offered to the children: 'As seen in the open questions, the range of learning goals was very broad. Many content topics from primary school mathematics were mentioned' (Benz 2012a, p. 259). The studies suggest the need to make the content areas explicit to kindergarten educators (Thiel 2010; Benz 2012a, b):

> These answers show very distinctly that it is absolutely important to clarify what early mathematics in kindergarten should mean, respectively which mathematical contents should be included or should be identified in early mathematics teaching. (Benz 2012b, p. 223)

Moreover, the so-called situational approach was and still is widespread and is represented in policy document. For example, 'educational work (…) is guided by the interests, needs and situations of the individual children' (KMK 2013, p. 101). In the radical form, this approach allows kindergarten educators to react and respond to children's questions and ideas, but direct intervention by educators is frowned upon. Although the radical principles are never actually followed, German kindergarten educators are still quite suspicious of any formal learning activities. This goes hand in hand with a deep distrust of learning material which are not 'natural' (like wooden blocks, organic materials from the gardens, etc.) and therefore not suitable for children. Developing an ICT concept for early mathematics learning in Germany has to take these facts into account.

Theoretical Framework

The research is framed by two major themes, our understanding of reasonable approaches to mathematics in kindergarten and of ICT use. Both aspects are guiding our content-related design and technical implementation. Therefore, these two aspects are discussed in the next sections.

Approaches to Early Mathematics

Getting into contact with the world as well as mathematical contents needs no technical support but a rich environment and other people with interest in both the content and the natural curiosity of young children. It is important to explore mathematical objects and situations, to talk about findings and hypotheses and to imitate more experienced people's actions and words. The richness of the environment a child grows up in is dependent on both spontaneous and specially designed interaction at kindergarten as well as at home (Anders et al. 2012). Educators, parents and other children need to be aware of mathematically fruitful situations and discussions (van Oers 2004; Gasteiger 2010) to foster the competencies of the younger ones:

> However, from our observation in classrooms involved in play, is clear that both creative construction and sensitive instruction are necessary elements for a developmentally productive organization of play and the development of mathematical thinking. (van Oers 2014, p. 121)

Interactions can be based on co-constructive, discovery learning in 'natural learning situations', i.e. everyday or play situations (Gasteiger 2012). Nevertheless, it may be important to guide learning in terms of offering activities with specific content and to prepare and enrich the environment children live in. Access to next levels of understanding in the zone of proximal development (Vygotsky 1978) is facilitated only by this enrichment. An ICT environment functions, at best, as a road map moving children through appropriate mathematical content areas which are in alignment with an early years' mathematics curriculum. In this approach, the 'power of knowledge differential' is utilised:

> Teachers know the convention of reasoning and representation that are involved in the patterns of mathematical thinking. Students initially may not have this awareness. There is also thus a power differentially involved. However, effective instruction can facilitate student's making of construction that lie within the canons of mathematically accepted knowledge, and yet there is room for creativity and enjoyment. (Presmeg 2014, p. 11)

Early years' mathematics should include various topics. Neither a sole emphasis on counting nor a unilateral training on one-to-one correspondence will support the learning of a viable image of mathematics. Rather, at least working with arithmetical and geometrical interesting problems or situations is important to foster children's mathematical learning. Content areas should not be chosen arbitrarily but closely connected to curriculum in school (Gasteiger and Benz 2012). Bredekamp (2004, p. 82) highlighted, 'perhaps, a bigger question (…) is how to distinguish between what children can learn and what they should learn'.

In our design and research project MaiKe, we take into account the wide range of competencies which are considered to promote a successful school beginning, that is, different content areas like number and operations, geometry and spatial sense, measurement, pattern, etc. described in the learning paths by NAEYC and NCTM (2010) or the big ideas by the Erikson Institute (Brownell et al. 2014). German mathematics education literature also focuses on these ideas (Benz et al. 2015; Steinweg 2008; Wittmann 2009).

Furthermore, special attention is paid to predictive competencies, which have an empirically proven impact on outcomes of second grade (Dornheim 2008). In particular, Dornheim (2008) identifies the following competencies to be predictive: counting, simultaneous perception (subitising), flexible counting (forward, backward, in steps) and part-whole relations (e.g. first additions like 2+1, 3+2), one-to-one relation, seriation and certain knowledge about numerals. Elements of 'spatial sense' (i.e. redrawing, bilateral symmetry, pattern) are predictive as well with respect to very young children (3- to 4-year-olds).

In summary as a basic principle, learning environments in early mathematics need to take into account 'connectivity' and 'compatibility' to school mathematics and predictive competencies. This principle precludes training of singular skills but the focus is on fundamental ideas of mathematics (Gasteiger and Benz 2012). Accordingly, MaiKe app design has to allow for interaction with mathematical key ideas entirely.

Tablet Use in Kindergarten

There is increasing availability of ICT like personal computers, mobile phones or smartphones and tablets for adults, children and also kindergarten children. However, as described earlier, we are aware that electronic media cannot be a substitute for real-world experiences in our project. Our assumption was that young children are playing with apps in their everyday life, and this contributed to the design project MaiKe. Following the view of NAEYC and NCTM (2010) that 'play does not guarantee mathematical development, but it offers rich possibilities' (p. 8), if the youngest ones play ICT apps, they should have the chance to benefit from ones which offer appropriate mathematical learning possibilities.

Additionally, design projects on ICT environments benefit from the invention of tablet and smartphone apps. Tablets no longer depend on hardware like mouse or keyboard. Therefore, the handling has become much easier for young children. Touchpad swiping might even make it possible for toddlers to 'use' tablets (Krauthausen 2012, p. 153).

In the popular press, ICT use is often seen as leading to children's isolation and impeding their interaction with others. Research studies have shown that the opposite is more likely:

> Computers serve as catalysts of social interaction. (…) Children prefer to work with friends rather than alone, and they display more positive emotion and interest when working together. (…) They show increased collaborative work, including spontaneous helping and teaching, and they discuss and build on one another's ideas…. (Clements and Sarama 2002, p. 341)

Consequently, mathematically sound learning apps like MaiKe may contribute to mathematical collaboration and interaction between peers or adults and children:

> Significant benefits are more likely when teachers follow up by engaging children in reflecting on and representing the mathematical ideas that have emerged in their play. Teachers enhance children's mathematics learning when they ask questions that provoke clarifications, extensions, and development of new understandings. (NAEYC & NCTM 2010, p. 8)

To provide opportunities for reflexion and discussion about mathematical content is yet another reason to select fundamental mathematical ideas as a basis for the app.

The number of existing early mathematics apps is enormous. Analysis of about 40 apps has shown striking results concerning educational and mathematical criteria (Steinweg and Weth 2014). Many apps focus only on a small range of content and seem not to distinguish between early years and school mathematics, even mathematical incorrect tasks can be found. Several apps offer colourful illustrations (mostly not really helpful for solving the tasks), are heavy on text and do not allow access for non-readers. Others offset the low or non-existing reading abilities of the users by using loud narrators speaking in high-pitched children's voices that are not easy to listen to. The search for an appropriate balance between learning and entertainment—often summarised as edutainment—is not over yet (Krauthausen 2012, p. 162).

One decidedly promising approach of design can be seen in the Israeli 'SlateMath for Kids' project. Founded by a mathematician and computer scientist, the content is mathematically correct and seems to be in line with Piagetian tasks, such as one-to-one correspondence. Unfortunately, most of them are heavy on text and cannot be handled by children (non-readers) on their own. The application is originally designed 'to help teachers teach math' (Kupferman and Schocken 2013, p. 10). Therefore, the application is not appropriate for German kindergarten context, described earlier, or home market. However, SlateMath design principle of using everyday situations as learning situations seems valuable:

> Our experience shows that Matific [i.e. application 'SlateMath for Kids'] endears math on children as young as 4 year-olds by helping them master *common tasks* that unfold in *common settings*: counting animals, hanging balloons, decorating cakes, and so on. There is no need to have a dinosaur slap its tail on the smaller of two numbers when there are many interesting ordinary scenarios in which order comparisons come to play. (Kupferman and Schoken 2013, p. 5)

The design should not focus on special effects and storytelling but rather on essential content in line with key ideas of mathematics. Ideally, app environments 'complement and expand what can be done with other media' (NAEYC and NCTM 2010, p. 9). This indicates that it is sensible to provide objects and tasks, which are common in real-world mathematical learning environments and vice versa.

The media set the limits of the technical implementations of mathematical learning environments. First of all, objects are never objects but images. Apps provide virtual experiences only. However, mathematical objects—even shapes—are abstract and in a way never touchable real-world objects. Mathematical learning environments always depend on indirect experiences, likewise an ICT environment. Second, touchpad swiping allows for sorting, matching and composing of images of objects. There is no way to prevent approaches by trial and error. Thus, an ICT environment needs to make use of this approach to initiate learning possibilities.

Being aware of these limitations, the design of an app could contribute only one of the many possible components of mathematical learning environments. The app may function as both a compilation of sound mathematical content and an initial impulse for further (and deeper) engagement with the content in direct interaction with other children, parents and kindergarten educators.

Design

We regard MaiKe[1] app as an electronic playground for 4- to 6-year-olds, offering chances to get to know important mathematical key ideas and to explore and improve mathematical competencies ('Mathematik entdecken' literally means 'discovering mathematics'). The app is designed to offer possibilities, which could lead to mathematical discourses and real-world interactions. In the following, the technical realisation in terms of screen design and handling and underlying principles are described.

Structure of MaiKe

The application's structure guides the children through six different worlds. Within each and every world, ten games always addressing different mathematical content are offered. Thus, the worlds are not bound to one certain content, but—like a minimalistic spiral curriculum—the children face a specific content in new games now and again during their journey through the worlds.

The level of difficulty of the tasks is increasing. For example, a game in the first world asks to complete a one to four number line by placing one missing object, and in a similar game in the fourth world, the child is asked to complete a one to ten number line with three blank spaces.

MaiKe's start screen allows access to the first world and the first game. Regardless of the success of playing the first game, the next game will be accessible after completing the first activity and so forth. Games in the first and second world consist of six tasks. The number of tasks needed to finish a game increases to ten as progress is made through the worlds. In total, 480 tasks are provided.

Design Principles

During the technical implementation process, there are three major principles leading:

(I) Unconditional access
(II) Mathematical sound representations
(III) Mathematical correctness

[1] MaiKe can be retrieved from Google Play Store: https://play.google.com/store/apps/details?id=de.unierlangen.maike. Accessed 22 June 2015.

Fig. 1 MaiKe game screen areas

Unconditional Access

Our idea is to keep it simple and self-explanatory. We want the children to have the opportunity to play MaiKe without any support by adults. This has to be emphasised due to the fact that there might not be any 'teaching' in German kindergarten.

Every child should have the chance to have access to MaiKe. Therefore, neither reading skills nor knowledge of numbers as numerals is conditional. Each activity is designed to be self-explanatory and features no verbal or text-based descriptions.

The handling required by children is straightforward and easy, based on simple swiping. The game screen is always split into visually separated areas (Fig. 1). In the pale-coloured area, objects can be moved (moving objects area ②). In the dark-coloured, static objects area ①, the matching moving objects have to be dragged to. Most of the tasks provide more movable objects than needed or matching. The screen has to be cleaned up before the next task automatically pops up. A dustbin is always provided in the left corner of the screen (dustbin area ③).

Mathematical Sound Representations

MaiKe makes a distinction between graphical illustrations, such as the start screen, and representations of educational material, such as in the pale-coloured area. The mathematical environments given in the games—where mathematical tasks have to be solved—provide representations or objects which are needed to solve the tasks. No further objects, identification figures (hero of a story or the like) or background illustrations are given.

Illustrations as a kind of motivation are only used in the starting screens of the six worlds. This means they appear only after and before 'doing mathematics'.

Finishing a game is rewarded by progressively completing the world illustration as in a jigsaw puzzle (beach landscape, fairy-tale world, dinosaurs, etc.). The scale of completeness shows both how many games have already played and the correctness of the offered solutions. Therefore, the illustrations serve as a feedback as well.

Mathematical Correctness

As a number of applications in early mathematics are not mathematically correct (an alarming rate of games ignores mathematical relations of geometrical shapes; some even struggle with arithmetic), mathematical correctness is regarded as a major MaiKe design principle. Moreover, the activities offered needed to be ‘a real mathematical activity’ in accordance with the definition by Wittmann (2010, p. 186; translation by the author):

> A real mathematical activity shares the characteristics:
> 1. ‘Elements’ offered have mathematical properties and mathematical relations.
> 2. The elements are handled by mathematical rules.
> 3. Activities are purposeful and always aim for patterns or solutions by using patterns.

MaiKe Contents: Examples of Realisation

This section provides some examples of the different content areas which are used to illustrate MaiKe’s technical implementation design principles.

Numbers and Operations

Cardinal and ordinal aspects of numbers are the major branches of arithmetic that children need to get into contact with (Wittmann 2009). Neither aspect precedes the other, and so MaiKe integrates the idea of the ordinal number line as well as the concept to identify the quantity of sets (cardinality) from the very beginning.

Benz (2014) shed light on the importance of perception, i.e. being able to perceive and understand a structured set of dots or objects. She also showed the individuality of the perception process of children in her study. The goal of quantity perception lies in the replacement of counting processes by simultaneous perception. Sets of up to three or four objects can be captured at a glance. Bigger samples need to be structured and perceived in a kind of part-whole concept. One promising approach is for children to take advantage of the power of five (Krauthausen 1995; Wittmann 1998). MaiKe activities invite the children to compare structured and nonstructured sets, whilst the structure always respects the power of five. Additionally, whenever possible, games provide access to analogue structured didactic material (field of ten).

Fig. 2 Mapping eggs in cartons with a finger number

Furthermore, throughout MaiKe, the first game in each content area integrates common everyday objects. Therefore, the first explorations of cardinal aspects of numbers also choose real-world objects, i.e. cartons of eggs (Fig. 1). Children are asked to map structured and nonstructured sets of eggs. Deep understanding of cardinality includes the competence to interpret the quantity of sets in various representations and easily swap one into another mentally. Therefore, egg cartons are increasingly compared with and finally replaced by the analogue structured didactical material, i.e. dots in a field of ten. In a further game, another analogue structured representation is introduced, finger numbers (Fig. 2). Numeral representations are offered only at the higher levels (worlds), because reading numerals needs to be based on understanding countable sets.

Providing countable sets might be questioned from an educational point of view. Although children are not forced to use the possibility to count—using individual counting strategies (Fuson 1988)—there is the chance to increasingly recognise the structure whilst playing and to make use of it and to attempt simultaneous perception (subitising).

The ordinal aspect of numbers, representing the counting order, is introduced through die representations (Fig. 3). Within the process, it is possible to count dots on a die if needed. However, some of the children may already 'know' the dot pattern and perceive the number by simultaneously building on their real-world experiences playing board games (Gasteiger 2012, 2013). If the children use their skills or start exploring dice pattern within this activity, they may gain the ordinal aspect of numbers as the predecessor and successor need to be detected. In addition, MaiKe introduces the conventional order of the number line from left to right.

Besides cardinal and ordinal number aspects, MaiKe provides set correspondence (one-to-one relations) and part-whole relation, which can be linked to early addition and subtraction tasks without symbolic equations or activities. Within the last world of games, there is a chance to try out writing numerals as well.

Fig. 3 Completing a number line of dice

Geometry

MaiKe focuses on plane and spatial geometry, symmetry and basic shapes. In particular, reconstructional or redrawing activities of line patterns, building blocks or shapes and sorting of shapes are given. In this way, geometrical aspects that foster spatial awareness can be explored.

Maier and Benz (2014) have shown that German children—more than English ones—have difficulties in identifying a right-angled triangle as a triangle. MaiKe avoids restrictions to the typical basic shapes equilateral triangle, square and circle but offers activities involving general quadrilaterals, ellipses, etc. in accordance with Clements and Sarama's suggestions (2007):

> Concepts of two-dimensional shapes begin forming in the pre-K years and stabilize as early as age six (…), so early experiences are important. This learning will be more effective if it includes a full range of examples and distractors to build valid and strong concept images…. (p. 230)

Shapes are to be classified in accordance with the geometrical relationships, for example, a square is a rectangle and a rectangle is a quadrilateral (Fig. 4). Young children understand shapes as entities—nevertheless, intuitive understanding of structures is possible according to van Hiele (1976). The iconic representation of shapes allows discovering some characteristics and differences. Navigating through the different geometrical games provides the opportunity to become increasingly aware of this mathematical structure.

Fig. 4 Finding every quadrilateral

Fig. 5 Seriation of real-world objects by (area) size

Quantities and Measurement

Exploring quantities like length, area, volume, mass (weight) and time mainly depends on real-world objects and hands-on activities. The technical environment is limited to activities concerning length or area (although, strictly speaking, it only appears to be area but the two-dimensional projection of volume). A typical challenge is a seriation problem in which a given number of objects have to be ordered by size. The underlying structure is the order relation of magnitudes (Fig. 5).

Mapping different sets (spoons and cups) by size, sorting by size and completing an order are different activities in this mathematical content.

Pattern

Mathematics as the science of patterns (Devlin 1997) cannot be restricted to patterns in sequences. Of course, structured dot patterns, the power of five, the order relation of magnitudes and the classification of geometrical shapes are based on mathematical patterns and structures too. Patterns define the beauty of mathematics:

> The mathematician's patterns, like the painter's or the poet's, must be beautiful, the ideas, like the colours or the words, must fit together in a harmonious way. Beauty is the first test; there is no permanent place in the world for ugly mathematics … It may be very hard to define mathematical beauty, but that is just as true of beauty of any kind—we may not know quite what we mean by a beautiful poem, but that does not prevent us from recognising one when we read it. (Devlin 1997, p. 6)

Seeking for patterns, symmetry and regularities and appreciating the beauty of mathematics are fundamental activities to understand mathematics:

> Pattern is less a topic of mathematics than a defining quality of mathematics itself. Mathematics 'makes sense' because its patterns allow us to generalize our understanding from one situation to another. Children who expect mathematics to 'makes sense' *look* for patterns. (Brownell et al. 2014, p. 84)

Sequential patterns of real-world objects (Fig. 6), tiles of equal shapes characterised by colour (Feynman 1995) or sequences of different shapes can be an initial starting point for recognising patterns and seeking awareness of patterns.

Geometrical ornaments are infinite mathematical objects. Due to the design principle of mathematical correctness, each pattern has neither a beginning nor an ending accordingly (Fig. 7).

Fig. 6 Pattern of real-world objects (sharks and dolphins)

Fig. 7 Pattern of shapes

First Case Studies

The design project aims to provide an app which meets the requirements of mathematical education research in the early years. It is, therefore, important to evaluate the implementation of the app in follow-up studies. Initially, some beta and part versions of MaiKe have been used in video-recorded case studies, mainly to test the unconditional access principle. Also, the children's reactions give feedback about different aspects of the apps that can be improved.

An example is provided of Ole (6 years old), who played one of the first versions of MaiKe shortly after he started school. The interview (author and interviewer is one and the same person; translation by the author) was not preceded by any instruction.

The start screen with the six different worlds is offered.
Int. Do you know how to start it? How to get there?
Ole Yes. Tapping.
Int. Yes, then do so.
The highlighted, first game (cardinality with solely egg cartons) is opened by tapping by Ole.
Int. What do one have to do here?
Ole Three (*touching a moveable carton with three eggs*).
Int. Where do they go?
Ole In the dustbin.
Int. Why?
Ole (*shyly laughing*) I don't know either.
Int. How many do we have here? (*pointing at the non-moveable egg carton on the right*)
Ole 1, 2, 3. Ah! There! (*swiping the moveable egg carton with three eggs on the right one*)
Two belongs to the dustbin (*swiping the egg carton with two eggs into the bin*).
Also to the dustbin. Four (*swiping the egg carton with four eggs into the bin*).

Ole finishes the first game and starts another one (ordinality with dice).
Ole 1, 3, 4. Then the 2 has to go there (*pointing at the gap in the row of dice given*).
Int. Try it out.

Ole is easily able to open the games, intuitively by tapping. He then acts cautiously in the first game in order not to make any mistakes. Although he again intuitively tries to move the moveable objects only, he is unsure where to swipe them. The interviewer focuses his attention on the non-moveable object. After that hint, Ole starts sorting the egg cartons one by one without hesitation. Starting the next game about the dice in a row, Ole needs no hints from the interviewer at all, and, furthermore, he comments on the objects and the task voluntarily. It took Ole 15:40 min to finish all 60 tasks in the first world. For one game (six tasks each), he needed 1:30 min on average.

In this first encounter with tablets (perspective on technical handling), acceptance, perseverance and first cases of performing are further evaluation criteria. Further case studies and document analysis are ongoing. Bigger samples of case studies will be selected, and institutional (kindergarten) as well as non-institutional (home) settings are evaluated. The effect on the mathematical learning will be tested by standardised methods (e.g. van Luit et al. 2001).

Remarks

In this article, we describe the design of the MaiKe project including technical implementation principles, placing different games throughout the six worlds and differentiating degrees of difficulty of recurring games, based on mathematical key ideas. An appropriate design seems to be achieved at this point in time. Evaluation studies and further case studies are underway.

Of course, playing an app cannot guarantee learning outcomes but provide learning chances. MaiKe offers kindergarten children possibilities to get in contact with appropriate mathematical content, to become aware of mathematical activities and to improve their competencies. Furthermore, we hope for implicit in-service training of kindergarten educators and parents in regard to the implementation. MaiKe offers adults chances to become aware of mathematical content and activities suitable for kindergarten children and to overcome the earlier described widespread uncertainty of kindergarten educators on which content should be provided. To ensure high quality of kindergarten learning environments is only one side of the coin. Recent studies have shown that support by parents at home is at least as important:

> Hence, our findings emphasize that it is important to make high-quality preschool education accessible for all children. However, we also found evidence for parents' potential to promote the development of their children's early numeracy skills and found that parental support at home seems to be a precondition for academic stimulation at preschool. However, not all parents may know how to best support their children and may need assistance. (Anders et al. 2012, p. 242)

MaiKe may serve as one of many small components of this assistance needed, as a starting point to initiate mathematical activities and talking about mathematical contents at home. Every virtual situation in the game can be transformed into a real-world situation with building blocks, cartons of eggs and other toys. MaiKe app games may provide some rewarding ideas for 'playing mathematics' between children and adults, parents or kindergarten educators.

References

Anders, Y., Rossbach, H., Weinert, S., Ebert, S., Kuger, S., Lehrl, S., et al. (2012). Home and preschool learning environments and their relations to the development of early numeracy skills. *Early Childhood Research Quarterly, 27*(2), 231–244.

Benz, C. (2012a). Maths is not dangerous—Attitudes of people working in German kindergarten about mathematics in kindergarten. *European Early Childhood Education Research Journal, 20*(2), 249–261.

Benz, C. (2012b). Attitudes of kindergarten educators about math. *Journal für Mathematikdidaktik, 33*(2), 203–232.

Benz, C. (2014). Identifying quantities—Children's constructions to compose collections from parts or decompose collections into parts. In U. Kortenkamp et al. (Eds.), *Early mathematics learning. Selected papers of the POEM 2012 conference* (pp. 189–203). New York: Springer.

Benz, C. (2016). Reflection: An opportunity to address different aspects of professional competencies in mathematics education. In T. Meaney, T. Lange, A. Wernberg, O. Helenius, & M. L. Johansson (Eds.), *Mathematics education in the early years—Results from the POEM conference 2014*. Cham: Springer.

Benz, C., Peter-Koop, A., & Grüßing, M. (2015). *Frühe Mathematische Bildung [Early mathematics education]*. Berlin: Springer.

Bredekamp, S. (2004). Standards for preschool and kindergarten mathematics education. In D. Clements & J. Sarama (Eds.), *Engaging young children in mathematics—Standards for early childhood mathematics education* (pp. 77–82). Mahwah, NJ: Lawrence Erlbaum.

Brownell, J., Chen, J.-Q., & Ginet, L. (2014). *Big ideas of early mathematics: What teachers of young children need to know*. Boston: Pearson.

Clements, D., & Sarama, J. (2002). The role of technology in early childhood learning. *Teaching Children Mathematics, 8*(6), 340–343.

Clements, D., & Sarama, J. (2007). *Early childhood mathematics education research—Learning trajectories for young children*. New York: Routledge.

Devlin, K. (1997). *Mathematics: The science of patterns*. New York: Scientific American Library.

Dornheim, D. (2008). *Prädiktion von Rechenleistung und Rechenschwäche [Prediction of numeracy competence and numeracy difficulty]*. Berlin: Logos.

Feynman, R. P. (1995). What is science? In D. K. Nachtigall (Ed.), *Internalizing physics: Making physics part of one's life: eleven essays of Nobel laureates* (pp. 99–112). Paris: United Nations Educational, Scientific and Cultural Organization.

Fuson, K. C. (1988). *Children's counting and concepts of number*. New York: Springer.

Gasteiger, H. (2010). *Elementare mathematische Bildung im Alltag der Kindertagesstätte [Elementary mathematics education in daily routine kindergarten]*. Münster: Waxmann.

Gasteiger, H. (2012). Fostering early mathematical competencies in natural learning situations—Foundation and challenges of a competence-oriented concept of mathematics education in kindergarten. *Journal für Mathematik-Didaktik, 33*(2), 181–201.

Gasteiger, H. (2013). Förderung elementarer mathematischer Kompetenzen durch Würfelspiele—Ergebnisse einer Interventionsstudie [Fostering elementary mathematics competencies by dice games—Results of an intervention study]. In G. Greefrath et al. (Eds.), *Beiträge zum Mathematikunterricht 2013* (pp. 336–339). Münster: WTM.

Gasteiger, H., & Benz, C. (2012). Mathematiklernen im Übergang—kindgemäß, sachgemäß und anschlussfähig [Learning mathematics during transition—Child appropriate, topic appropriate, and connective]. In S. Pohlmann-Rother & U. Franz (Eds.), *Kooperation von KiTa und Grundschule* (pp. 104–120). Köln: Wolters Kluwer.

Hundeland, P. S., Erfjord, I., & Carlsen, M. (2013). Use of digital tools in mathematical learning activities in the kindergarten: Teachers' approaches. In B. Ubuz et al. (Eds.), *Proceedings of the eighth congress of European research in mathematics education* (pp. 2108–2117). Ankara: Middle East Technical University.

KMK—Secretariat of the Standing Conference of the Ministers of Education and Cultural Affairs of the Länder in the Federal Republic of Germany. (2013). The education system in the Federal Republic of Germany 2011/2012—Early childhood education and care. http://www.kmk.org/fileadmin/doc/Dokumentation/Bildungswesen_en_pdfs/dossier_en_ebook.pdf. Accessed 22 June 2015.

Krauthausen, G. (1995). Die 'Kraft der Fünf' und das denkende Rechnen [The power of five and thoughtful numeracy]. In G. Müller & E. Wittmann (Eds.), *Mit Kindern rechnen* (pp. 87–108). Frankfurt & Main: Der Grundschulverband e.V.

Krauthausen, G. (2012). *Digitale Medien im Mathematikunterricht der Grundschule [Digitale media in primary school mathematics education]*. Berlin: Springer.

Kupferman, R., & Schocken, S. (2013). The Matific approach to early-age math education. http://media.wix.com/ugd/ea1b59_bcd7e5b280154a92b7558754f64e4b86.pdf. Accessed 22 June 2015.

Lange, T., & Meaney, T. (2013). iPads and mathematical play: A new kind of sandpit for young children. In B. Ubuz et al. (Eds.), *Proceedings of the eighth congress of European research in mathematics education* (pp. 2138–2147). Ankara: Middle East Technical University.

Maier, A. S., & Benz, C. (2014). Children's constructions in the domain of geometric competencies (in two different instructional settings). In U. Kortenkamp et al. (Eds.), *Early mathematics learning: Selected papers of the POEM 2012 conference* (pp. 173–187). New York: Springer.

NAEYC (National association for the education of young children) & NCTM (National council for teachers of mathematics). (2002/updated 2010). Early childhood mathematics: Promoting good beginnings. http://www.naeyc.org/files/naeyc/file/positions/psmath.pdf. Accessed 22 June 2015.

Presmeg, N. (2014). A dance of instruction with construction in mathematics education. In U. Kortenkamp et al. (Eds.), *Early mathematics learning: Selected papers of the POEM 2012 conference* (pp. 9–17). New York: Springer.

Steinweg, A. S. (2008). Zwischen Kindergarten und Schule—Mathematische Basiskompetenzen im Übergang [Between kindergarten and school—Basic mathematics competencies in the transition period]. In F. Hellmich & H. Köster (Eds.), *Vorschulische Bildungsprozesse in Mathematik und in den Naturwissenschaften* (pp. 143–159). Bad Heilbrunn: Klinkhardt.

Steinweg, A. S., & Weth, T. (2014). Auch das noch? Tablets im Kindergarten [Really? Tablets in kindergarten]. In J. Roth & J. Ames (Eds.), *Beiträge zum Mathematikunterricht 2014—Band 2* (pp. 1167–1170). Münster: WTM.

Thiel, O. (2010). Teachers' attitudes towards mathematics in early childhood education. *European Early Childhood Education Research Journal, 18*(1), 105–115.

van Hiele, P. M. (1976). Wie kann man im Mathematikunterricht den Denkstufen Rechnung tragen? [How can thinking-steps be addressed in mathematics education?]. *Educational Studies in Mathematics, 7*(1–2), 157–169.

van Luit, J., van de Rijt, B., & Hasemann, K. (2001). *Osnabrücker Test zur Zahlbegriffsentwicklung [Osanbruecker number-concept-test]*. Göttingen: Hogrefe.

van Oers, B. (2004). Mathematisches Denken bei Vorschulkindern [Mathematical thinking of preschool-children]. In W. E. Fthenakis & P. Oberhuemer (Eds.), *Frühpädagogik international. Bildungsqualität im Blickpunkt* (pp. 313–330). Wiesbaden: VS Verlag für Sozialwissenschaften.

van Oers, B. (2014). The roots of mathematising in young children's play. In U. Kortenkamp et al. (Eds.), *Early mathematics learning: Selected papers of the POEM 2012 conference* (pp. 111–123). New York: Springer.

Vygotsky, L. S. (1978). *Mind in society*. Cambridge, MA: Harvard University Press.

Wittmann, E. (1995). Mathematics education as a 'design science'. *Educational Studies in Mathematics, 29*(4), 355–374.

Wittmann, E. (1998). Standard number representations in the teaching of arithmetic. *Journal für Mathematikdidaktik, 19*(2/3), 149–178.

Wittmann, E. (2009). *Das Zahlenbuch—Handbuch zur Frühförderung [The number book—Handbook of early support]*. Stuttgart: Klett.

Wittmann, E. (2010). Grundsätzliche Überlegungen zur frühkindlichen Bildung in der Mathematik [Fundamental thoughts about early education in mathematics]. In M. Stamm & D. Edelmann (Eds.), *Frühkindliche Bildung, Betreuung und Erziehung. Was kann die Schweiz lernen?* (pp. 177–195). Zürich: Rüegger.

"I Spy with My Little Eye": Children Comparing Lengths Indirectly

Johanna Zöllner and Christiane Benz

Abstract In preschool settings, learning takes place in informal situations, which can make it very challenging for preschool teachers to identify learning possibilities or "teachable moments". Because children in preschool often deal with length in their daily life, many teachable moments can be identified in situations when children are comparing and measuring length. Fostering children's competencies in the area of length in informal natural learning situations needs preschool teachers to have pedagogical content knowledge about comparing and measuring in order to "see" or perceive these competencies in children's activities. This knowledge is the basis for identifying natural learning situations. Competencies required for comparing lengths indirectly will be analysed in this chapter.

Introduction

Many government documents in different countries as well as theoretical and empirical studies in the field of mathematics education propose that natural or informal learning situations or teachable moments should be identified and used in preschool in order to foster children's early mathematical competencies (Gasteiger 2010; Ginsburg et al. 2008). Natural and informal learning situations for mathematics learning, such as comparing and measuring length, occur in children's free play, in developing their own games and in their daily routine. However, to compare and measure lengths, children need to use different components of the concept of length.

In order to recognise and foster children's competencies concerning comparing and measuring length, preschool teachers need pedagogical content knowledge so that they can recognise competencies in children's activities. Then preschool teachers are able to help children develop procedural and conceptual knowledge of length. In this chapter, the pedagogical content knowledge about comparing length indirectly will be described. The indirect comparison of length was chosen because many competencies in regard to a concept of length are faced when children are

J. Zöllner (✉) • C. Benz
University of Education Karlsruhe, Karlsruhe, Germany
e-mail: zoellner@ph-karlsruhe.de; benz@ph-karlsruhe.de

T. Meaney et al. (eds.), *Mathematics Education in the Early Years*,
DOI 10.1007/978-3-319-23935-4_20

comparing indirectly. In order to illustrate the competencies needed for an indirect comparison, a meta-analysis of empirical and theoretical studies is discussed to concretised activities in the children's play and everyday life situations. After analysing and presenting pedagogical content knowledge concerning comparing and measuring length, implications for the daily life in preschool settings are discussed.

Components of an Indirect Comparison

Indirect comparison was chosen in order to illustrate pedagogical knowledge because it requires many partial competencies, which are linked with comparing and measuring length. These partial competencies are also necessary in qualitative comparisons as well as in quantitative measuring. Many authors have described partial competencies of a concept of length (like understanding of the attribute, additivity, concept of units, counting the units, knowing the standardised units, origin, measuring competencies with standardised tools, estimating, etc.) (e.g. Battista 2006; Boulton-Lewis et al. 1996; Carpenter 1971; Carpenter and Lewis 1976; Clements and Sarama 2009; Lehrer 2003; McDonough and Sullivan 2011; Nührenbörger 2002; Piaget et al. 1974; Schmidt and Weiser 1986). Lack of space means that they cannot all be described in detail here. Through indirect comparison, the network of some partial components and the relationship between these individual competencies can be described.

The different options children have to compare length indirectly will be discussed initially. Then, four partial competencies which are necessary to compare indirectly are described in detail: comparing directly, using tools on the basis of conservation, conducting unit iteration and using the idea of unit and proportionality. These four competencies are chosen, because they show how partial competencies of the concept of length do not develop in levels in a linear manner. Rather, they show it is a developing net, in which many competencies determine each other or are interdependent. In describing these four competencies, further connected competencies will be referred to in order to reveal the netlike structure for describing the concept of length.

Different Options to Compare Indirectly

The different options for comparing indirectly demand plenty of competencies. For example, when children are building streets with building blocks, they may want to compare which of two streets is longer. To do this, they need to compare two different lengths, which cannot be placed next to each other because they are in different positions (or built at different times); consequently, they have to perform an indirect comparison (Franke and Ruwisch 2010).

The first possibility to compare the length of the streets is to perceive the streets holistically and to relate them to each other (Battista 2006; Boulton-Lewis et al. 1996). When preschool children perceive and compare holistically, they often do not use any reasoning for judging the relation between the lengths: "Holistic visual recognition does not require reasoning" (Heuvel-Panhuizen and Elia 2011, p. 631). The children simply *see* their solution to the task of comparing and thus do not question their solution.

The children can also estimate, for which they need much knowledge and experience of length. Children of preschool age may not have the necessary experiences to have developed these competencies; therefore, this solution process is not discussed further.

If children use a medium for the comparison, there are different options for comparing indirectly. By analysing the different options the children use, preschool teachers could "spy with their little eye" the different competencies in the children's activities. For example, the children could use different objects as a medium for measuring: a stick, a bar, a rod, their own body parts, a measuring tape, a folding yardstick or a metre stripe. Any object could be used as medium for measuring; however, objects which have a clear linear shape seem to be more suitable. Depending on which medium the children choose, different approaches to accomplishing the indirect comparison between the lengths of two streets, described in the following as a and b, unfold:

1. If a child chooses a medium (m) that is shorter than one of the streets (a) but longer than the second street (b), a can be compared with m: $a>m$. Then, m will be compared with b: $m>b$ (see Fig. 1). In this case, a transitive conclusion can be made: When $a>m$ and $m>b$, then $a>b$.
2. However, if a child chooses a medium that is longer than both of the two streets, the length of the one street (a) could be marked on the medium (m). The child could also memorise one point of the scale, if it chooses a standardised measuring tool. Alternatively, the child could mark the end points with fingers and thus

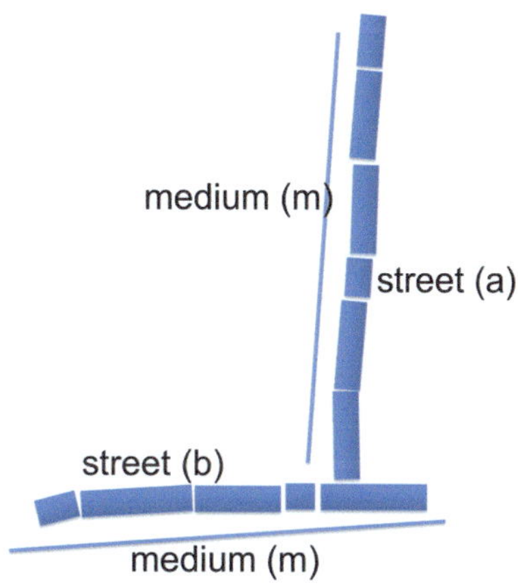

Fig. 1 One possibility to build the streets and to compare them with the same medium (m)

representing the length of street (a) on the medium (m). The distance between the starting point and the end point could be described as m'. Consequently, $a=m'$. When the child places the medium next to the second street (b), the distance m' is compared with the length of b. With this comparison, the child could draw a conclusion that if $a=m'$ and $m'<b$, then $a<b$.

Another possibility emerges; if $a=m'$ and $m'=b$, then, with the help of the transitivity of the equivalence relation, it can be concluded that $a=b$.

3. If a child chooses a medium (m) that is so short that it has to be used repeatedly on the two streets (a and b), the relation of length between the two streets will be determined by the comparison of the numbers of repetitions. For example, the medium (m) could be put three times on a, and the same medium (m) could be put five times on b, leading to the conclusion that $a=3\times m$ and $b=5\times m$. From this, it can be concluded that $b=a+2m$ and consequently $b>a$. A similar procedure is possible when instead of one medium several media with the same length are used. The procedure described here, both with one and with several media, corresponds to *unit iteration* (see below).
4. The chosen medium could also be shorter than the two streets, but may only be placed once completely along the streets. In this case, the child has to put the medium one time on a and one time on b and has to regard the rest of the street a' and b' separately. It is the relationship between a' and b' which gives some information about the relation between the lengths of the two streets. The child has to compare the "leftover lengths" of the streets, the lengths a' and b' with the help of m. The relation of the lengths a and b (the whole streets) leads to the following assumption: $a=m+a'$ and $b=m+b'$; this means the length of the medium and length of the rest of the street are equal to the whole length of the street. If the rest of the street a is shorter than the rest of street b ($a'<b'$), then also the whole of the street a is shorter than street b ($a<b$), because $m+a'<m+b'$. This procedure is also necessary in case 3, if the difference between the lengths after the repeated usage of the medium is shorter than m.
5. If there are several media with different lengths available (m, m_1, m_2, m_3), which are in each case shorter than a, the children could put these (without gaps and without overlapping) on street a, so that their sum equals the length of a ($a=m+m_1+m_2+m_3$). Now, the same media (m, m_1, m_2, m_3) could be put on street b. Because of the relation of the sum of the media, the relation between a and b can be inferred.

All these procedures can be conducted both with standardised measuring instruments and with non-standardised media. If the children use standardised tools like a ruler or a measuring tape, they could use the property of such tools and match the length of the street directly to a measuring number, for example, 3 m. If the children also measure the length of the second street, they can compare the numbers with each other. In this case, the concrete procedure is dependent on the length of the standardised medium as described in Sections 1–5. However, it is also possible that the children choose a standardised measuring instrument, but use it as a non-standardised medium (e.g. like a stick) (Zöllner and Benz 2013).

Which approach the children use is dependent on the available materials and on the idea or the knowledge the children have about comparing indirectly. Certainly, it should be a long-term goal in dealing with length that each one of these possibilities could be a choice for children to use.

Different Competencies Which Are Used by Comparing Indirectly

In describing the different possibilities on how to accomplish an indirect comparison, it is clear that a lot of competencies are necessary for the different options. Several of these partial competencies are discussed in more detail in the following section. For a preschool teacher, who observes the children when comparing indirectly and who wants to supportively and promotionally take part in the play, it is helpful to be aware of these partial competencies and their meaning in developing children's concept of length. To do this, the teacher needs to give suitable prompts or provide adequate materials to challenge children in that moment.

Comparing Directly

Using the example above, the children want to compare the lengths of two streets with each other. Although the children cannot do a direct comparison, they still need the competence to do a direct comparison, because in each of the possible procedures, the children use a direct comparison at least once. For example, using the possibility, which is described in the first option ($a<m$; $b>m$), the children could choose to use a rope for their indirect comparison. They would initially compare the length of the rope with the first street. The goal of the direct comparison is a qualitative statement such as "one of the streets is longer than the rope" or "the rope and the street are of the same length". In order to come to a valid statement, the child needs to place the rope on or alongside the street in such a way that the rope has the same shape as the street. The child could now—because of the position of the end points of the street and the rope—notice (visual or tactile-kinaesthetic) the relation of length between the street and the rope (Bright 1976 cited in Nührenbörger 2002). The concrete application of the direct comparison is dependent on the attributes of the medium and the objects that are to be compared. If the child wants to compare solid objects that are consistent in their length, for example, sticks, bars or boxes, the child can "simply" place them next to each other. Depending on the position of the end points, the child can decide which of the objects is longer. However, if the objects are flexible or differently shaped (such as rope, flexible paper stripes, telescopes, range of jars), the child has to determine that the objects that should be compared are in the same shape.

Comparing directly demands several competencies:

- Awareness that only the linear property of length of the objects is important: All other properties of the objects that are to be compared have to be ignored, only the length counts in the comparison. Depending on the concrete problem, it may be important to distinguish between the concepts: height, length, width, depth and so on.
- Additivity and partitioning: The children have to realise when comparing directly that the longer object could be disjointed in two parts: one part that is as long as the first object and the other part equalling the overlapping piece (Griesel 1996; Nührenbörger 2002).
- Knowledge and usage of suitable adjectives of length: In order to describe the result of the direct comparison, the children have to understand the appropriate adjectives of length (such as longer, shorter, higher, etc.) and be able to use them.
- Conservation: The child should be aware that the length of the objects does not change by changing their shape (see below).

Studies reveal that children in the preschool age are in general—in spite of all these required competencies—able to do a direct comparison (Heuvel-Panhuizen and Elia 2011). Results of infant research lead to the assumption that children have inherently a "compare scheme", in regard to both discrete and continuous quantities. They are disposed to recognise imprecise differences between continuous quantities (Krajewski 2013). Thus, direct comparison and conservation mutually develop each other. Insight into conservation is only possible if children are able to do a direct comparison. Contrariwise, the children have to realise that the displacement of an object that is to be compared does not lead to a change in length.

Using Tools on the Basis of Conservation

Conservation is the knowledge that a quantity does not change, if nothing is taken away or added to it (Lefrancois 1994). The work of Piaget placed conservation as of central importance for general cognitive development as well as for the development of the concept of length: "to measure a length means to first displace another, whose sustainment is secured during the displacement" (Piaget et al. 1974, p. 93). The displacement is a movement, and this is "a congruent transformation of the figures of the space, i.e. the length AB of a displaced object stays AB" (p. 119).

The majority of recent descriptions of the concept of length of children also include conservation (Clements and Stephan 2004; Cross et al. 2009; Nührenbörger 2002; Battista 2006). However, there is disagreement over Piaget's idea that the conservation of length does not develop until the age of 7 or 8 years[1].

[1] In this chapter, Piaget's research method and results are not critically evaluated. For a detailed discussion of his work, see, for example, Elkind (1967).

Fig. 2 Conservation in the equivalence format

The understanding of conservation was investigated in more detail in several follow-up studies. One aspect is especially relevant for the description of the concept of length. Elkind (1967) and Acredolo (1982) differentiate between the operationalisation of the *conservation in the equivalence format* and the *conservation in the identity format.*

Conservation in the equivalence format describes the idea of conservation, using the work of Piaget. For example, two sticks are compared concerning their length, and their lengths are determined to be equal. Then, one of the sticks is moved. Now, the children are asked to determine again whether their lengths are equal. Here, the equivalence of the length of the two sticks is in the focus (see Fig. 2).

In contrast to this, the *conservation in the identity format* only uses one single stick. The position of this stick is altered, and the children discover that the length of the stick does not change when it is moved.

Elkind (1967) discovers that children are often led by their perception when solving tasks concerning the equivalence format. They possibly realise the conservation on a cognitive level, but their perception leads them to the assumption that one of the sticks is longer. According to Elkind, the children realise the conservation in the identity format much earlier. This is due to the fact that the perception does not mislead them. In regard to this, Hiebert (1981) states that children use a medium for indirect comparisons, moving it, for example, from one object to another, without thinking that it might have changed its length, suggesting that they have an intuitive understanding of the conservation in the identity format.

When comparing the length of two streets, Hiebert seems correct as hardly any child could contemplate possible changes in the length, when the medium is moved. Children usually implicitly assume that the length of the used objects does not change through displacement. In this respect, one can agree with Piaget that conservation is a basic insight in the beginning of the development of a concept of length; however, it is the conservation in the identity format. Tasks concerning the conservation in the equivalence format are also important as children need to realise that in some situations perception alone is not sufficient for problem solving. Rather, it is an interplay between perception and cognitive processes which leads to more effective solutions.

Being aware of these differences and finesses, the preschool teacher can guide the children's attention to the ideas about conservation in a suitable moment. In this way, the children gradually become aware that the length of an object does not change if it is moved. In difficult situations, when the perception of the children would lead to other assumptions, the teachers can make use of the knowledge that the children have already acquired in other concrete actions.

Conducting Unit Iteration

Unit iteration is a central partial competence for the concept of length. This can be found in many descriptions of length (Clements and Sarama 2009; Lehrer 2003; Barrett et al. 2003; Kamii 2006). If children choose one medium or several media of the same length (e.g. matches), which are shorter than the two streets, a unit iteration is needed. The children either put one match repeatedly next to the two streets or they use several matches and put them one after another next to the two streets. The length of one single match becomes the unit. The length of the streets is divided through this unit into several equal parts. When all these parts of the length are counted, the number of units constitutes the measuring result. Therefore, unit iteration can be seen as a quantitative measuring process: "When we measure we associate the length of an object with a number" (Barrett et al. 2003, p. 19).

Comparing the length of the two streets *a* and *b* does not have to be compulsory result in a counting comparison. There may be a one-to-one correspondence between the partial lengths of *a* and *b*. Then, the unit iteration could lead to a qualitative comparison of the kind "*a* is longer than *b*".

Unit iteration integrates several partial competencies:

- Conservation: The children must be aware that the length of a match does not change when it is moved.
- Partitioning/additivity: The two streets are divided into several parts of the same length. The total length is equivalent to the sum of the partial lengths.
- Understanding the units and proportionality (see below).
- Placing matches end to end, without gaps or overlapping: The comparison is only valid if there is no gap between the single units and no overlapping.
- Comparison of the number of the equally long partial lengths for both streets.

Carpenter (1971) discovered in research with year one pupils that they successfully used unit iteration. Another study (Zöllner in preparation) revealed that preschool children also are able to do this. However, they use the unit iteration only if there is no other medium available for them (especially no longer standardised medium). Hiebert (1984) also states that preschool children do use unit iteration. However, he describes that they may have difficulties placing the units without gaps or overlapping on or next to the object.

However, it may also be that the children have no understanding of units yet. This is the next competence that is described.

Using the Idea of Unit and Proportionality

Understanding of units and proportionality plays a central role, especially in unit iteration. It is also important for an understanding of the measuring process with standardised measuring tools. It should be mentioned here that a measuring process can also be carried out by children although they do not have a differentiated idea of measuring (Zöllner and Benz 2013).

Measuring includes the choice of a suitable unit, decomposing the length in equal parts (procedural activity: unit iteration) and the counting of the single units (Benz et al. 2015). Some children who use iteration do not pay attention to the need for equal partial lengths (units), especially in their initial encounters with measuring. A common example is measuring distances with the help of steps where children may not pay attention to the need for steps to be of the same length. A reliable comparison with unit iteration can only emerge, if the comparative length is subdivided into equally long partial lengths.

If the units are counted, the unit iteration corresponds de facto to a measuring process. The statement of a measuring result, for example, "the street is as long as 15 matches" or "the street is 2 m long", contains for this reason a measuring number and a unit. The measuring number is dependent on the measuring unit. The understanding of this fact is called proportionality (Mitchell 2011; Lehrer 2003). The study of Schmidt and Weiser (1986) reveals that children who compare two units of length concentrate on the measuring number and are making correct judgements if the measuring unit is the same. They can correctly answer the question: "What is longer—5 smurf metres or 3 smurf metres?" In so doing, they perform a *transfer within* (Osborne 1976, p. 19) through recognising the numbers as the length. Carpenter and Lewis (1976) describe that first- and second-grade children perform this transfer within, even when the same unit is not used. For half of the year one and year two pupils in their study, the object is longer which has more units to be counted or otherwise where the measuring number is higher, even if the units were different (e.g. 3 m is shorter than 5 cm, because $3<5$). However, if changing the setting, again more children realise the proportionality: Carpenter and Lewis (1976) show two equally long stripes to children. On one of the stripes, four equally long paper stripes are placed. Then the children are asked how many of these shorter paper stripes would be necessary to fill the second stripe. All the results, where the children answered more than four, were counted correctly. In this setting, Carpenter and Lewis (1976) observe that 35 of 51 children could solve the task correctly and that they referred in their explanations to the length.

In comparing two measurement amounts (e.g. 3 m and 5 cm), two ideas are possibly competing with each other, similar to the conservation in the equivalence format. On the one hand, children have already had the experience that $3<5$ and so apply this knowledge to the length without regarding the units. On the other hand, the children become aware that one needs less units to put on a length if these units are longer. For children in preschool age or in year one or year two, this knowledge seems not to be tested and consolidated in many cases compared to the awareness that $3<5$:

> Children develop the notion of the inverse relationship before they realize that equal quantities are still equal even though they have measured a different number of units. Thus, it appears that children do not develop the notion of the inverse relationship between unit size and number of units through experience measuring with different sized unit. (Carpenter and Lewis 1976, p. 57)

For a comprehensive use of units and proportionality, several competencies are needed, which are described earlier: conservation, direct comparison, additivity and partitioning.

Each partial competence of the concept of length is connected with the other competencies through many interrelations. Nevertheless, there are basic competencies that serve as a condition for the further development, for example, the ability to do a direct comparison of the length of objects. Some more competencies are described briefly below.

Meaning for the Everyday Life of Preschool Teachers

In this section, important competencies needed for comparing indirectly are discussed in regard to how preschool teachers can make use of them. Benz (2012) highlights the need for preschool teachers to identify meaningful mathematical situations and to use them to further develop children's mathematical competencies. The usefulness of situations about measuring and comparing can be "seen" more easily by preschool teachers, if they have the necessary and differentiated knowledge about the individual competencies and how they are connected to the development of the concept of length. For example, Bush (2009) cites a teacher who expressed her inability to support the children in their development of their concept of length:

> I need to teach length to my students but I don't really know where to start. I don't really know what's important for them to understand and what I should focus my teaching on. (Bush 2009, p. 29)

When preschool teachers want to identify and support mathematical competencies concerning measuring and comparing in play situations, the importance of pedagogical content knowledge about the concept of length becomes apparent. For example, if children want to make a "street map" of their streets that are built with buildings blocks, they could use two cars in order to determine the width of the street. Here, the preschool teacher should be aware that supporting children to use unit iteration could be unhelpful. An intervention of the preschool teacher with the hint that there should be no gap between the cars would be useless, because every street is in reality clearly wider than two cars, because they have to pass without touching each other. However, if the children want to compare the length of their streets and if they are not able to do this with a direct comparison, it is absolutely reasonable to use the cars as an arbitrary medium and to put them without gaps next to each other and then to compare the number of the cars that are used. A discussion about the choice of the cars, what would happen if we would only use lorries for the comparison and so on, could put more focus on the use of units of the same length and on proportionality.

In dealing with length, other competencies are also acquired that are important in other areas. For example, conservation in both forms also plays an important role in arithmetic. The number of a quantity does not change, if nothing is added or taken away. The part-whole relationship of numbers is given a central role for a development of mathematical understanding. Dornheim (2008) states that the missing of an understanding of the part-whole relationship indicates arithmetic disability. These ideas can be initiated through the additivity or respectively the partitioning, which play both an important role in direct comparison and unit iteration.

Because the competences are connected like a net, in daily situations often many different competences are used. Therefore, it is necessary that preschool teachers know the competences so that they are able to recognise and foster these competencies. Furthermore, the communication with other experts but also with the parents about the experiences and activities of the children and the consequences for the mathematical development is facilitated.

References

Acredolo, C. (1982). Conservation—nonconservation: Alternative explanations. In C. J. Brainerd (Ed.), *Children's logical and mathematical cognition* (pp. 1–31). New York: Springer.

Barrett, J. E., Jones, G., Thornton, C., & Dickson, S. (2003). Understanding children's developing strategies and concepts for length. In D. H. Clements & G. Bright (Eds.), *Learning and teaching measurement* (pp. 17–30). Reston, VA: National Council of Teachers of Mathematics. 2003 Yearbook.

Battista, M. T. (2006). Understanding the development of students' thinking about length. *Teaching Children Mathematics, 13*(3), 140–146.

Benz, C. (2012, July 8–15). *Learning to see—Representation, perception and judgement of quantities of numbers as a task for in-service teacher education in preschool education.* Paper presented at the 12th international congress on mathematical education, Seoul, Korea.

Benz, C., Peter-Koop, A., & Grüßing, M. (2015). *Frühe mathematische Bildung. Mathematiklernen der Drei-bis Achtjährigen [Early mathematical education].* Heidelberg: Springer.

Boulton-Lewis, G. M., Wilss, L. A., & Mutch, S. L. (1996). An analysis of young children's strategies and use of devices for length measurement. *Journal of Mathematical Behaviour, 15*, 329–347.

Bush, H. (2009). Assessing children's understanding of length measurement: A focus on three key concepts. *Australian Primary Mathematics Classroom, 14*(4), 29–32.

Carpenter, T. P. (1971). The role of equivalence and order relations in the development and coordination of the concepts of unit size and number in selected conservation type measurement problems. *Report from the Project on Analysis of Mathematics Instruction* (Tech. Rep. No. 178). Madison, WI: University of Wisconsin.

Carpenter, T. P., & Lewis, R. (1976). The development of the concept of a standard unit of measure in young children. *Journal for Research in Mathematics Education, 7*(1), 53–58.

Clements, D. H., & Sarama, J. (2009). *Learning and teaching early math. The learning trajectories approach.* New York: Routledge.

Clements, D. H., & Stephan, M. (2004). Measurement in pre-k to grade 2 mathematics. In H. D. Clements, J. Sarama, & A. DiBiase (Eds.), *Engaging young children in mathematics. Standards for early childhood mathematics education* (pp. 299–320). Mahwah, NJ: Erlbaum.

Cross, C. T., Woods, T. A., & Schweingruber, H. (2009). *Mathematics learning in early childhood: Paths toward excellence an equity.* Washington, DC: National Academy of Sciences.

Dornheim, D. (2008). *Prädiktion von Rechenleistung und Rechenschwäche. Der Beitrag von Zahlen-Vorwissen und allgemein-kognitiven Fähigkeiten [Prediction of arithmetic ability and dyscalculia. The contribution of numerical knowledge and general cognitive abilities].* Berlin: Logos.

Elkind, D. (1967). Piaget's conservation problems. *Child development, 38*, 15–27.

Franke, M., & Ruwisch, S. (2010). *Didaktik des Sachrechnens in der Grundschule [Didactics of practical arithmetic in primary school]* (2nd ed.). Heidelberg: Spektrum.

Gasteiger, H. (2010). *Elementare mathematische Bildung im Alltag der Kindertagesstätte. Grundlegung und Evaluation eines kompetenzorientierten Förderansatzes [Elementary mathematics education in everyday life in kindergarten].* Münster: Waxmann.

Ginsburg, H. P., Lee, J. S., & Boyd, J. S. (2008). Mathematics education for young children: What it is and how to promote it. *Social Policy Report, 22*(1), 3–22.
Griesel, H. (1996). Grundvorstellungen zu Größen [Basic ideas for measurement]. *Mathematik lehren, 78*, 15–19.
Heuvel-Panhuizen, M. van den, & Elia, I. (2011). Kindergartners' performance in length measurement and the effect of picture book reading. *ZDM: The international Journal on Mathematics Education, 43*(5), 621–635.
Hiebert, J. (1981). Results and implications from national assessment. *Arithmetic Teacher, 26*(6), 38–43.
Hiebert, J. (1984). Why do some children have trouble learning measurement concepts? *Arithmetic Teacher, 31*(7), 19–24.
Kamii, C. (2006). Measurement of length: How can we teach it better? *Teaching Children Mathematics, 13*(3), 154–158.
Krajewski, K. (2013). Wie bekommen Zahlen einen Sinn? Ein entwicklungspsychologisches Modell zur zunehmenden Verknüpfung von Zahlen und Größen [How do numbers get a meaning? A developmental psychological model concerning the relationship of numbers and measurement]. In M. Aster & J. H. Lorenz (Eds.), *Rechenstörungen bei Kindern. Neurowissenschaft, Psychologie, Pädagogik [Arithmetic disability of children, neuroscience, psychology and pedagogy]* (2nd ed., pp. 155–180). Göttingen: Vandenhoeck & Ruprecht.
Lefrancois, G. R. (1994). *Psychologie des Lernens [Psychology of learning]*. Berlin: Springer.
Lehrer, R. (2003). Developing understanding of measurement. In J. Kilpatrick, G. W. Martin, & D. Schifter (Eds.), *A research companion to principles and standards for school mathematics* (pp. 179–192). Reston, VA: National Council of Teachers of Mathematics.
Mitchell, A. E. (2011). *Interpreting students' explanations of fraction tasks, and their connections to length and area knowledge* (doctoral dissertation, Australian Catholic University).
McDonough, A., & Sullivan, P. (2011). Learning to measure length in the first three years of school. *Australian Journal of Early Childhood, 36*(3), 27–35.
Nührenbörger, M. (2002). *Denk- und Lernwege von Kindern beim Messen von Längen. Theoretische Grundlegung und Fallstudien kindlicher Längenkonzepte im Laufe des 2. Schuljahres [Children's ways to think and to learn in situations of measurement with length]*. Hildesheim: Franzbecker.
Osborne, A. R. (1976). The mathematical and psychological foundations of measure. In R. A. Bradbard & D. A. Lesh (Eds.), *Number and measurement. Papers from a research workshop* (pp. 19–46). Columbus, OH: ERIC Clearinghouse of Science, Mathematics and Environmental Education.
Piaget, J., Inhelder, B., & Szeminska A. (1974). *Die natürliche Geometrie des Kindes [The child's conception of geometry]* (R. Heipke Trans.). Stuttgart: Klett (Original work published 1948).
Schmidt, S., & Weiser, W. (1986). Zum Maßzahlverständnis von Schulanfängern [About the knowledge of measure of children starting school]. *Journal für Mathematikdidaktik, 7*(2/3), 121–154.
Zöllner, J. (in preparation). *Das Längenkonzept von Kindergartenkindern [The concept of length of young children]*.
Zöllner, J., & Benz, C. (2013, February 6). *How four to six year old children compare length indirectly*. Paper presented at the eight congress of European research in mathematics education (Cerme8), Antalya, Turkey. Online: http://cerme8.metu.edu.tr/wgpapers/WG13/WG13_Zollner.pdf

The Role of Conceptual Subitising in the Development of Foundational Number Sense

Judy Sayers, Paul Andrews, and Lisa Björklund Boistrup

Abstract Evidence indicates that children with a well-developed number sense are more likely to experience long-term mathematical success than children without. However, number sense has remained an elusive construct. In this chapter, we summarise the development of an eight-dimensional framework categorising what we have come to call foundational number sense or those non-innate number-related competences typically taught during the first years of schooling. We also show, drawing on grade one lessons from Hungary and Sweden, how focused instruction on conceptual subitising, the teaching of children to identify and use easily recognisable groups of objects to structure children's understanding of number, facilitates children's acquisition of a range of foundational number sense-related competences.

Introduction

Over the last 15 years since the publication of Clements' (1999) seminal paper, various scholars, particularly in the USA, have been encouraging teachers to attend to the development of young learners' conceptual subitising (e.g. Clements and Sarama 2009; Conderman et al. 2014). Conceptual subitising is the ability to recognise quickly and without counting relatively large numerosities by partitioning these large groups into smaller groups that can be individually subitised (Clements and Sarama 2007; Geary 2011). Various claims, discussed below, have been made with respect to the efficacy of conceptual subitising-focused instruction. In a related vein, our own recent work has focused on a conceptualisation of foundational number sense (FoNS), which we describe as those number-related competences expected of a typical first-grade student that require instruction (Back et al. 2014). FoNS is characterised by eight components, described below. The purpose of this chapter, drawing on excerpts from grade one lessons taught by a case study teacher in each of Hungary and Sweden, is to examine the extent to which conceptual subitising-focused

J. Sayers (✉) • P. Andrews • L. Björklund Boistrup
Stockholm University, Stockholm, Sweden
e-mail: judy.sayers@mnd.su.se

T. Meaney et al. (eds.), *Mathematics Education in the Early Years*,
DOI 10.1007/978-3-319-23935-4_21

activities have the propensity to facilitate students' acquisition of the various FoNS components and, in so doing, examine the warrant for their claimed efficacy.

What Is Subitising?

Subitising refers to being instantly and automatically able to recognise small numerosities without having to count (Clements 1999; Jung et al. 2013; Moeller et al. 2009; Clements and Sarama 2009). Children as young as three are typically able to subitise numerosities up to three (Fuson 1988; Moeller et al. 2009), while most adults are able instantly to recognise without counting the numerosity represented by the dots on the face of a die (Jung et al. 2013). This process, innate to all humans, is typically known as perceptual subitising (Gelman and Tucker 1975) and forms an element of the preverbal number sense described below. In short, perceptual subitising is recognising a numerosity without using other mathematical processes (Clements 1999).

Conceptual Subitising

However, a second form of subitising, conceptual subitising (Clements 1999), which is not unrelated to FoNS, has been shown to have considerable implications for teaching and learning. Conceptual subitising relates to how an individual identifies 'a whole quantity as the result of recognizing smaller quantities… that make up the whole' (Conderman et al. 2014, p. 29). More generally, it can be summarised as the systematic management of perceptually subitised numerosities to facilitate the management of larger numerosities (Obersteiner et al. 2013). For example, when a child is confronted by two dice, one showing three and another showing four, each is perceptually subitised before any sense of seven can emerge.

Subitising can be construed as having a synonymity with the spatial structuring of amounts (Battista et al. 1998). In this case, the ability to recognise and manipulate numbers spatially, through the use of, for example, dice, dominoes and ten frames, plays a significant role in the development of children's understanding of both number and arithmetic (Hunting 2003; Mulligan and Mitchelmore 2009; Van Nes and De Lange 2007; Van Nes and Van Eerde 2010). Indeed, research has shown that conceptual subitising can be taught through mathematical tasks that provide structured images of numbers (Clements 2007; Mulligan et al. 2006), including fingers to represent small numbers (Penner-Wilger et al. 2007).

In addition to being a powerful tool in the development of children's general understanding of numbers (Jung 2011; Penner-Wilger et al. 2007), conceptual subitising has been linked positively to a variety of particular learning outcomes such as counting and counting speed (Benoit et al. 2004) and an understanding of cardinality (Baroody 2004; Butterworth 2005; Jung 2011). Conceptual subitising underpins children's understanding of the equivalence of different decompositions or partitions of numbers (Hunting 2003; Van Nes and De Lange 2007), commutativity of addition (Cowan and Renton 1996) and the part-whole knowledge (Jung et al. 2013; Young-Loveridge 2002) necessary for understanding that 8+6=14 because 5+5=10, 3+1=4 and 10+4=14 (Van Nes and Doorman 2011).

Importantly, poor performance on both perceptual subitising (Landerl et al. 2004) and conceptual subitising (Mulligan et al. 2006) may be linked to later mathematical difficulties. In particular, they may be handicapped in their learning of arithmetic (Clements 1999).

What Is Number Sense?

Number sense has a 'traditional emphasis in early childhood classrooms' (Casey et al. 2004, p. 169) and is a key component of many early years' mathematics curricula (Howell and Kemp 2005; Yang and Li 2008). However, it has, for many years, remained definitionally elusive (Gersten et al. 2005). As Griffin (2004) noted:

> What is number sense? We all know number sense when we see it but, if asked to define what it is and what it consists of, most of us, including the teachers among us, would have a much more difficult time. Yet this is precisely what we need to know to teach number sense effectively. (p. 173)

Three Conceptions of Number Sense

Our constant comparison analysis (Strauss and Corbin 1998) of the literature, a process which has been described extensively in Andrews and Sayers (2015), has identified three distinct conceptions. The first, an innate or preverbal number sense (Butterworth 2005; Ivrendi 2011; Lipton and Spelke 2005), comprises an understanding of small quantities that allows for comparison. For example, children at 6 months can discriminate numerosities with a 1:2 ratio (Feigenson et al. 2004), while children at 4 can subitise the numerosity of sets containing up to five items (Gelman and Tucker 1975). These numerical discriminations are thought to underpin the acquisition of verbal counting skills (Gallistel and Gelman 2000) and arithmetic (Zur and Gelman 2004). This preverbal number sense develops in the early years as an innate consequence of human and other species' evolution and importantly is independent of formal instruction (Dehaene 2001; Feigenson et al. 2004).

Our second perspective relates to what we have labelled foundational number sense (FoNS). FoNS comprises those number-related understandings that require instruction and which typically occur during the first years of school (Ivrendi 2011; Jordan and Levine 2009). It is something 'that children acquire or attain, rather than simply possess' (Robinson et al. 2002, p. 85) and reflects, inter alia, elementary conceptions of number as a representation of quantity or a fixed point in the counting sequence (Griffin 2004).

We return to FoNS shortly, but first we summarise our third perspective, which we have labelled applied number sense. Applied number sense refers to those core number-related understandings that permeate all mathematical learning (Faulkner 2009; Faulkner and Cain 2013; National Council of Teachers of Mathematics 1989). Applied number sense refers to the 'basic number sense which is required by all adults regardless of their occupation and whose acquisition by all students should be a major goal of compulsory education' (McIntosh et al. 1992, p. 3).

Defining Foundational Number Sense

Over the last 2 years, we have been developing a simple-to-operationalise framework for analysing the FoNS-related opportunities that teachers provide for their students. To achieve these objectives, we exploited the constant comparison analysis advocated by grounded theorists. In brief, peer-reviewed research papers focused on grade one students' (typically 6–7-year-olds) acquisition of number-related competence were identified. These were read and FoNS-related categories identified. With each new category, previous articles were re-examined for evidence of it. This approach, drawing on literature from psychology, mathematics education, learning difficulties and generic education, placed *rote counting to five* and *rote counting to ten*, two narrow categories discussed by Howell and Kemp (2005), within the same broad category of systematic counting. In some respects, this remains a work in progress. In its first manifestation (Back et al. 2014), seven components were identified and evaluated against case study teaching in England and Hungary. Two teachers' lessons, focused explicitly on number sequence-related learning, showed that the framework operationalised six of the seven categories but indicated, also, differences in the ways in which the various components interacted when different excerpts were analysed. More recently, we have presented a stronger explanatory narrative for the eight-component FoNS framework (Andrews and Sayers 2015; Sayers and Andrews 2015) and provide, through an examination of number line-related exemplary teaching in Poland and Russia (Andrews et al. 2015), further evidence of the analytical power of the FoNS framework. These eight components, each summarised briefly, are:

Number Recognition

FoNS-aware children are able to recognise number symbols and know their associated vocabulary and meaning (Malofeeva et al. 2004). They can both identify a particular number symbol from a collection of number symbols and name a number when shown that symbol (Clarke and Shinn 2004; Gersten et al. 2005; Van de Rijt et al. 1999; Yang and Li 2008). Children who experience difficulty with number recognition experience later mathematical problems generally (Lembke and Foegen 2009) and particularly with subitising (Koontz and Berch 1996; Stock et al. 2010). Alternatively, children who recognise numbers are more able to manage multi-digit arithmetic than those who cannot (Desoete et al. 2012; Krajewski and Schneider 2009). Such skills are better predictors of later mathematics achievement than either general measures of intelligence or earlier achievement scores (Geary et al. 2009), with effects lasting as late as adolescence (Geary 2013).

Systematic Counting

FoNS-aware children can count systematically (Berch 2005; Clarke and Shinn 2004; Gersten et al. 2005; Griffin 2004; Van de Rijt et al. 1999) and understand ordinality (Ivrendi 2011; Jordan et al. 2006; LeFevre et al. 2006; Malofeeva et al. 2004; Van Luit and Schopman 2000). FoNS-aware children count to 20 and back or count upwards and backwards from an arbitrary starting point (Jordan and Levine 2009; Lipton and Spelke 2005), knowing that each number occupies a fixed position in the sequence of all numbers (Griffin et al. 1994). The skills of symbolic number ordering underpin later arithmetical competence in general (Gersten et al. 2005; Passolunghi et al. 2007; Stock et al. 2010) and mental arithmetical competence in particular (Lyons and Beilock 2011).

Awareness of the Relationship Between Number and Quantity

FoNS-aware children understand the relationship between number and quantity. In particular, they understand not only the one-to-one correspondence between a number's name and the quantity it represents but also that the last number in a count represents the total number of objects (Jordan and Levine 2009; Malofeeva et al. 2004; Van Luit and Schopman 2000). The correspondence between a number's name or symbol and the quantity represented is, essentially, a human invention and thus requires instruction if students are to understand it (Geary 2013). Children who have difficulty with this mapping process tend to experience later mathematical difficulties (Kroesbergen et al. 2009; Mazzocco et al. 2011).

Quantity Discrimination

FoNS-aware children understand magnitude and can compare different magnitudes (Clarke and Shinn 2004; Griffin 2004; Ivrendi 2011; Jordan et al. 2006; Jordan and Levine 2009; Yang and Li 2008). They deploy language like 'bigger than' or 'smaller than' (Gersten et al. 2005), understanding that eight represents a quantity that is bigger than six but smaller than ten (Baroody and Wilkins 1999; Lembke and Foegen 2009). Magnitude-aware children have moved beyond counting as 'a memorized list and a mechanical routine, without attaching any sense of numerical magnitudes to the words' (Lipton and Spelke 2005, p. 979). Moreover, magnitude awareness has been shown to be a predictor, independently of ability or age, of more general mathematical achievement (Aunio and Niemivirta 2010; De Smedt et al. 2009, 2013; Desoete et al. 2012; Holloway and Ansari 2009; Nan et al. 2006; Stock et al. 2010).

An Understanding of Different Representations of Number

FoNS-aware children understand that numbers can be represented differently (Ivrendi 2011; Jordan et al. 2007; Yang and Li 2008) and that these 'act as different points of reference' (Van Nes and Van Eerde 2010, p. 146). The better children understand a number line, for example, the higher their later arithmetical achievement (Siegler and Booth 2004; Booth and Siegler 2006, 2008). The better a child understands a partition as a representation of a number, the better developed is that child's later understanding of numerical structures (Thomas et al. 2002) and arithmetical skills (Hunting 2003). The more competent a child is with regard to the use of fingers in both counting and early arithmetic skills that can be taught effectively (Gracia-Bafalluy and Noël 2008), the more competent that child is in later years (Fayol et al. 1998; Jordan et al. 1992; Noël 2005). Significantly, the use of finger strategies increases as socio-economic status increases, justifying targeted interventions (Jordan et al. 1992; Levine et al. 1992). The use of manipulatives, particularly linking blocks, facilitates counting and the identification of errors (Van Nes and Van Eerde 2010). Thus, the better the connections between different representations, the more likely a child is to become arithmetically competent (Mundy and Gilmore 2009; Richardson 2004; Van Nes and De Lange 2007; Van Nes and Van Eerde 2010).

Estimation

FoNS-aware children are able to estimate, whether it be the size of a set (Berch 2005; Jordan et al. 2006, 2007; Kalchman et al. 2001; Malofeeva et al. 2004; Van de Rijt et al. 1999) or an object (Ivrendi 2011). Estimation involves moving between representations—sometimes the same, sometimes different—of number, for example, placing a number on an empty number line (Booth and Siegler 2006). However, the skills of estimation are dependent on the skills of a child to count (Lipton and Spelke 2005). Estimation is thought to be a key determinant of later arithmetical competence, particularly in respect of novel situations (Booth and Siegler 2008; Gersten et al. 2005; Holloway and Ansari 2009; Libertus et al. 2011; Siegler and Booth 2004).

Simple Arithmetic Competence

FoNS-aware children can perform simple arithmetical operations (Ivrendi 2011; Jordan and Levine 2009; Malofeeva et al. 2004; Yang and Li 2008), skills which underpin later arithmetical and mathematical fluency (Berch 2005; Dehaene 2001; Jordan et al. 2007). Indeed, simple arithmetical competence, which Jordan and Levine (2009) describe as the transformation of small sets through addition and subtraction, has been found to be, at grade one, a stronger predictor of later mathematical success than measures of general intelligence (Geary et al. 2009; Krajewski and Schneider 2009). However, drawing on their experiences of combining physical objects, children's ability to solve nonverbal problems develops before the ability to solve comparable word problems (Levine et al. 1992).

Awareness of Number Patterns

FoNS-aware children understand and recognise number patterns and, in particular, can identify a missing number (Berch 2005; Clarke and Shinn 2004; Gersten et al. 2005; Jordan et al. 2006, 2007). Such skills reinforce the skills of counting and facilitate later arithmetical operations (Van Luit and Schopman 2000). Importantly, failure to identify a missing number in a sequence is one of the strongest indicators of later mathematical difficulties (Chard et al. 2005; Clarke and Shinn 2004; Gersten et al. 2005; Lembke and Foegen 2009).

In summary, our systematic analysis of the literature identified eight distinct but not unrelated characteristics of FoNS. The fact that they are not unrelated is important because number sense:

> relies on many links among mathematical relationships, mathematical principles…, and mathematical procedures. The linkages serve as essential tools for helping students to think about mathematical problems and to develop higher order insights when working on mathematical problems. (Gersten et al. 2005, p. 297)

In other words, without the encouragement of such links, there is always the risk that children may be able to count competently but not know, for example, that four is bigger than two (Okamoto and Case 1996).

Implications of FoNS Related Learning in Conceptual Subitising Tasks

The quality of a child's FoNS has substantial implications. On the one hand, poorly developed number sense has been implicated in later mathematical failures (Jordan et al. 2009; Gersten et al. 2005), while on the other, research has shown that the better a child's number sense, the higher his or her later mathematical achievements both in the shorter term (Aubrey and Godfrey 2003; Aunio and Niemivirta 2010) and the longer term (Aubrey et al. 2006; Aunola et al. 2004). Moreover, without appropriate intervention, which research shows can be effective (Van Luit and Schopman 2000), children who start school with limited number sense are likely to remain low achievers throughout their schooling (Aubrey et al. 2006). Basic counting and enumeration skills are predictive of later arithmetical competence in England, Finland, Flanders, the USA, Canada and Taiwan, respectively (Aubrey and Godfrey 2003; Aunola et al. 2004; Desoete et al. 2009; Jordan et al. 2007; LeFevre et al. 2006; Yang and Li 2008), indicating a cross-culturally common phenomenon. In similar vein, the ability to identify missing numbers and discriminate between quantities are also predictors of later success (Chard et al. 2005; Clarke and Shinn 2004; Jordan et al. 2009), as is competence with number combinations (Geary et al. 2000, 2009; Locuniak and Jordan 2008). With respect to conceptual subitising, as indicated above, it is known that various approaches to the teaching of conceptual subitising can support the development of children's understanding of both number and arithmetic (Hunting 2003; Mulligan and Mitchelmore 2009; Van Nes and De Lange 2007). However, it is not known how such activities impact on other FoNS-related competences, and thus this is the focus of this chapter.

Methods and Results

Having derived the eight components, our aim was to evaluate their efficacy as a means of identifying FoNS-related opportunities provided by different teachers in different classroom and cultural contexts. Such an evaluative process should facilitate both refinement of the instrument and confirmation of its sensitivity to cultural nuances. The data came from grade one lessons that had been video recorded as part of other projects but which were made available to us for the purpose of FoNS-related analyses. The use of pre-existent data for secondary analyses is well established in social science research (Heaton 2008; Murphy 2014) and allows for new insights (Brewer 2012). In this case, the fact that data were not collected with FoNS-related analyses in mind means that teachers' actions were not constrained by prior knowledge of our interests. Importantly, both sets had been collected in similar ways—case study examinations of a grade one teacher's practice—and subjected to similar ethical considerations (Dale et al. 1988). For example, both teachers, construed locally (in their context) as effective, were video recorded in ways that would optimise capturing their actions and utterances, and both had been recorded over several lessons to minimise the likelihood of showpiece lessons.

From the perspective of analysis, each video, with transcripts, was repeatedly scrutinised for evidence of FoNS components by two of the three authors independently. These analyses were then compared and agreements reached with respect to which FoNS-related components were being encouraged at different times. A key principle underpinning the development of the framework was the desire to create a simple-to-operationalise tool that would allow for the episodes of a lesson to be multiply coded in order to support both simple analyses based on the frequencies of particular events and more sophisticated analyses based on the interactions of those events (see Andrews and Sayers 2015).

In the following, we examine how two such case study teachers, Klara from Hungary and Kerstin from Sweden, both pseudonyms, worked with their grade one classes (6–7-year-olds). Our focus is on the ways in which different FoNS components can be seen in excerpts we construe as being focused on the development of children's conceptual subitising. We do not offer any evaluative commentary, as our intention was solely to examine how activities focused on conceptual subitising yielded FoNS-related learning opportunities.

The Hungarian Excerpts

Klara's First Excerpt

In an early lesson, Klara was observed to use domino templates and counters. Working individually, children were asked to use the domino template to represent two numbers that added to six. She then collated responses and, with each suggestion, placed a prepared domino on the board. As she worked, she encouraged children to be clear in their descriptions by insisting on their using the terms left and right in relation to the domino they described.

The final arrangement can be seen in Fig. 1. Each domino in each equivalent pair was placed adjacent to the other, although this was not consistently managed, with Klara asking why the double three was lonely. This elicited the response that three on the left and three on the right is the same as three on the right and three on the left. That is, any physical transformation of the double-three domino would give the same domino.

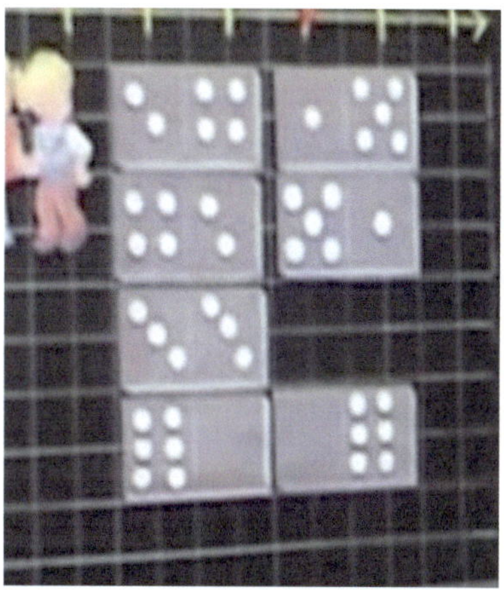

Fig. 1 Seven dominoes showing representations of six

Commentary on Klara's First Excerpt

In this task, the dice provided children with familiar arrays to support their subitising of integers up to six. In presenting it, Klara did not explicitly mention subitising. However, it is our view that the use of such arrays encourages children to see common subitised numerosities. Interestingly, in comparison with the similar task undertaken by Kerstin, the failure to list the dominoes in numerical order missed an opportunity for greater pattern work, which seemed a rare oversight on Klara's part from analysis of other lessons. From the FoNS perspective, particularly in the invitation to combine dots to make six, the activity had a focus on number patterns and simple addition. Moreover, the dominoes themselves provided evidence of different representations of number, and the act of linking numbers to the dots was evidence of Klara relating numbers to quantity.

Klara's Second Excerpt

The second excerpt began with Klara announcing that her class was to work on a task involving a bus trip with the possibility of ten children going. She revealed the picture shown in Fig. 2 and, pointing to the zero at the top, asked how many children

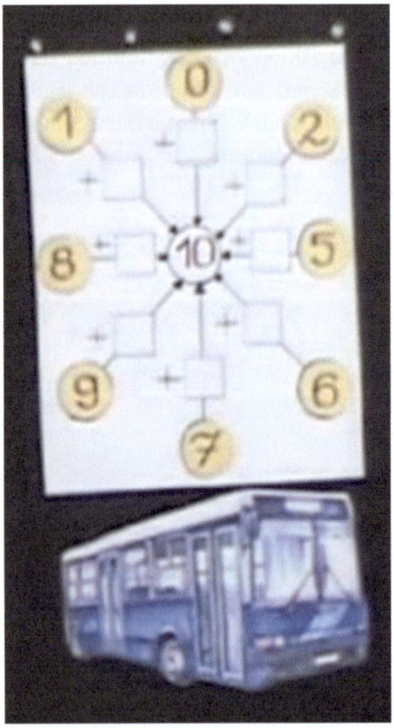

Fig. 2 Initial bus trip problems

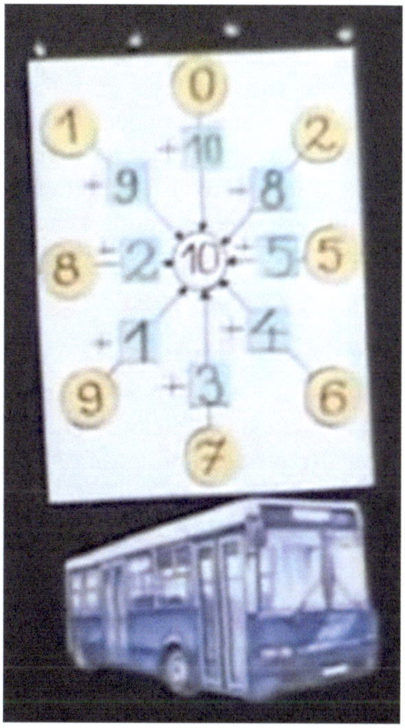

Fig. 3 Completed bus trip problems

would be able to join the trip if none were already on the bus. She received the answer of ten and placed the number 10, on a blue background, beneath the figure zero. After a few minutes, the picture, as shown in Fig. 3, was completed with eight different complements to ten having been added to the diagram. In so doing, she had encouraged her students to see every pair of complements with two pairs, $1+9$ and $9+1$ and $2+8$ and $8+2$, being repeated. Throughout the process, Klara encouraged her students to use their hands. In particular, she discussed how the fingers of two hands can be used to represent ten before demonstrating, with respect to two children already on the bus, that if both hands are held open and two fingers are closed—representing the two children on the bus—the remaining fingers represent the number of additional students allowed to travel. Thus, three fingers on one hand and five on the other, drawing on subitised numerosities, facilitate a structural awareness of eight as the sum of five and three.

Commentary on Klara's Second Excerpt

Klara did not make explicit the relationship between finger use and subitising, but we consider it to be there. The explicit act of, say, closing two fingers and then observing that three plus five fingers remain while not disallowing the possibility of counting as a strategy clearly encourages children to focus on the structural relations and the immediate recognition of subitised numerosities. With regard to FoNS, at least four categories were evident. The picture around which Klara structured the different tasks encouraged number recognition and the relationship of number to quantity. Students' use of fingers was an exploitation of different representations of number, while simple addition, possibly subtraction, was the explicit focus. There were also opportunities for those children who preferred to do so to exploit systematic counting as an additive strategy.

Klara's Third Excerpt

Klara revealed two pictures adjacent to each other, one showing eight girls and the other five boys, before posing the question, how many children would there be in total? Before moving to the solution, Klara, in response to students' answers to her questions, wrote underneath the pictures

$$8_{gy} + 5_{gy} = ?_{gy}.$$

In this instance, 'gy' was an agreed abbreviation of gyerekek (children). After this, she turned her attention to the eight girls, which she represented by placing eight red counters on a 'board', as shown in Fig. 4. Next, turning to the five boys, she asked, how many would be needed to complete the 'board'? She received an answer of two and added them to the board in blue. Next, asking how many boys remain, she was told three and completed the picture as in Fig. 5. Finally, she asked

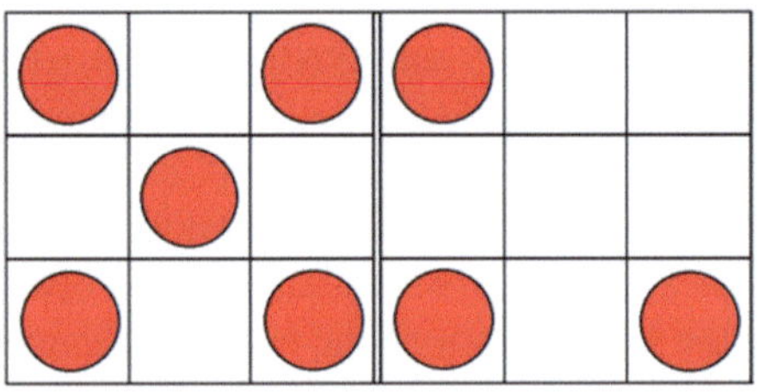

Fig. 4 Klara's placing of eight *red* counters on a 'board'

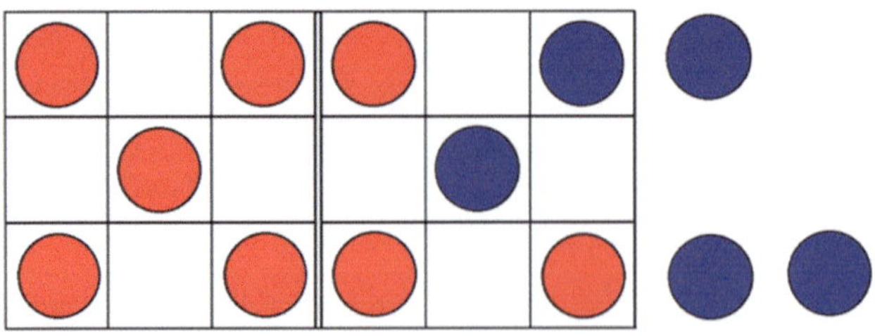

Fig. 5 Klara's complete representation of 8 + 5 = 13

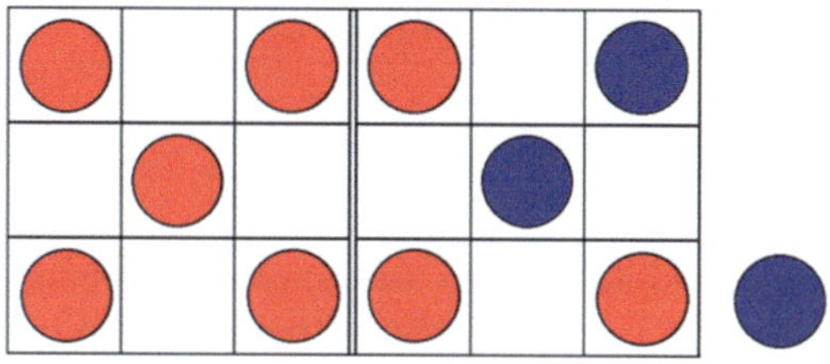

Fig. 6 The completed board for $8+3=11$

how many children were there in total and was told 13. With further probes, she elicited the result that the 13 was a result of ten plus three.

Following this, Klara revealed the next variation of the task, which, as shown in Fig. 6, was the 'board' for eight plus three. Klara's questions led to her class to agree that the 'board' showed a representation of $8+2+1$, which was $10+1$ or 11.

This discussion led to the following being written beneath the image

$$8+2+1=11$$

and

$$8+3=11$$

This process was repeated, exactly as above, for $8+4$ and $8+6$. In all three cases, the class chanted the process. For example, '8 plus 2 equals 10; 10 plus 1 equals 11'.

Commentary on Klara's Third Excerpt

The third excerpt was more focused on conceptual subitising than the earlier one. Klara's focus, in her use of the board, was the representation of ten as two fives. In so doing, she drew on familiar subitised numerosities. That is, her students were familiar with and able to recognise the properties of fiveness and were comfortable with ten as the juxtaposition of two fives. Thus, the addition of five to eight, drew, essentially, on familiar subitised numerosities of five or less. In so doing, she addressed several components of FoNS, not least of which was the explicit focus on simple addition. Her formulation beneath the two pictures encouraged number recognition. The use of the board and its associated counters provided another representation of number and an explicit link between number and quantity. It could also be argued that the use of counters, and the familiar representation of five, was an encouragement for students to explore arithmetical procedures as structural patterns.

The Swedish Excerpts

Kerstin's First Excerpt

Each child in the class was given a small bowl containing six small pebbles and a sheet of paper, laid landscape on the desktop. The paper was halved by means of a pen or pencil laid vertically down the centre of the sheet. Kerstin asked her students to take their six pebbles and, in any way they liked, place some on one side of the divide and the others on the other. The only rule is that all six pebbles must be used.

While they were doing this, Kerstin attached a metre rule to her whiteboard to create two distinct halves in the same way as her students. She wrote six at the top before inviting her students for different partitions of six. A child volunteered three and three. Kerstin placed, towards the vertical middle of her board, three discs to one side of her line and three to the other.

The process continued, and a second child suggested two and four. Kerstin placed these above the previous. A third child suggested five and one, and it now becomes clear that Kerstin was placing her counters in such a way that each ordered pair had a well-defined place on her board, with left zero, right six at the top coming down to left six, right zero at the bottom. The fourth child suggested six and zero, the next four and two, the next zero and six, before the final child offered one and five. Thus, seven sets of counters had been placed systematically on the board, with each pair, representing a partition of six, in a well-defined position. At this stage, Kerstin drew a horizontal line across the board to separate each pair to create the effect shown in Fig. 7.

Fig. 7 Kerstin's completed diagram

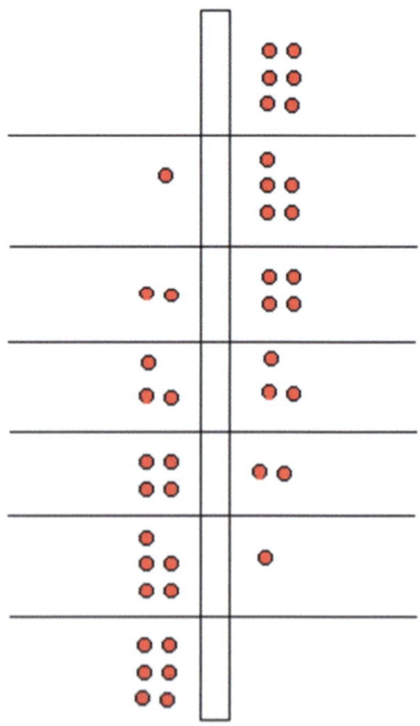

Kerstin asked her class how many ways they could make six in this way and received the answer, 'sju', seven.

Commentary on Kerstin's First Excerpt

We would argue that this activity clearly encouraged students to engage in conceptual subitising. The breaking down of six into different additive pairs, most of which were amenable to perceptual subitising, allowed students to instantly see six. From the perspective of FoNS, several categories were evident. Firstly, the use of pebbles served to remind children that numbers represent quantities. Secondly, the various partitions of six allowed children to see different representations of the same number. Thirdly, if only implicitly, the same act of partitioning encouraged children to engage with simple addition. Fourthly, the manner in which Kerstin arranged the solutions on the board highlighted two forms of number pattern. On the one hand, there was the clear distinction between the patterns formed by even and odd integers, although, of course, this could also be construed as another perspective on the representation of numbers. On the other hand, the sequencing of the solutions highlighted the fact that as one set of numbers decreases, the other increases. Fifthly, the arrangement of the partitions on the board could be construed as an encouragement for children to see numbers as having well-ordered places in the sequence of all numbers as part of a drive to facilitate their counting competence.

Kerstin's Second Excerpt

Kerstin invited the class to play a game in pairs. One child would take the six pebbles and, behind his or her back, distribute them between his and her two hands. He or she would then reveal one hand's contents, and the other child had to say what was in the closed hand. Then the pair would swap roles and repeat the process. Thus, many opportunities were given for children to rehearse the partitions of and complements to six. During this time, Kerstin circulated the room, asking student pairs questions like, if I have two in one hand, how many do I have in the other? At the end of this episode, Kerstin alerted her students to the symmetry of the arrangements on the board by pointing out the connections between four plus two and two plus four and the same for one and five and five and one. She also reminded her students, by moving counters from right to left, that each row of her table summed to six.

Commentary on Kerstin's Second Excerpt

As with her first activity, it seems that this task was focused on the development of conceptual subitising. When circulating the room, Kerstin's questions, focused on the complements to six, drew on children's mental representations of perceptually subitised numerosities like two and four. When summarising the task and its relationship to what was on the board, her explicit matching of, say, four with two and two with four further supported six as a conceptually subitised construct. With respect to FoNS-related opportunities, the game allowed children to consolidate the connection between number and quantity and, of course, cardinality as a component of systematic counting. It presented different representations of number and, implicitly, exploited simple arithmetic or counting to locate missing numbers. Also, in alerting her children to the symmetry of the relationship, she was encouraging an awareness of pattern, even if she was not exploring missing values.

Kerstin's Third Excerpt

At the start of this third excerpt, Kerstin distributed a worksheet (Fig. 8) to each child. On the worksheet were eight drawings of pairs of hands. One hand was open and showed some pebbles, and the other hand was closed to represent the hidden pebbles. As can be seen, the pictures alternated with respect to which hands were open and which closed. Students were invited to find the missing number of pebbles and write the answer in beneath the relevant hand. There were four pairs for six and four for seven.

While children worked, Kerstin circulated and helped those in need. This typically involved her modelling a situation with pebbles in her own hands. On later occasions, particularly when moving working on tasks involving seven, students were encouraged to model the situation as in Fig. 9.

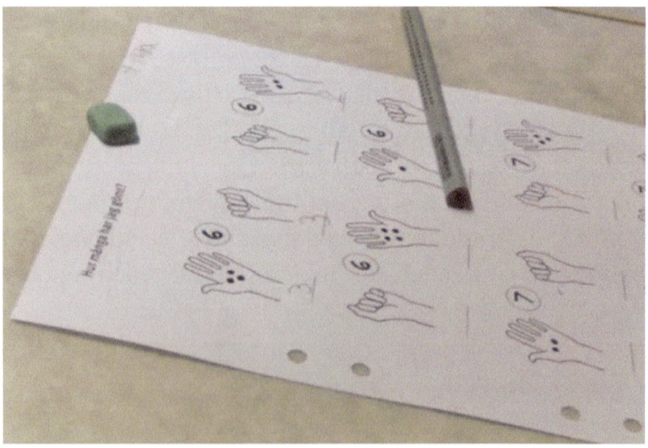

Fig. 8 Kerstin's worksheet

Fig. 9 The subitising model encouraged by Kerstin

Commentary on Kerstin's Third Excerpt

While it is conceivable that some students may have employed a counting-on strategy, this final activity offered an opportunity for students to consolidate their conceptual subitising of six. This could have been achieved either by means of a mental representation of perceptually subitised numerosities or the use of pebbles as in Fig. 9. With respect to seven, a number less amenable to perceptual subitising, Kerstin's encouragement to use the pebbles provided an explicit connection to conceptual subitising strategy. From the perspective of FoNS, the exercise consolidated the connection between number and quantity and cardinality as a component of systematic counting. It also presented different representations of number and, implicitly, exploited simple arithmetic or counting to locate missing numbers. Also, students were encouraged to engage in number recognition.

Discussion

In this chapter, we examine how activities focused on conceptual subitising have the potential to facilitate children's acquisition of the various components of foundational number sense (FoNS). Interestingly, *fons* is the Latin word for fount or spring, which seems apposite for such an important underpinning mathematical understanding.

Our analyses, summarised in Table 1, indicate that in both cases, Klara from Hungary and Kerstin from Sweden, a significant proportion of the eight FoNS components were identified in each excerpt. In neither set of excerpts was there evidence of quantity discrimination or estimation, although it is probably unrealistic, or even unreasonable, to assume all FoNS components to be addressed in every sequence of activities, as some are more likely to lend themselves to particular components than others. However, the number of FoNS components identified in each excerpt—consistently between four and five—indicated that claims made for the efficacy of teaching which focused on conceptual subitising (Clements 1999; Conderman et al. 2014; Sadler 2009) are not without warrant.

It is also interesting to note that in neither case was conceptual subitising an explicit intention—neither teacher was aware of the term—nor were teachers expecting to address FoNS categories of learning. However, both teachers had similar broad learning outcomes in knowing and understanding combinations of numbers. It is also interesting to note that despite substantial differences in the management of their lessons—Klara spent all her lesson orchestrating whole class activity with only occasional expectations of students working individually, while Kerstin spent the great majority of her time managing and supporting students working individually—the FoNS components addressed in their respective excerpts were remarkably similar.

Finally, in an earlier chapter, in which FoNS categories were applied to tasks and instruction focused on mathematical sequences, similar results were obtained for Klara, in Hungary, and Sarah, in England (Back et al. 2014). Evidence from the two analyses suggests that instructional tasks focused on sequences and instructional tasks focused on conceptual subitising appear rich in their potential for realising a range of FoNS components. Without wishing to overstate the significance of these results, results yielded by small case studies, it is worth suggesting that a potentially exciting line of future enquiry would be to identify other topics and ways of teaching them with similar potential for FoNS developments.

Implications

A crude interpretation of the data in Table 1 could be that the mean number of categories applied to a teacher's episodes offers a measure of didactical or instructional complexity. Indeed, were such figures available for representative samples of teachers from each country, then such conclusions may have validity and, importantly, reflect those 'culturally determined patterns of belief and behaviour, frequently beneath articulation, that distinguish one set of teachers from their culturally different colleagues' (Andrews and Sayers 2013, p. 133). However, this chapter offers

Table 1 FoNS-related summary of the various excerpts

		Excerpts					
		Klara			Kerstin		
		1	2	3	1	2	3
FoNS components	Number recognition		X	X			X
	Systematic counting		X		X	X	X
	Relating number to quantity	X	X	X	X	X	X
	Quantity discrimination						
	Different representations	X	X	X	X	X	X
	Estimation						
	Simple arithmetic	X	X	X	X	X	X
	Number patterns	X		X	X	X	

two case studies showing that Klara paid more attention to systematic counting than Kerstin, differences that can only allude to the sensitivity of the FoNS framework to culturally located differences. Thus, if the FoNS framework is to be shown to be useful to both researchers and teachers internationally, there is a need to:

- Evaluate the framework at scale to confirm its cultural sensitivity and propensity for generating culturally located models of FoNS-related practice.
- Develop curriculum support tools for teachers to plan an explicit incorporation of FoNS categories in their teaching.
- Develop a diagnostic tool for teachers to assess individual grade one children's FoNS-related understanding.

Finally, in this and other teacher-focused evaluations of the FoNS framework, no evidence emerged of teachers addressing the development of children's estimation skills, prompting the question, why was this the case?

Acknowledgements We are grateful to our colleague Jenni Back for sharing her Hungarian data.

References

Andrews, P., & Sayers, J. (2013). Comparative studies of mathematics teaching: Does the means of analysis determine the outcome? *ZDM: The International Journal on Mathematics Education, 45*(1), 133–144.

Andrews, P., & Sayers, J. (2015). Identifying opportunities for grade one children to acquire foundational number sense: Developing a framework for cross cultural classroom analyses. *Early Childhood Education Journal*. doi:10.1007/s10643-014-0653-6.

Andrews, P., Sayers, J., & Marschall, G. (2015). *Developing foundational number sense: Number line examples from Poland and Russia*. Paper presented to the ninth congress of European research in mathematics education (CERME9), Prague.

Aubrey, C., Dahl, S., & Godfrey, R. (2006). Early mathematics development and later achievement: Further evidence. *Mathematics Education Research Journal, 18*(1), 27–46.

Aubrey, C., & Godfrey, R. (2003). The development of children's early numeracy through key stage 1. *British Educational Research Journal, 29*(6), 821–840.

Aunio, P., & Niemivirta, M. (2010). Predicting children's mathematical performance in grade one by early numeracy. *Learning and Individual Differences, 20*(5), 427–435.

Aunola, K., Leskinen, E., Lerkkanen, M.-K., & Nurmi, J.-E. (2004). Developmental dynamics of math performance from preschool to grade 2. *Journal of Educational Psychology, 96*(4), 699–713.

Back, J., Sayers, J., & Andrews, P. (2014). The development of foundational number sense in England and Hungary: A case study comparison. In B. Ubuz, Ç. Haser, & M. Mariotti (Eds.), *Proceedings of the eighth congress of the European Society for Research in Mathematics Education* (pp. 1835–1844). Ankara: Middle East Technical University on behalf of ERME.

Baroody, A. (2004). The developmental bases for early childhood number and operations standards. In D. Clements & J. Sarama (Eds.), *Engaging young children in mathematics: Standards for early childhood mathematics education* (pp. 173–219). Mahwah, NJ: Lawrence Erlbaum.

Baroody, A., & Wilkins, J. (1999). The development of informal counting, number, and arithmetic skills and concepts. In J. Copley (Ed.), *Mathematics in the early years* (pp. 48–65). Reston, VA: National Council of Teachers of Mathematics.

Battista, M., Clements, D., Arnoff, J., Battista, K., & Van Auken Borrow, C. (1998). Students' spatial structuring of two-dimensional arrays of squares. *Journal for Research in Mathematics Education, 29*, 503–532.

Benoit, L., Lehalle, H., & Jouen, F. (2004). Do young children acquire number words through subitizing or counting? *Cognitive Development, 19*(3), 291–307.

Berch, D. B. (2005). Making sense of number sense. *Journal of Learning Disabilities, 38*(4), 333–339.

Booth, J., & Siegler, R. (2006). Developmental and individual differences in pure numerical estimation. *Developmental Psychology, 42*(1), 189–201.

Booth, J., & Siegler, R. (2008). Numerical magnitude representations influence arithmetic learning. *Child Development, 79*(4), 1016–1031.

Brewer, E. W. (2012). Secondary data analysis. In J. Goodwin (Ed.), *Sage secondary data analysis* (pp. 165–176). London: Sage.

Butterworth, B. (2005). The development of arithmetical abilities. *Journal of Child Psychology and Psychiatry, 46*(1), 3–18.

Casey, B., Kersh, J., & Young, J. (2004). Storytelling sagas: An effective medium for teaching early childhood mathematics. *Early Childhood Research Quarterly, 19*(1), 167–172.

Chard, D., Clarke, B., Baker, S., Otterstedt, J., Braun, D., & Katz, R. (2005). Using measures of number sense to screen for difficulties in mathematics: Preliminary findings. *Assessment for Effective Intervention, 30*(2), 3–14.

Clarke, B., & Shinn, M. (2004). A preliminary investigation into the identification and development of early mathematics curriculum-based measurement. *School Psychology Review, 33*(2), 234–248.

Clements, D. (1999). Subitizing. What is it? Why teach it? *Teaching Children Mathematics, 5*(7), 400–405.

Clements, D. (2007). Curriculum research: Toward a framework for "research-based curricula". *Journal for Research in Mathematics Education, 38*(1), 35–70.

Clements, D., & Sarama, J. (2007). Early childhood mathematics learning. In F. Lester (Ed.), *Handbook of research on teaching and learning mathematics* (pp. 461–555). Greenwich, CT: Information Age.

Clements, D., & Sarama, J. (2009). *Learning and teaching early math: The learning trajectories approach*. New York: Routledge.

Conderman, G., Jung, M., & Hartman, P. (2014). Subitizing and early mathematics standards: A winning combination. *Kappa Delta Pi Record, 50*(1), 18–23.

Cowan, R., & Renton, M. (1996). Do they know what they are doing? Children's use of economical addition strategies and knowledge of commutativity. *Educational Psychology, 16*(4), 407–420.

Dale, A., Arber, S., & Procter, M. (1988). *Doing secondary analysis*. London: Unwin Hyman.

De Smedt, B., Noël, M.-P., Gilmore, C., & Ansari, D. (2013). How do symbolic and non-symbolic numerical magnitude processing skills relate to individual differences in children's mathematical skills? A review of evidence from brain and behavior. *Trends in Neuroscience and Education, 2*(2), 48–55.

De Smedt, B., Verschaffel, L., & Ghesquière, P. (2009). The predictive value of numerical magnitude comparison for individual differences in mathematics achievement. *Journal of Experimental Child Psychology, 103*(4), 469–479.

Dehaene, S. (2001). Précis of the number sense. *Mind & Language, 16*(1), 16–36.

Desoete, A., Ceulemans, A., De Weerdt, F., & Pieters, S. (2012). Can we predict mathematical learning disabilities from symbolic and non-symbolic comparison tasks in kindergarten? Findings from a longitudinal study. *British Journal of Educational Psychology, 82*(1), 64–81.

Desoete, A., Stock, P., Schepens, A., Baeyens, D., & Roeyers, H. (2009). Classification, seriation, and counting in grades 1, 2, and 3 as two-year longitudinal predictors for low achieving in numerical facility and arithmetical achievement? *Journal of Psychoeducational Assessment, 27*(3), 252–264.

Faulkner, V. (2009). The components of number sense. *Teaching Exceptional Children, 41*(5), 24–30.
Faulkner, V., & Cain, C. (2013). Improving the mathematical content knowledge of general and special educators: Evaluating a professional development module that focuses on number sense. *Teacher Education and Special Education: The Journal of the Teacher Education Division of the Council for Exceptional Children, 36*(2), 115–131.
Fayol, M., Barrouillet, P., & Marinthe, C. (1998). Predicting arithmetical achievement from neuro-psychological performance: A longitudinal study. *Cognition, 68*(2), B63–B70.
Feigenson, L., Dehaene, S., & Spelke, E. (2004). Core systems of number. *Trends in Cognitive Sciences, 8*(7), 307–314.
Fuson, K. (1988). *Children's counting and concept of number*. New York: Springer.
Gallistel, C., & Gelman, R. (2000). Non-verbal numerical cognition: From reals to integers. *Trends in Cognitive Sciences, 4*(2), 59–65.
Geary, D. (2011). Cognitive predictors of achievement growth in mathematics: A 5-year longitudinal study. *Developmental Psychology, 47*(6), 1539–1552.
Geary, D. (2013). Early foundations for mathematics learning and their relations to learning disabilities. *Current Directions in Psychological Science, 22*(1), 23–27.
Geary, D., Bailey, D., & Hoard, M. (2009). Predicting mathematical achievement and mathematical learning disability with a simple screening tool. *Journal of Psychoeducational Assessment, 27*(3), 265–279.
Geary, D., Hamson, C., & Hoard, M. (2000). Numerical and arithmetical cognition: A longitudinal study of process and concept deficits in children with learning disability. *Journal of Experimental Child Psychology, 77*(3), 236–263.
Gelman, R., & Tucker, M. (1975). Further investigations of the young child's conception of number. *Child Development, 46*(1), 167–175.
Gersten, R., Jordan, N., & Flojo, J. (2005). Early identification and interventions for students with mathematics difficulties. *Journal of Learning Disabilities, 38*(4), 293–304.
Gracia-Bafalluy, M., & Noël, M. P. (2008). Does finger training increase young children's numerical performance? *Cortex, 44*(4), 368–375.
Griffin, S. (2004). Building number sense with number worlds: A mathematics program for young children. *Early Childhood Research Quarterly, 19*(1), 173–180.
Griffin, S., Case, R., & Siegler, R. (1994). Rightstart: Providing the central conceptual prerequisites for first formal learning of arithmetic to students at risk for school failure. In K. McGilly (Ed.), *Classroom lessons: Integrating cognitive theory and classroom practice* (pp. 24–49). Cambridge, MA: MIT Press.
Heaton, J. (2008). Secondary analysis of qualitative data: An overview. *Historical Social Research/Historische Sozialforschung, 33*(3) (125), 33–45.
Holloway, I., & Ansari, D. (2009). Mapping numerical magnitudes onto symbols: The numerical distance effect and individual differences in children's mathematics achievement. *Journal of Experimental Child Psychology, 103*(1), 17–29.
Howell, S., & Kemp, C. (2005). Defining early number sense: A participatory Australian study. *Educational Psychology, 25*(5), 555–571.
Hunting, R. (2003). Part-whole number knowledge in preschool children. *Journal of Mathematical Behavior, 22*(3), 217–235.
Ivrendi, A. (2011). Influence of self-regulation on the development of children's number sense. *Early Childhood Education Journal, 39*(4), 239–247.
Jordan, N., Huttenlocher, J., & Levine, S. (1992). Differential calculation abilities in young children from middle- and low-income families. *Developmental Psychology, 28*(4), 644–653.
Jordan, N., Kaplan, D., Locuniak, M., & Ramineni, C. (2007). Predicting first-grade math achievement from developmental number sense trajectories. *Learning Disabilities Research & Practice, 22*(1), 36–46.
Jordan, N., Kaplan, D., Nabors Oláh, L., & Locuniak, M. (2006). Number sense growth in kindergarten: A longitudinal investigation of children at risk for mathematics difficulties. *Child Development, 77*(1), 153–175.

Jordan, N., Kaplan, D., Ramineni, C., & Locuniak, M. (2009). Early math matters: Kindergarten number competence and later mathematics outcomes. *Developmental Psychology, 45*(3), 850–867.

Jordan, N., & Levine, S. (2009). Socioeconomic variation, number competence, and mathematics learning difficulties in young children. *Developmental Disabilities Research Reviews, 15*(1), 60–68.

Jung, M. (2011). Number relationships in a preschool classroom. *Teaching Children Mathematics, 17*(9), 550–557.

Jung, M., Hartman, P., Smith, T., & Wallace, S. (2013). The effectiveness of teaching number relationships in preschool. *International Journal of Instruction, 6*(1), 165–178.

Kalchman, M., Moss, J., & Case, R. (2001). Psychological models for the development of mathematical understanding: Rational numbers and functions. In S. Carver & D. Klahr (Eds.), *Cognition and instruction: Twenty-five years of progress* (pp. 1–38). Mahwah, NJ: Lawrence Erlbaum.

Koontz, K., & Berch, D. (1996). Identifying simple numerical stimuli: Processing inefficiencies exhibited arithmetic learning disabled children. *Mathematical Cognition, 2*(1), 1–23.

Krajewski, K., & Schneider, W. (2009). Early development of quantity to number-word linkage as a precursor of mathematical school achievement and mathematical difficulties: Findings from a four-year longitudinal study. *Learning and Instruction, 19*(6), 513–526.

Kroesbergen, E., Van Luit, J., Van Lieshout, E., Van Loosbroek, E., & Van de Rijt, B. (2009). Individual differences in early numeracy. *Journal of Psychoeducational Assessment, 27*(3), 226–236.

Landerl, K., Bevan, A., & Butterworth, B. (2004). Developmental dyscalculia and basic numerical capacities: A study of 8-9-year-old students. *Cognition, 93*(2), 99–125.

LeFevre, J.-A., Smith-Chant, B., Fast, L., Skwarchuk, S.-L., Sargla, E., Arnup, J., et al. (2006). What counts as knowing? The development of conceptual and procedural knowledge of counting from kindergarten through Grade 2. *Journal of Experimental Child Psychology, 93*(4), 285–303.

Lembke, E., & Foegen, A. (2009). Identifying early numeracy indicators for kindergarten and first-grade students. *Learning Disabilities Research & Practice, 24*(1), 12–20.

Levine, S., Jordan, N., & Huttenlocher, J. (1992). Development of calculation abilities in young children. *Journal of Experimental Child Psychology, 53*(1), 72–103.

Libertus, M. E., Feigenson, L., & Halberda, J. (2011). Preschool acuity of the approximate number system correlates with school math ability. *Developmental Science, 14*(6), 1292–1300.

Lipton, J., & Spelke, E. (2005). Preschool children's mapping of number words to nonsymbolic numerosities. *Child Development, 76*(5), 978–988.

Locuniak, M., & Jordan, N. (2008). Using kindergarten number sense to predict calculation fluency in second grade. *Journal of Learning Disabilities, 41*(5), 451–459.

Lyons, I., & Beilock, S. (2011). Numerical ordering ability mediates the relation between number-sense and arithmetic competence. *Cognition, 121*(2), 256–261.

Malofeeva, E., Day, J., Saco, X., Young, L., & Ciancio, D. (2004). Construction and evaluation of a number sense test with Head Start children. *Journal of Educational Psychology, 96*(4), 648–659.

Mazzocco, M. M., Feigenson, L., & Halberda, J. (2011). Preschoolers' precision of the approximate number system predicts later school mathematics performance. *PLoS One, 6*(9), e23749.

McIntosh, A., Reys, B., & Reys, R. (1992). A proposed framework for examining basic number sense. *For the Learning of Mathematics, 12*(3), 2–8.

Moeller, K., Neuburger, S., Kaufmann, L., Landerl, K., & Nuerk, H. C. (2009). Basic number processing deficits in developmental dyscalculia: Evidence from eye tracking. *Cognitive Development, 24*(4), 371–386.

Mulligan, J., & Mitchelmore, M. (2009). Awareness of pattern and structure in early mathematical development. *Mathematics Education Research Journal, 21*(2), 33–49.

Mulligan, J., Mitchelmore, M., & Prescott, A. (2006). Integrating concepts and processes in early mathematics: The Australian pattern and structure mathematics awareness project (PASMAP).

In J. Novotná, H. Moraová, M. Krátká, & N. Stehlíková (Eds.), *Proceedings of the 30th annual conference of the International Group for the Psychology of Mathematics Education* (Vol. 4, pp. 209–216). Prague, Czech Republic: PME.
Mundy, E., & Gilmore, C. (2009). Children's mapping between symbolic and nonsymbolic representations of number. *Journal of Experimental Child Psychology, 103*(4), 490–502.
Murphy, D. (2014). Issues with PISA's use of its data in the context of international education policy convergence. *Policy Futures in Education, 12*(7), 893–916.
Nan, Y., Knösche, T., & Luo, Y. J. (2006). Counting in everyday life: Discrimination and enumeration. *Neuropsychologica, 44*(7), 1103–1113.
National Council of Teachers of Mathematics. (1989). *Curriculum and evaluation standards for school mathematics*. Reston, VA: NCTM.
Noël, M. P. (2005). Finger gnosia: A predictor of numerical abilities in children? *Child Neuropsychology, 11*(5), 413–430.
Obersteiner, A., Reiss, K., & Ufer, S. (2013). How training on exact or approximate mental representations of number can enhance first-grade students' basic number processing and arithmetic skills. *Learning and Instruction, 23*, 125–135.
Okamoto, Y., & Case, R. (1996). Exploring the microstructure of children's central conceptual structures in the domain of number. In R. Case, Y. Okamoto, G. Sharon, A. McKeough, C. Bleiker, B. Henderson, K. Stephenson, R. Siegler, & D. Keating (Eds.), *The role of central conceptual structures in the development of children's thought* (pp. 27–58). Wiley on behalf of the Society for Research in Child Development.
Passolunghi, M., Vercelloni, B., & Schadee, H. (2007). The precursors of mathematics learning: Working memory, phonological ability and numerical competence. *Cognitive Development, 22*(2), 165–184.
Penner-Wilger, M., Fast, L., LeFevre, J., Smith-Chant, B., Skwarchuk, S., Kamawar, D., et al. (2007). The foundations of numeracy: Subitizing, finger gnosia, and fine-motor ability. In D. McNamara & J. Trafton (Eds.), *Proceedings of the 29th Annual Cognitive Science Society* (pp. 1385–1390). Austin, TX: Cognitive Science Society.
Richardson, K. (2004). Making sense. In D. Clements & J. Sarama (Eds.), *Engaging young children in mathematics: Standards for early childhood mathematics* (pp. 321–324). Mahwah, NJ: Lawrence Erlbaum.
Robinson, C., Menchetti, B., & Torgesen, J. (2002). Toward a two-factor theory of one type of mathematics disabilities. *Learning Disabilities Research & Practice, 17*(2), 81–89.
Sadler, F. (2009). Help! They still don't understand counting. *Teaching Exceptional Children Plus, 6*(1), 1–12.
Sayers, J., & Andrews, P. (2015). *Foundational number sense: Summarising the development of an analytical framework.* Paper presented to the Ninth Congress of European Research in Mathematics Education (CERME9), Prague.
Siegler, R., & Booth, J. (2004). Development of numerical estimation in young children. *Child Development, 75*(2), 428–444.
Stock, P., Desoete, A., & Roeyers, H. (2010). Detecting children with arithmetic disabilities from kindergarten: Evidence from a 3-year longitudinal study on the role of preparatory arithmetic abilities. *Journal of Learning Disabilities, 43*(3), 250–268.
Strauss, A., & Corbin, J. (1998). *Basics of qualitative research: Techniques and procedures for developing grounded theory*. London: Sage.
Thomas, N., Mulligan, J., & Goldin, G. (2002). Children's representation and structural development of the counting sequence 1-100. *The Journal of Mathematical Behavior, 21*(1), 117–133.
Van de Rijt, B., Van Luit, J., & Pennings, A. (1999). The construction of the Utrecht early mathematical competence scale. *Educational and Psychological Measurement, 59*(2), 289–309.
Van Luit, J., & Schopman, E. (2000). Improving early numeracy of young children with special educational needs. *Remedial and Special Education, 21*(1), 27–40.
Van Nes, F., & De Lange, J. (2007). Mathematics education and neurosciences: Relating spatial structures to the development of spatial sense and number sense. *The Montana Mathematics Enthusiast, 4*(2), 210–229.

Van Nes, F., & Doorman, M. (2011). Fostering young children's spatial structuring ability. *International Electronic Journal of Mathematics Education, 6*(1), 27–39.

Van Nes, F., & Van Eerde, D. (2010). Spatial structuring and the development of number sense: A case study of young children working with blocks. *The Journal of Mathematical Behavior, 29*(2), 145–159.

Yang, D. C., & Li, M. N. (2008). An investigation of 3rd-grade Taiwanese students' performance in number sense. *Educational Studies, 34*(5), 443–455.

Young-Loveridge, J. (2002). Early childhood numeracy: Building an understanding of part-whole relationships. *Australian Journal of Early Childhood, 27*(4), 36–42.

Zur, O., & Gelman, R. (2004). Young children can add and subtract by predicting and checking. *Early Childhood Research Quarterly, 19*(1), 121–137.

Part V
Professional Development

Teachers' Interpretation of Mathematics Goals in Swedish Preschools

Laurence Delacour

Abstract The purpose of this chapter is to investigate how two preschool teachers interpret the mathematical goals outlined in the Swedish preschool curriculum. The ways in which these preschool teachers transform, clarify and concretise the mathematics goals are analysed. The data indicates a tendency towards two different approaches of interpreting and implementing mathematics: a comprehensive approach and an academic approach. Based on these preschool teachers' interpretation and implementation of the mathematics goals, the consequences in the form of qualification, socialisation and subjectification will be discussed.

Introduction

The Organisation for Economic Co-operation and Development (OECD 2013) Programme for International Student Assessment (PISA) shows that children in both Europe and the United States have difficulty succeeding in mathematics. Similarly, researchers have recognised that early childhood education and care (ECEC) that includes mathematics has an impact on children's learning outcomes and can increase their opportunities in later life (Duncan et al. 2007; Barber 2009; Doverborg and Pramling Samuelsson 2011). Consequently, ECEC has become a policy priority in many countries (Taguma et al. 2012), and a stronger focus on mathematics in preschool is being highlighted in government policies and curricula.

In Sweden, the government indicates that society now places greater demands on mathematical understanding and skills, and, therefore, Sweden must meet these new social demands in order to benefit child development (Government Offices 2010). At the same time, the government states that Swedish preschools have not made full use of children's desire to learn (Government Offices 2010). In order to increase the quality of Swedish ECEC, Sweden selected one of the five policy levers that the

L. Delacour (✉)
Malmö University, Malmö, Sweden
e-mail: laurence.delacour@mah.se

T. Meaney et al. (eds.), *Mathematics Education in the Early Years*,
DOI 10.1007/978-3-319-23935-4_22

OECD publication Starting Strong III (OECD 2012) presents as having a positive effect on early child development and learning. Sweden selected "Policy Lever 2: Designing and implementing curriculum and standards" as the lever to focus on. It is considered that a curriculum framework can help teachers to focus on the progression of their pedagogical goals and on the children's development (Siraj-Blatchford et al. 2002). For that reason, a revised preschool curriculum was introduced in Sweden in 2011 (The National Agency for Education 2010) in which the goals for children's mathematical development were made clearer in both scope and content. Consequently, preschools must now strive to ensure that each child develops, expresses and uses his or her understanding of space, shape, location, direction, sets, quantity, order, number concepts, measurement, time and change. Furthermore, preschools must strive to ensure that each child develops his or her ability to reflect and test solutions; distinguish, express, examine and present mathematical concepts; follow interrelationships; and develop their mathematical skill in putting forward and following reasoning (The National Agency for Education 2010).

How children are to create, explore and use mathematics is not specified in the Swedish curriculum, as it is a goal-oriented document without suggestions on how to teach. However, the curriculum states that Swedish preschools should build on democratic foundations, be open to different ideas and encourage children to learn (The National Agency for Education 2010). It is therefore interesting to investigate how some preschool teachers interpret and implement these goals and what consequences their interpretations can have on children's learning outcomes.

Interpretation of the Curriculum

Recent national and international studies show how preschool curricula are interpreted by teachers and other educators. Jonsson (2011) studied how the preschool curriculum in Sweden could be identified in activities for children between 1 and 3 years of age and how preschool teachers handle the demands placed on children's learning. Her study shows that preschool teachers seem to perceive preschool play and school topics as equally important. Preschool teachers tend to adapt the activity content from moment to moment, based on their interpretations of individual child's interests and needs—what Jonsson calls the "present's didactics". For them, learning occurs when children interact, discuss, experience, imitate, explore and test. The teacher acknowledges the children's behaviour and thereby creates opportunities but does not teach using the formal approaches of school.

Aubrey and Durmaz (2012) investigated the relationship between the new curriculum in England and young children's mathematics in practice, how teachers understand and interpret the curriculum and how 5-year-old children react to the mathematical situations offered. They found that practices varied greatly. All teachers were in favour of the new curriculum, which they felt helped them to achieve a balance between teacher-initiated and child-initiated activities and make them more play based and child centred.

Nevertheless, Aubrey and Durmaz (2012) found that there were differences between urban and rural preschools. The urban preschool studied worked mostly with number-concept understanding. Mathematical situations were mainly teacher initiated, and communication occurred mostly from the preschool teacher towards the children, with the children not being offered any opportunities to talk, question and interact. The groups were often large. The rural preschool offered children very topic-oriented play, as well as various games and outdoor practical activities in different group constellations (e.g. pairs of children, groups of four children and larger groups). Besides number perception, shape and position were also covered. Mathematical situations were often child initiated, and communication took place between children and adults and between children and children. The children had opportunities to investigate.

In China, a new mathematics curriculum for preschool was introduced in 2001, and tremendous changes have taken place, according to Ma (2012). Ma's study is based on an analysis of 13 mathematical situations in preschools in China. These 13 cases had been previously assessed by teachers and researchers as "excellent" examples of how good education is conducted. The results show the importance of children's holistic development; the importance of giving children opportunities to explore, collaborate and link mathematics to real life; and how teachers use a variety of resources to teach. Ma saw that in these situations, children were given a great deal of space in the interactions between teachers and the children.

A study by Lembrér and Meaney (2014) shows how the "schoolification" of the Swedish preschool curriculum, with its increased emphasis on mathematical goals, could affect the kinds of activities that preschool teachers offer to children. They suggest that teachers mostly plan activities in which children are socialised to become members of society and to reproduce currently accepted norms and values. Lembrer and Meaney argue that if teachers permitted mathematical activities to be influenced and directed by the children themselves, while still basing them on norms and values, the children could be positioned as both being (showing capability) and becoming (progressing on a journey towards adulthood). This approach would contribute to a socialisation that better prepares children for the uncertain future they will meet.

An idea that seems to be prevalent among preschool teachers is that a preschooler learns by participating in everyday situations. Therefore, many educators consider that children are learning all the time and from everything (Björklund 2007; Doverborg and Pramling Samuelsson 1999). Some educators posit that children discover mathematics in a natural way when they play games, build with blocks or tidy up toys. However, the problem with these approaches is that it is usually children who already have some knowledge and interest who get the most out of such activities (Ahlberg 2000). Children learn independently to a certain extent but must be challenged to think one step further and to see things from different perspectives. The foundations of mathematics can be laid through activities such as building with sand, playing with water and various containers or doing puzzles. However, in order to be able to build on these foundations and understand and put words to mathematical concepts, children need a teacher:

> Free play can provide a useful foundation for learning, but a foundation is only an opportunity for building a structure. Adult guidance is necessary to build a structure on the foundation of children's informal mathematics. (Hildebrandt and Zan 2002, p. 2)

In Sweden, Doverborg and Pramling Samuelsson (2011) made a longitudinal study in which they compared two groups of children at two preschools. In Group 1, the preschool teacher consciously worked on mathematics with the children, while in Group 2, the preschool teacher focused mostly on language. The teacher in Group 1 worked systematically to develop number concepts with eight children aged 2–3. Eight months later, there was a big difference between the two groups; and 3 years later, the group that had worked with mathematics could solve problems more easily and had a greater understanding of why mathematics was important. Doverborg and Pramling Samuelsson argue that if children are motivated to use mathematics to solve everyday problems, they will be motivated to learn more advanced mathematical concepts later.

However, the way in which mathematics is communicated between teachers and children will affect the outcome. Alrø and Johnsen-Høines (2010) argue that the quality of communication between teachers and children affects the quality of mathematical learning. They consider that an investigative approach to mathematics education is important in order for children to develop innovative skills and critical thinking. Alrø and Høines observe in their study that it can be difficult for children to take responsibility for their answers and be active if they only participate in evaluating talk aimed at determining what is right or wrong.

According to Björklund (2016), children can understand and learn mathematical concepts more easily when the teacher makes relationships visible to the children, based on the children's own initiatives, in their explorative play. Teacher scaffolding, with different feedback strategies such as open-ended questions, seems to be more important than instructions in supporting children's learning (Bäckman 2016).

Thulin (2011) highlights how science content was communicated between preschool teachers and children in a Swedish preschool. She relates her findings to the demands of the current changing educational mission and the increased science goals in the revised curriculum. According to Thulin, in order to follow the curriculum and communicate about specific concepts, it helps if the teacher focuses on content. Otherwise, the desired content—in this case, the science concepts—will be submerged beneath other goals such as social training.

Deliyianni et al. (2009) state that formal mathematics teaching at early ages can get in the way of children thinking for themselves; however, according to Chen and McCray (2014), *instructions* (i.e. experiences aligned primarily with the teacher's goals) and *constructions* (i.e. processes young children actively engage in to acquire concepts and skills) should be integrated. Because both elements are needed in order to help children experience greater gains in learning outcomes, the teacher's role does not need to conform to an irreducible and contradictory dichotomy of a *comprehensive approach* versus an *academic approach* (Taguma et al. 2012).

Research Questions

This chapter discusses preschool teachers' interpretations of the mathematical goals specified in the curriculum for preschool. This work comes from a larger study about the implementation of the revised Swedish curriculum (Delacour 2013). The research questions for this chapter are:

- How do some preschool teachers interpret and implement the mathematical goals in the Swedish preschool curriculum?
- What consequences do these preschool teachers' transformations of the mathematical goals have on the children's qualification, socialisation and subjectification?

When teachers talk about how they interpret the objectives for mathematics, they also give examples to illustrate how they transform these objectives into practice. *Qualification* refers to the need for children to gain understanding, insight, knowledge and skills about mathematical concepts (Biesta 2011). *Socialisation* refers to the insertion of children into the existing order and the transmission of particular norms and values. *Subjectification* refers to the opportunity for children to be and to become a subject and to be able to solve mathematical problems in their own way.

The Revised Curriculum for Preschool

In regard to early childhood services, most OECD countries have curricula for guidance on what should be done. However, approaches for the care and education of the youngest children vary from country to country. Countries in the social pedagogy tradition try to maintain an open and holistic curriculum and use a *comprehensive approach* that centres on the child. Countries in which early education has been closely associated with primary school tend to privilege readiness for school and use an *academic approach* (Taguma et al. 2012). The question of the "correct curriculum approach" is the subject of considerable debate among ECEC policymakers.

For the holistic development of the child, where general knowledge, social and emotional well-being and communication are taken into consideration, the teacher must broaden the scope and use a *comprehensive approach* (Bertrand 2007; OECD 2006). In contrast, an *academic approach* focuses on important educational goals but risks limiting the possibility of a child-centred environment that is characterised by self-initiated activity, creativity and self-determination (Eurydice 2009; Prentice 2000). The *comprehensive approach*, on the other hand, risks losing a focus on important mathematical goals (Pianta et al. 2009; Bertrand 2007). Opinions vary on which curriculum provides the best OECD quality, setting up a clear dichotomy between separate academic and comprehensive approaches or/and a *mixed model* that combines different curriculum approaches (Taguma et al. 2012).

According to Pramling and Pramling Samuelsson (2011), the revised curriculum content in Sweden places equal value on social and cognitive learning. According to the curriculum, learning should start from children's development, experiences, interests and circumstances and should take place in a playful manner (The National Agency for Education 2010). This description can be connected to the *comprehensive* approach and to a holistic view of children's development. Although the focus on caring is unchanged in the new curriculum, more emphasis is placed on learning and knowledge (Roth 2011). An academic approach seems to be given a higher priority than before, although the holistic development of the child is still in focus, despite the revision of the goals.

Sweden has a decentralised management system which means that goals are decided at the national level, while the selection of content is determined at the municipal and local levels (Roth 2011). The purpose of decentralisation is to give more freedom to teachers. Therefore, it is up to the teachers to combine the different approaches because how children are to create, explore and use mathematics is not specified in the curriculum. The objectives are formulated as goals for preschools to strive towards, with no specific goals for children to achieve.

Theoretical Framework

The focus of this study is on the transformed curriculum, based on preschool teachers' narratives about their interpretations of the national objectives for mathematics and on the examples they give to illustrate their narratives. Transformation means to reshape and adapt, and it is the teachers who are the main actors in the transformation of the curriculum (Linde 2006). This process involves breaking down, clarifying and concretising the national goals.

According to Biesta (2011), the curriculum document describes what knowledge children should encounter as considered by politicians as being necessary for functioning in society. However, the curriculum is transformed in various ways; what preschool teachers choose to focus on will depend to some extent on the views they have of the preschool's role in society (Linde 2006). The mathematical objectives in the revised curriculum are just one of many factors that will affect what happens in preschools. According to Linde, some teachers may focus more on childcare, while others focus more on education or on learning.

When the teacher chooses to focus on mathematical goals and on learning, they prioritise the *qualification* of the children. The curriculum describes what mathematical qualifications are necessary for preschool children to gain. For example, the guidelines state that the children should be provided with experiences to sort and classify objects and make comparisons.

Teachers do not perceive the curriculum in the same way (Uljens 2011), and it takes a long time to achieve change. Teachers select and exclude portions of curriculum according to their background, education and experience. For example, if teachers believe that care and education are the most important parts of the preschool's role, they will focus on the parts of the curriculum that refer to how learning should start from children's development, experiences, interests and circumstances and occur in a playful manner. Thus, they will focus on the socialisation of the children. Socialisation relates to meaning creation and investigation rather than acquiring knowledge (Biesta 2011). Biesta argues that socialisation is an insertion of newcomers into existing orders and a transmission of particular norms and values. When preschool teachers introduce mathematical concepts in a playful manner and take the children's interests into consideration, they want the children to be socialised into liking mathematics and realising how important it is (Biesta 2011).

Differences in how teachers interpret the curriculum may come from their views on children and childhood rather than their perception of the subject. Teachers have their own practices and their own ways of planning and implementing their activities. Their personal experiences, knowledge and ambition affect what parts of the curriculum they choose to work with (Linde 2006). Some teachers think that preschool should focus more on socialisation, in order to help children to integrate into society; others focus more on qualification, in order to prepare children for school; still others believe that some other concept, such as children's freedom, should be in focus (Biesta 2011). Children's influence and freedom are important parts of the Swedish curriculum (The National Agency for Education 2010), and the preschool teacher's way of dealing with this part of the curriculum will affect the subjectification of the children.

According to Biesta (2011), subjectification refers to the opportunity to be and to become a subject and focuses on individual freedom, one's own voice and uniqueness. Education should not only bring children into existing sociocultural and political regimes; it should also help them to free themselves from such regimes (Biesta 2011). Educational institutions give children a number of representative voices through qualification and socialisation. Children are given the opportunity to express their views as members of particular communities, traditions and discourses. There are no training programmes that can help children to find their own voice when the situation demands it, states Biesta. Adults can, however, prevent this uniqueness from emerging if they prevent children from meeting otherness and difference. When children encounter situations that may interfere with their "normal" way of being, and which require a committed and responsible response, their personal uniqueness has the opportunity to appear. Gaining access to common, representative voices is necessary; but education should also contribute to the subjectification process and enable children to become more independent in thought and action (Biesta 2011).

According to Biesta (2011), a good education can be achieved through a balance of the three different elements: qualification, socialisation and subjectivity. When teachers focus on qualification through the mathematical goals, in order to privilege readiness for school, they use an academic approach.

When teachers focus on the other part of the curriculum, where the Swedish tradition of social pedagogy has maintained an open and holistic curriculum, they use a comprehensive approach that centres on the child. General knowledge, social and emotional well-being and communication are taken into consideration in the comprehensive approach (Bertrand 2007; OECD 2006), where self-initiated activity, creativity and self-determination characterise this child-centred environment. A child-centred environment enables children to become more independent in thought and action, which Biesta named subjectification. When teachers use both an academic approach and a comprehensive approach, they are able to balance qualification, socialisation and subjectivity. The revision of the Swedish curriculum for preschool can be interpreted as part of the current tendency to strike a balance between these different approaches (Delacour 2013).

Linde (2006) argues that the time teachers have at their disposal and the number of children they have in their class or group are the result of government decisions, which in turn affect how teachers interpret the curriculum and what they choose to do. Other factors that affect how the curriculum is interpreted include the material that is available, the preschool's environment and how the group of children interacts.

Method

In order to understand how some preschool teachers break down, clarify and concretise the mathematical goals in the revised curriculum, two interviews were conducted separately with four preschool teachers, in which the teachers were asked one by one to explain and give examples of how they transformed the curriculum and what activities they did with the children. In the first interview, the teachers had the curriculum in front of them, read through the mathematical goals and explained how they transformed the goals in practice. In the second interview, the teachers told how they planned a specific mathematical activity. The teachers work at two different preschools. All four have an interest in mathematics, although mathematics was not included in the course work during these teachers' initial teacher education.

Lotta and Susan work together at one preschool, with a group of 4- and 5-year-old children. Lotta has worked in different preschools for 32 years, while Susan has worked for 9 years at two different preschools. Susan attended mathematics courses and Reggio Emilia courses at the university after completing her teaching degree.

Malin and Åsa work together at the other preschool, also with 4- and 5-year-old children. Åsa attended mathematics courses at the university after completing her teaching degree. Previously, Åsa worked as a teacher in compulsory school.

The teachers were interviewed individually so that they would not be influenced by their colleagues and could express their own points of view. The interviews were semi-structured in order to be flexible, to allow the follow-up of ideas and to ask supplementary questions (Bryman 2011).

The preschools are located in two small communities in the same municipality. There are no major differences between the preschools in terms of staff composition, group size or children's sociocultural and economic backgrounds.

Analysis of the Data

To analyse and interpret the data, a hermeneutic approach (Bryman 2011) was used. The data was divided into five categories, which came out of the information provided by the interviewees: theme, interest, question/answer, reflection/motivation and feedback/evaluation. Each category is discussed below.

Theme: Mathematics may or may not be the theme of a given activity, although it could be included within a different theme, such as the environment. Determining the theme contributed to determining which approach was used—comprehensive or academic. For example, when a preschool teacher chose to use mathematics within an environmental theme, she used a comprehensive approach, because she tried to include other parts of the curriculum such as motor skills, creative ability and free play. In this case, mathematics was only one part of what the children were doing. On the other hand, when the teacher chose mathematics as a theme, they focused mostly on the mathematical goal and used an academic approach. In these cases, the preschool teacher's motivation was about qualification: helping the children to gain understanding, insight, knowledge and skills.

Interest: Mathematics instruction can either be based on the children's interest or on the teacher making the children interested in something. When preschool teachers plan their practice based on the children's interest, they can open up unexpected and creative ways for children to discover mathematics. Such an environment is child centred and the activities are self-initiated. The teachers used a comprehensive approach in these cases, as the children have the possibility to influence what would happen—allowing subjectification to occur. On the other hand, when preschool teachers think that they have to arouse the children's interest in mathematics, they do not see the children as capable. In this case, the teacher's role is to prepare children for school, based on the knowledge they believe the children need. In such situations, teachers use an academic approach, and the socialisation is on the children becoming.

Question/answer: This category identifies who formulates and answers the questions asked during the activity. When children are allowed to formulate their own questions and answers, they are being given the opportunity to acquire their own way of thinking and solving problems—leading to subjectification (Biesta 2011). When preschool teachers formulate questions whose answers are related to mathematical concepts and formal education, they transfer predetermined norms and knowledge to the children. Since their focus is on preparation for school, they tend to use an academic approach, where children are qualified and socialisation is on becoming.

Reflection/motivation: This category involves children either trying out and reflecting on mathematical concepts in groups or following the teacher's instructions and justifying their answers to the teacher. When groups reflect on mathematical concepts together, with each child coming up with his or her own reflection, the children become more independent in thought and action—leading to subjectification. When the children are given instructions by the teacher, based on the mathematics concepts that the children should be qualified for, the teachers highlight what they consider to be necessary for the children to function in society. The children reproduce established knowledge, as a result of the teacher using an academic approach.

Feedback/evaluation: This category connects to how teachers provided feedback to the children and evaluated their behaviour, responses and understanding of mathematical concepts. Based on what the teacher evaluates, approves and gives feedback on, the children respond to the teacher's expectations. If the correct answer is valued, the children will follow the norms and values of the society. If the children's creativity is valued, they will be able to produce their own responses.

The five categories are outlined in Table 1.

Teachers' Different Approaches

In the analysis of the interviews, the preschool teachers' interpretations of the mathematics objectives differed, leading to two separate ways of transforming the objectives in practice and indicating a tendency towards two different ways of communicating mathematics to the children. These approaches affect how the curriculum is transformed (Linde 2006).

In the preschool where Lotta and Susan work, the ways of working with the children are inspired by Reggio Emilia. The children's interests are important, and children are encouraged to formulate their own questions and find their own answers. The teacher seems to communicate mathematics based on the children's interest. The environment is child centred, and the tendency is towards the comprehensive approach (CA from this point forward).

Table 1 Categories used in the analysis

	"Comprehensive approach"	"Academic approach"
Theme	The teacher bases her work on a theme (e.g. the environment) that includes mathematics	The teacher chooses mathematics as a theme and focuses on mathematics skills and knowledge, not connected to other situations or concepts
Interest	The activities are based on children's interests, and the teacher pays attention to mathematical concepts when they arise in the situation	The teacher decides what could interest the children and what they need to learn
Questions/answer	The teacher allows the children to formulate their own questions/answers	The teacher helps the children to find the correct answer
Reflection/motivation	The children are given the opportunity to discover mathematics by themselves and share with the group	The teacher gives considerable information and the children must justify their answers
Feedback/evaluation	Children's initiatives are valued	The correct answer is valued

In the other preschool, the managers of the preschools have decided that all the preschools would work with mathematics. The teachers plan their mathematical activities according to what they believe the children need to learn. This involves a more academic approach (AA from this point forward).

The Comprehensive Approach

From the CA, the teacher's approach is comprehensive, with a holistic development of the child in which general knowledge, social and emotional well-being and communication are taken into consideration. Lotta gave an example of how she worked with mathematics within a theme:

Lotta: Vi hade "Olles Skidfärd" från Elsa Beskow som grund för verksamheten. Utifrån den vi byggde skidor med mjölkkartong och så fick dom göra skidan lika långa som dom själva var.

Lotta: We had "Ollie's Ski Trip" by Elsa Beskow as the basis for the activities. Based on this, they built skis with milk cartons and made the skis as long as themselves.

From this, it can be construed that the children learnt the concept of length as well as how to use it to build skis, based on listening to a fairy tale. They used their motor skills, their imaginations and their bodies.

The preschool teachers discussed how the children needed to feel mathematics with their bodies, for example, by walking a tightrope, feeling distance and seeing forms, colours and numbers:

Susan: När dom är unga är det viktigt från början att dom får en känsla för matematik. Det du lär med kroppen stannar i knoppen.

Susan: When they are young, it is important, from the beginning, to get the feeling of mathematics. What you learn with the body stays in your mind.

Susan clearly saw her role as providing the children with experiences in which learning takes place through action. The teacher believes in a holistic method, where both body and mind are important for the understanding of mathematical concepts. Consequently, the mathematical situations that these teachers offer to the children are based on learning by doing.

The teachers give the children the opportunity to think for themselves and are interested to hear what the children are thinking. They do not socialise the children to conform to norms about "right" or "wrong"—and in this way, they allow subjectification to occur:

Lotta: Att man får dom att tänka till och kanske att själva komma med en fundering eller kanske inte ett svar men det blir en dialog lite oss emellan. … det finns inte rätt eller fel utan att barnen själva får finna en kommentar eller ett resonemang kring hur dom tänker.

Lotta: We help them to think and perhaps to come up with a thought or maybe not an answer but a little dialogue between us … there is no right and wrong without the children themselves finding a comment or a discussion about how they think.

The teachers at this preschool discuss and pay attention to mathematical concepts as and when they appear in an activity. Lotta and Susan's attitude towards mathematics learning in preschool seems to be that only the mathematical goals that the children are interested in must be taken into consideration:

Susan: Det var barnen som pratade om former … och vi utgick ifrån det
Susan: It was the children who talked about shapes … and we picked up on it.

Susan said that the curriculum goals are not goals that must be achieved and that since preschool is not compulsory, play is more important and the children should learn at their own pace. These preschool teachers started from the children's interests and actions when they communicated mathematics. When the teacher followed the children's interest, thoughts and experiences, she saw herself as researching along with the children. What will be discovered was not predetermined, opening up the situation for innovative discovery and subjectification. The children were socialised as being. Knowledge of mathematical concepts, or qualification, did not seem to be highlighted, as it was left to the children to choose to talk about mathematics.

Lotta: Det är lite också utefter deras intresse och man känner av lite.
Lotta: It's a little about their interest, and you are to be little sensitive.
Susan: Vi delar dom I grupper allt eftersom intresset från deras sida … grundmål är att barnens frågor och tankar är viktiga, deras erfarenheter. Och det visar sig genom att vara medupptäckare.
Susan: We divide the children into groups based on their interests … the goal is that children's questions and thoughts are important, as are their experiences. And we discover together.

The children formulated their own questions and answers. These teachers seemed to expect the children to become actors in their own learning, by taking the initiative to seek their own answers, consider their friends' different solutions and recognise several possible ways to solve problems. The children's self-confidence in mathematics was to be strengthened, but not just by evaluating their responses as being either right or wrong. The children seemed to be given the opportunity to meet otherness and differences and thus be able to develop their unique voice:

Susan: Det är inte: mitt är det rätta! För ibland är det barn som söker det: Och vad är det nu? Vad ska det vara? Nej, det är utifrån vad dom ser. Det är inte jag som styr. … Hur tänker du här? Eller hur blev det så? Dom är fria att komma fram till sina egna lösningar.
Susan: It's not a matter of, "I am right!" For sometimes it's about what children are searching for. And what is it now? What should it be? It is not me that

controls it. No, it's based on what they see. … How do you think? Or how did it happen? They are free to arrive at their own solutions.

Lotta told how they were working with the idea of the environment. The children learned about sorting garbage, recycling cans and so on. She and Susan wanted the children to have fun and to listen to each other's different solutions. The teachers encouraged the children to formulate their own questions and answers in order to show the complexity of mathematics. Mathematical concepts became the focus if they appeared in the situation. The children are assigned agency in the form of participation, codetermination and influence. The teacher focuses on the children's socialisation by connecting mathematics to a theme. For example, Lotta wanted the children to have a holistic view of mathematics and to understand how mathematics can be used to improve the environment. The aim in this kind of situation is to give children a positive and enjoyable experience of mathematics in the spirit of community, respect and cooperation; making this indicates that a comprehensive approach was used. Children were viewed as having an innate potential that can be expressed and made available for the construction of a better society. The teachers intended for the children to take responsibility for their own education.

These preschool teachers seemed to accept that the children had the ability to seek knowledge and to learn from each other. Consequently, they expected the children to take an active part in the planning of the activity by showing and sharing interests with each other. They saw children as capable of seeking knowledge and as having their own reasoning—a way of thinking that is connected to Biesta's (2011) concept of subjectification.

The differences in how preschool teachers interpreted the curriculum may stem more from their views on children and childhood. For example:

Susan: Vi bollar med varandra och försöker att barnen bollar tillsammans för dom är nyfikna, dom vill lära.

Susan: We communicate with each other and try to get kids to communicate together because they are curious, they want to learn.

Lotta: Dom är duktiga på att själva hitta lite lösningar på problem. Att vi tänker så och att barnen själva får finna en kommentar eller ett resonemang kring hur dom tänker och det sker en utveckling verkligen.

Lotta: They are good at finding the solutions to problems … we think so, and the children themselves may come up with a comment or a discussion of how they think and that's real development.

These two preschool teachers generally used the CA approach. It was the children's behaviour and interests that determined the mathematical concepts that were addressed in a situation. The preschool teachers often thought about the mathematical concepts they wanted to communicate, but they let the children control the content and introduce different concepts. The teachers preferred the children to see mathematical concepts as being anchored to something familiar and understandable to them, communications in which the children used their bodies, their minds, their imagination and each other in order to understand the world around them. The teachers, therefore, chose to use mathematics within a theme. When the children explored and focused on mathematical concepts, they could find their own answers, reflect and draw their own conclusions. The children reflected on what was happening and explained and justified to each other. The children were given plenty of time to discover at their own pace. The teacher gave feedback to the children when they took initiative and found their own solutions, which improved the children's opportunities for subjectification:

Lotta: Men så spännande det är. Man kan hitta massor med lösningar. Ni är jätte duktiga på att komma på olika lösningar ju.

Lotta: But how exiting! It's possible to find lots of solutions. You are very good at finding different solutions.

The Academic Approach

In the transformation of mathematical goals through the academic approach (AA), mathematics at the second preschool was expressed as an easier form of school mathematics. The main objective was to prepare children for school:

Åsa: När jag jobbade på lågstadiet var det mycket matematik. … så jag hade rätt mycket med mig men det var för lite äldre barn men jag har haft mycket nytta av å plocka ner på barns nivå här.

Åsa: When I worked in school, there was a lot of maths. It was for slightly older kids, but I got a lot of benefit from it and here I take it down to the younger children's level.

Åsa seemed to prioritise readiness for school and used an academic approach that can be connected to qualification. Her view of the preschool's goals was different from Lotta's and Susan's. Within the AA, an understanding of abstract concepts is central with children needing to pay attention to differences and justify their answers. Mathematics is about learning to recognise and name shapes and about understanding fractions and patterns. Although these preschool teachers worked with concrete materials, the mathematical concepts were sometimes abstract. According to these teachers, when children followed every step of a process, they formed mental images and thus were able to think abstractly:

Åsa: Vi har börjat arbeta abstrakt för vi upptäckte att de har blivit duktiga … det är faktiskt många som kan tänka abstrakt för vi har arbetat så mycket med det konkreta. De har fått mentala bilder …

Åsa: We started with the abstract concept now because we found that they have become skilled. … Surprisingly, many do actually understand the abstract, because we have worked in such a very concrete way before. They have mental images.

Malin: Nu har vi nya barn så vi får börja om från början så det är viktigt att verkligen visa saker: ja vi har ett helt äpple och för att göra det riktigt vi visar och sätta den ihop och delar den och sätta den ihop igen och till slut sitter det där ... en kvart... I Påskas kunde några barn förstå en åttonde del när vi delade lera ... därför de hade fått följa hela processen att dela den.

Malin: Now we have new children so we get to start over, so it's also important to really show things: Yes, we have a whole apple and to make it real, we show it and put it together and take it apart and put together again and so finally it sits there ... a quarter ... At Easter, some children were able to understand "one sixteenth" when they shared clay ... because they had experienced the whole process of dividing it into parts.

The teachers talked about learning mathematics as a goal in itself. Mathematics was woven into everything. To focus on mathematics over a longer period was a decision that came from management. The theme of "mathematics" was concluded with an exhibition at which all the community preschools showed what they had done with the children. As Linde (2006) suggests, teacher's choices of which part of the curriculum they want to work with can be influenced by management's decisions. When the focus stays mainly on mathematical concepts over a long period of time, with a final exhibition to showcase the work, the teacher is unlikely to lose focus on important mathematical goals. Thus, the children will be well prepared to understand school mathematics. For example, Malin prepared situations focused on mathematical concepts. She was aware that other goals could interact with mathematics, but her focus was mainly on helping the children to understand the concepts she had planned:

Malin: Jag brukar alltid ha en grundtanke I min planering. Jag får in språk, jag får in sociala samspel eller samarbete, jag får in mer men det är ändå matematiken jag har som fokus. Jag fokuserar på former, geometriska former ... Vi gick på jakt, triangel jakt, rektangel, kvadrat, cirkel. Vi var ute, vi pratade om former och vi gjorde diagram på tavlan om var vi hittade mest former

Malin: I am always thinking about mathematics in my planning. I get into the language, I get into social interaction and collaboration, I get into more, but it is still mathematics that I have in mind ... I focus on shapes, geometric shapes ... we went hunting, hunting for triangles, rectangles, squares, circles. We were out in the neighbourhood, we talked about shapes, and we made diagrams on the board about where we found most of the shapes.

In their preparation and teaching, these preschool teachers focused on certain mathematical concepts; they communicated these by initiating situations in which the children come in contact with the concepts by following the teachers' instructions and trying things:

Åsa: Jag ska jobba mycket med siffran 2... vi kommer att fortsätta jobba med cirkel och gå runt. ... Och då kommer dom att få hålla upp den längsta pinnen ... och den kortaste, och sen får dom hålla upp den tjockaste och den smalaste ...

Åsa: I will work a lot with the number 2…we'll continue to work with the circle and walk around. …And then they'll hold up the longest stick…and the shortest, and then I'll get them to hold up the thickest and the thinnest…

The preschool teacher's task here is to evoke the children's interest in mathematics. The mathematical concepts are chosen based on what the teachers believe that the children need, prioritising readiness for school:

Åsa: Vi kan inte sitta och vänta att barnen ska visa intresse för nånting utan vi måste få dom intresserad … När jag jobbade på lågstadiet hade barnen jätte jätte svårt dels för mönsterbildning och dels för bråk och jag tänker att så mycket som vi har jobbat med sjättedelar och fjärdedelar. Jag hoppas att det kommer att märkas i skolan.

Åsa: We can't just sit and wait for the children to show an interest in something; we have to get them interested …When I worked in school, children had a really, really hard time with patterns and fractions, and I think that as we have worked so much with sixths and fourths, I hope there will be a noticeable difference in school.

When a preschool teacher keeps to her lesson plan and helps the children to find the correct answer, her focus is on readiness, which is connected to qualification and becoming:

Åsa: När dom ska lyfta upp höger pinne och det är svårt så räcker dom upp vänster handen så sager jag så: "Nej, nej, höger, höger, höger, höger, höger, höger, höger" och så gör dom så, "rätt, rätt" (visar med rösten och kroppen hur hon hjälper barnen på ett roligt sätt).

Åsa: When they are going to lift up the right stick and it is hard and so they raise the left hand, I say: "No, no, right, right, right, right, right, right, right" and they do so, "right, right" [showing with her voice and body how she helps children in a fun way].

In the AA, preschool teachers provide a great deal of information and determine the pace of the situation. This approach helps the children to maintain focus on the mathematical concept that the teacher has planned, in order to prepare them for school:

Åsa: Jag tycker om använda mycket på en gång så jag lägger ner en massa legobitar och säger: "allt är om 4". Så dom ska välja vilka fyra legobitar dom vill och sen ska dom lägga ner dom så det blir en så lång rad som möjligt. Vi mätar längden och sen dom lägger ner dom och vänder legobittarna så det blir så kort… sen dom gör det så högt som möjligt.

Åsa: I like to use a lot at one time, so I put out a lot of Lego pieces and said: "Everything is about 4." So they had to choose which four Lego pieces they wanted, and then they would lay them down so it would be as long a line as possible. We measured the length and then they would lay them down and turn the Lego pieces so it was as short ... then they had to make it as high as possible.

Malin and Åsa encourage the children to be better and better at understanding and using the mathematical concepts in the curriculum. They also try to challenge the children by offering more advanced activities:

Malin: Man känner dom som är lite längre fram, som behöver lite mer utmaning. ... där kan man pressa dem lite.

Malin: You know those who are a little further on, who need to be challenged a bit more. ... Where you can push them a little.

Åsa: Ska vi ha en tävling ... och priset idag är en smörgås om ni svarar rätt. Och så frågar jag [barnen] då, då kan jag ha rätt så avancerat[frågor].

Åsa: We should have a contest ... and the prize today is a sandwich if you answer correctly. And so I ask [the children] then, then I can have quite advanced [questions].

These teachers have a tendency to transform the curriculum into goals to achieve—a setting where children's knowledge and skills are encouraged and rewarded.

Åsa and Malin consider socialisation to be about ensuring that the children experience mathematics as fun and interesting while also recognising mathematics as important to their future achievement in school. Åsa and Malin told the children how good they were and that they already knew things that school children had not yet learned. These preschool teachers tried to create a climate where children were able to think, dare, try or even guess. Children were expected to listen to each other and show respect for each other's answers. The teachers described the purpose of creating such an environment as being about helping children to challenge themselves.

Åsa: Har du ett klimat där det är tillåtet att tycka, tänka, våga, gissa så utmanar barnen sig själva och varandra ... det är jätte viktigt med uppdrag och använda rösten och använda ansiktsuttryck. De tycker det är kul när man är kontrollant.

Åsa: If we try to have a climate where it is allowed to think, think, dare, guess, so the children challenge themselves and each other ... it is really important with these tasks to use your voice and use facial expressions. They think it's fun when you are the controller.

This teacher views her role as teacher as guiding and teaching predetermined knowledge to the children; however, she believes that the children will learn more easily if they have fun. In the interviews, both teachers suggested that they used body language, facial expressions and other ways of presenting situations in order to communicate mathematics in a fun and exciting way. Lotta and Susan, on the other hand, also think that the children should have fun, but their view is that the fun should come from the children themselves and that the teacher's role is to explore with the children. The final outcome is not decided in advance.

Like the teachers in Linde's (2006) study, the view that Åsa and Malin have of their role as teachers influences how they work; and even though Åsa and Malin are aware of the other parts of the curriculum, such as language, social interaction and collaboration, they worked with the AA in relationship to the mathematics goals. These preschool teachers selected specific mathematical concepts based on what they think the children need in order to be prepared for school. They start a situation by introducing a mathematical concept which is not linked to a separate theme. The preschool teachers using the AA are sensitive to how the children are doing and their capabilities. The teachers had the children focus on what they prepared and helped each child to understand and find solutions, often with the assistance of other children. The teachers considered their responses to be important, so that children did not feel singled out and lose interest in mathematics when they did not find the right answer. Furthermore, these teachers' interview responses suggested they gave many instructions and talked a lot. The children learned to use a representative voice, but they did not seem to have the opportunity to express their individual freedom, so that their own voice and uniqueness could flourish.

Conclusion

The analysis of the data suggests that these preschool teachers had different views about their roles as implementers of the curriculum and that they are communicating mathematics differently with the children at their preschools. As Linde (2006) suggests, preschool teachers' different backgrounds, experiences of mathematics, views about a good childhood, views on preschools' role and views of their roles as preschool teachers influence the activities they plan, what they will focus on and what they will reward. According to Linde, two of the factors that can affect how teachers interpret the curriculum are their previous experiences and what the management considers important and places priority on. In the case of Åsa and Malin, the decision to work with mathematics was made by their preschool bosses. They could not decide how long the preschools in their community would work with mathematics as a theme or that the theme would conclude with an exhibition. As well, Åsa had previously worked in a school and seen what children had difficulty with there. In contrast, Susan and Lotta's preschool was inspired by the Reggio Emilia pedagogic approach, where working with different themes according to the children's interest is prioritised.

In the preschool where Lotta and Susan work, the *comprehensive approach* is prominent as mathematical activities are based on the children's influence which positions the children as being (Lembrér and Meaney 2014). The children are given the opportunity to be more independent in thought and action—developing subjectification (Biesta 2011). However, this approach carries the risk that the teacher loses focus on the mathematical goals (Pianta et al. 2009; Bertrand 2007) and that the mathematical concepts are not visible for the children (Thulin 2011). General knowledge, social and emotional well-being and communication are respected, but children did not always get answers to their questions, and it was left to them to see if they could talk their way through the mathematical concepts.

In the preschool where Malin and Åsa work, the *academic approach* is prominent. These teachers privilege readiness for school and qualify the children to understand the mathematical concepts specified in the curriculum. However, when teachers are mainly focused on readiness for school, there is a risk that they will limit the children's self-initiated activities, creativity and self-determination (Eurydice 2009; Prentice 2000) and that the children will not have opportunities to think by themselves (Deliyianni et al. 2009). In order to help children to understand the mathematical concepts, these teachers evaluated children's answers in terms of right or wrong, making it difficult for the children to take responsibility for their answers and be active in their solutions.

Pramling and Pramling Samuelsson (2011) argue that placing the same value on social and cognitive learning is defining the revised curriculum content in Sweden, but this study shows that teachers do not place the same value on both social and cognitive learning in practice. Rather what they did depended on their backgrounds, experiences of mathematics, views on childhood, views on the role of preschools and views on their role as preschool teachers (Linde 2006).

How mathematics should be communicated with respect to the Swedish preschool tradition, in which play and children's interests are central, is a challenge for many teachers. This study indicates that thought and reflection are needed about how preschool teachers can help children to communicate using both a representative and a unique, innovative voice, by combining a holistic development that gives children general knowledge and social and emotional well-being with an academic approach that focuses on important educational goals. A greater awareness of the different ways of communicating mathematics can affect how children are qualified, socialised and subjectified and can support balancing these three essential components.

References

Ahlberg, A. (2000). *Att se utvecklingsmöjligheter i barns lärande. Matematik från början [To see developing opportunities in children's learning. Mathematics from the beginning].* Göteborg: Göteborgs universitet. Nationellt centrum för ma-tematikutbildning.

Alrø, H., & Johnsen-Høines, M. (2010). Critical dialogue in mathematics education. In H. Alrø, O. R. Christensen, & P. Valero (Eds.), *Critical mathematics education: Past, present and future* (pp. 11–22). Rotterdam: Sense.

Aubrey, C., & Durmaz, D. (2012). Policy-to-practice contexts for early childhood mathematics in England. *International Journal of Early Years Education, 20*(1), 59–77.

Barber, P. (2009, January 28 to February 1). Introduction. In V. Durand-Guerrier, S. Soury-Lavergne, & F. Arzarello (Eds.), *Proceedings of the sixth congress of the European Society for Research in Mathematics Education, Lyon, France* (pp. 2535–2536). Institut National de Recherche Pèdagogique. Retrieved from http://www.inrp.fr/editions/editions-electroniques/cerme6/working-group-14.

Bäckman, K. (2016). Children's play as a starting point for teaching shapes and patterns in the preschool. In T. Meaney, T. Lange, A. Wernberg, O. Helenius, & M. L. Johansson (Eds.), *Mathematics education in the early years—Results from the POEM2 conference, 2014*. New York: Springer.

Bertrand, J. (2007). Preschool programs: Effective curriculum. Comments on Kagan and Kauerz and on Schweinhart. In *Encyclopedia on early childhood development, Centre of Excellence for Early Childhood*. Montreal: Development and Strategic Knowledge Cluster on Early Child Development. http://www.child-encyclopedia.com/documents/BertrandANGxp.pdf.

Biesta, G. (2011). *God utbildning i mätningens tidevarv [Good education in an age of measurement]*. Stockholm: Liber.

Björklund, C. (2007). *Hållpunkter för lärande: småbarns möten med matematik [Hold Points for Learning: Toddlers meeting with mathematics]*. Dissertation, Åbo Akademis förlag, Åbo.

Björklund, C. (2016). Playing with patterns: Conclusions from a learning study with toddlers. In T. Meaney, T. Lange, A. Wernberg, O. Helenius, & M. L. Johansson (Eds.), *Mathematics education in the early years—Results from the POEM2 conference, 2016*. Cham: Springer.

Bryman, A. (2011). *Samhällsvetenskapliga metoder [Social scientific methods]*. Lund: Studentlitteratur.

Chen, J. Q., & McCray, J. (2014). Intentional teaching: Integrating the processes of instruction and construction to promote quality early mathematics education. In C. Benz, B. Brandt, U. Kortenkamp, G. Krummheuer, S. Ladel, & R. Vogel (Eds.), *Mathematics education perspective on early mathematics learning between the poles of instruction and construction* (pp. 257–274). New York: Springer. http://link.springer.com/chapter/10.1007/978-1-4614-4678-1_16.

Delacour, L. (2013). *Didaktiska kontrakt i Förskolepraktik. Förskollärares transformering av matematiska mål i ett läroplansdidaktiskt perspektiv [Didactic contract in Early Childhood Practice. Preschool teachers' transformation of mathematical objectives in a curriculum didactic perspective]*. Doctoral dissertation, Malmö University, Malmö.

Deliyianni, E., Monoyiou, A., Elia, I., Georgiou, C., & Zannettou, E. (2009). Pupils' visual representations in standard and problematic problem solving in mathematics: Their role in the breach of the didactical contract. *European Early Childhood Education Research Journal, 17*(1), 95–110.

Doverborg, E., & Pramling Samuelsson, I. (1999). *Förskolebarn i matematikens värld [Preschoolers in a mathematical world]*. Stockholm: Liber.

Doverborg, E., & Pramling Samuelsson, I. (2011). Early mathematics in the preschool context. In N. Pramling & I. Pramling Samuelsson (Eds.), *Education encounters: Nordic studies in early childhood didactics* (pp. 37–64). New York: Springer.

Duncan, G., Dowserr, C., Claessens, A., Magnusson, K., Huston, A., Klebanov, P., Pagani, L., Feinstein, L., Engel, K., & Gunn, B. (2007). School readiness and later achievement. *Developmental Psychology, 43*(6), 1428–1446.

Eurydice. (2009). *Early childhood education and care in Europe: Tackling social and cultural inequalities*. Brussels: Eurydice.

Government Offices, Department of Education, working group (U 2010: A). *Promemoria*, 2010-06-24, U2010/4443/S. Electronically available 2013-02-14.

Hildebrandt, C., & Zan, B. (2002). Using group games to teach mathematics. In R. DeVries (Ed.), *Developing constructivist early childhood curriculum: Practical principles and activities* (pp. 193–208). New York: Teachers College Press.

Jonsson, A. (2011). *Nuets Didaktik—Förskolans lärare talar om läroplan för de yngsta [The present's didactics—Preschool teacher talk about the curriculum for the youngest]*. Dissertation, Institutionen för pedagogik, kommunikation och lärande, Göteborgs universitet.

Lembrér, D., & Meaney, T. (2014). Socialisation tensions in the Swedish preschool curriculum: The case of mathematics. *Educare: Vetenskapliga Skrifter, 2014*(2), 89–106.

Linde, G. (2006). *Det ska ni veta! En introduktion till läroplansteori [That you shall know! An introduction to curriculum theory]*. Lund: Studentlitteratur.

Ma, Y. (2012). An analysis of the characteristics and strategies of the excellent teachers in mathematics lessons in primary school. In *Regular lecture 12th international congress on mathematical education, COEX, Seoul, Korea*. Available from http://www.icme12.org/sub/sub02_04.asp.

OECD. (2006). *Starting strong II: Early childhood education and care*. Paris: OECD. doi:10.1787/9789264035461-en.

OECD. (2012). *Starting strong III: A quality toolbox for early childhood education and care*. Electronically available, 14 Feb 2013. http://www.stofnanir.hi.is/rannung/sites/files/rannung/Starting%20Strong%20III.pdf.

OECD. (2013). *PISA 2012 results: What students know and can do: Students performance in mathematics, reading and science* (Vol. 1). Paris: OECD.

Pianta, R. C., Barnett, W. S., Burchinal, M., & Thornburg, K. R. (2009). The effects of preschool education: What we know, how public policy is or is not aligned with the evidence base, and what we need to know. *Psychological Science in the Public Interest, 10*(2), 49–88.

Pramling, N., & Pramling Samuelsson, I. (2011). *Educational encounters: Nordic studies in early childhood didactics*. Dordrecht: Springer.

Prentice, R. (2000). Creativity: A reaffirmation of its place in early childhood education. *The Curriculum Journal, 11*(2), 145–158.

Roth, A.-C. V. (2011). *De yngre barnens läroplanshistoria—didaktik, dokumentation och bedömning i förskola [The younger children's curriculum history—didactics, documentation and assessment in preschool* (2nd ed.). Lund: Studentlitteratur.

Siraj-Blatchford, I., Sylva, K., Muttock, S., Gilden, R., & Bell, D. (2002). *Researching effective pedagogy in the early years* (DfES Research Report 356). London: DfES.

Taguma, M., Litjens, I., & Makowiecki, K. (2012). *Quality matters in early childhood education and care: Finland*. Paris: OECD.

The National Agency for Education [Skolverket]. (2010). *Förskola i utveckling, bakgrund till ändringar i förskolans läroplan [Preschool in development, background to the changes in the preschool curriculum]*. Solna: Åtta 45.

Thulin, S. (2011). *Lärares tal och barns nyfikenhet. Kommunikation om na-turvetenskapliga innehåll i förskolan [Teacher talk and children's curiosity. Communication about science content in preschool]*. Dissertation, Växjö University Press, Växjö.

Uljens, M. (2011). *Didaktik (didactic)*. Lund: Studentlitteratur.

Reflection: An Opportunity to Address Different Aspects of Professional Competencies in Mathematics Education

Christiane Benz

Abstract One major challenge in early mathematics childhood education is to support children's constructive learning. For this, different professional competencies are necessary. Nearly 100 years ago, Dewey already pointed out the impact of reflection on professional development in education. Reflection is still seen as an essential component or a key element of professional development, because in the reflection process, different aspects of professional competencies are interweaved like pedagogical content knowledge and action-related competencies as well as other aspects like beliefs and emotions. In this paper, an innovative in-service and pre-service education bachelor course for early mathematics education is presented. It is designed to give both professionals and students the possibility to develop various professional competencies. One major component can be identified in reflection. Therefore, selected evaluation results of the reflective modules will be presented.

Introduction

For a long time, mathematics education has not been part of pre-service education in early childhood education in Germany and other countries. After changing curricula and educational policy, a need for designing new components of mathematics education in pre-service education arose. Moreover, there is also a need for developing in-service education for early childhood, because for many of the professionals currently working in kindergarten or preschool, early mathematical education was not part of their own pre-service education.

The long-term in-service project 'Children and Adults Explore Mathematics together'—which is linked to the innovative structures in pre-service education within the Karlsruhe Bachelor of Arts (BA) course 'Childhood Pedagogy'—acts an answer to these new demands. The different components of an in-service

C. Benz (✉)
University of Education Karlsruhe, Karlsruhe, Germany
e-mail: benz@ph-karlsruhe.de

T. Meaney et al. (eds.), *Mathematics Education in the Early Years*,
DOI 10.1007/978-3-319-23935-4_23

and pre-service education for early mathematics education are designed to give both professionals and students the possibility to connect various professional competencies.

In this paper, firstly, results of empirical studies are analysed regarding requirements that professionals are supportive in early mathematics education. Then, the important role of reflection in developing competencies—in order to meet these requirements—is emphasised. In the analysis of the long-term in-service project and the innovative structures of pre-service education courses, the focus is on the reflection component.

Theoretical and Empirical Background

Competencies to Support (Mathematical) Learning in Early Childhood Education

The results of the British EPPE study (Effective Provision of Preschool Education) which is linked to the project Research in Effective Pedagogy in the Early Years (REPEY) reveal the importance of social interaction for children's learning. The studies show that positive outcomes 'are closely associated with adult–child interactions … that involve some element of sustained shared thinking' (Siraj-Blatchford and Sylva 2004, p. 720).

> Sustained shared thinking occurs when two or more individuals 'work together' in an intellectual way to solve a problem, clarify a concept, evaluate an activity, extend a narrative etc. Both parties must contribute to the thinking and it must develop and extend the understanding. (Sylva et al. 2004, p. vi)

This highlights the important role of a supportive interaction between adults and children, which is characterised by the connection between instructive and constructive moments. Using instructive aspects, constructive learning is supported. As well, professionals' various attitudes and performances that promote children's learning were revealed (Siraj-Blatchford 2007): Supportive professionals observe children's activities systematically, give feedback during activities, ask and interact with children, instruct and provide playing and learning environments, offer group and individual activities, provide a choice of games, create a balance between activities initiated by adults and by children and possess knowledge of children's development. Strehmel (2008) describes similar competencies, which are fundamental for supporting children's learning in early childhood education. She highlights that a high quality of pedagogical processes will rely on giving stimuli and making suggestions for self-directed learning and being sensitive and careful to the children and responsive to the individual needs, interests and educational background—the latter ones require diagnostic competencies. As one of the main competencies, other researchers point out the ability to observe and interpret complex situations in pedagogical daily routine (Nentwig-Gesemann 2007) as well as the ability to understand

and design processes of interaction of adults and children (Kasüschke and Fröhlich-Gildhoff 2008). All descriptions contain different aspects of instruction and construction as complementary aspects. Through sensitive instructional approach, children's constructive learning will be supported.

Although these descriptions are not specific to mathematical learning competencies, many components are also described in models of teachers' competencies concerning mathematics education in school (Baumert and Kunter 2011). As well in regard to research about the competencies of professionals in supporting preschool mathematics education, similar components are identified:

> To implement early mathematics education in natural learning situations and to ensure that children with different levels of knowledge and skills can profit, early childhood educators need wide-ranging knowledge and competencies. First of all, they need content knowledge. They have to see the relations between mathematics in the early years and later on to guarantee coherent mathematical learning. (Gasteiger 2014, p. 278)

Further, Gasteiger points out that professionals need pedagogical content knowledge and—as a part of content knowledge—diagnostic knowledge as well as action competencies (Gasteiger 2014; see also Chen and McCray 2014). Referring to other studies (Baumert and Kunter 2006; Stipek et al. 2001), Gasteiger also highlights the influence of beliefs, attitudes and motivation. Synthesising the results of the different empirical studies and existing models of professional competencies concerning mathematics education, the following categories of competencies or orientations seem to be relevant for supporting children's early mathematical learning:

1. *Content knowledge, pedagogical content knowledge and knowledge of children's development* constitute the basis for fostering children's mathematical competencies. Professionals need this kind of knowledge in order to notice children's mathematical competencies in their activities in order to initiate sustained shared thinking processes or other kinds of supporting interactions between children and adults concerning mathematics education.
2. Professionals need *action competencies* in order to notice, initiate and design interactions, which support mathematical competencies. Because of the informal nature of preschool settings, identifying 'teachable moments' is quite challenging for preschool teachers (Ginsburg et al. 2008). This special ability can be seen as one major aspect of action competencies of preschool teachers.
3. The relevance and influence of attitudes, beliefs and motivational and volitional tendencies concerning action-related competencies are highlighted in the description by Weinert (2001):

> The theoretical construct of action competence comprehensively combines those intellectual abilities, content-specific knowledge, cognitive skills, domain-specific strategies, routines and subroutines, motivational tendencies, volitional control systems, personal value orientations and social behaviours into a complex system. (p. 51)

When designing in-service and pre-service education, these different aspects of professional competencies have to be considered.

One major challenge is to support professionals so that they can develop these different competencies. As Dewey (1910) and Schön (1983) point out, reflection can serve as a bridge between these different competencies. The role of reflection is discussed in the next section.

The Role of Reflection in Professional Development of Preschool Teachers

Dewey emphasises the importance of reflection for learning in general as well as for pre-service and in-service teacher education more than a hundred years ago. In his work *How We Think*, Dewey (1910) defines reflective thoughts as

> active, persistent, and careful consideration of any belief or supposed form of knowledge in the light of the grounds that support it, and the further conclusions to which it tends. (Dewey 1910, p. 9)

He highlights the importance of reflection for learning processes in general and identifies different components as steps in reflection. Referring to Dewey, Kolb (1984) describes in his experiential learning model not only different steps or stages but constitutes a circulation model of sequential components as following: concrete experience, reflective observation, abstract conceptualisation and active experimentation. The reflective observation constitutes the connection between practical and theoretical abilities. In other words, pedagogical content knowledge and knowledge of children's development and action-related competencies interweave in the phase of reflection.

The third aspect described above *attitudes, beliefs and motivational and volitional tendencies* is not included in the model of Dewey and Kolb. Nevertheless, the importance of emotions is taken into account in the model by Zull (2002, 2004). He sees the experiential model of Kolb as an overlay over the structure of the brain and points out the importance of emotions. 'Even if we experience something that has happened to us before, it is hard to make meaning of it unless it engages our emotions' (Zull 2002, p. 166). Barrett (2005) explains that Zull

> also points out that reflection is a search for connections (2002 p. 167) and suggests that we have to seriously consider the role of emotion if we want to foster deep learning (2002 p. 169). (Barrett 2005, p. 20)

It therefore seems that the different aspects of professional competencies in early childhood can be addressed in the process of reflection. Above all, reflection can play an important role in preschool teachers' practice to support early mathematics learning in preschool: 'the teachers must be able to reflect on children's co-constructing learning processes. This encloses the ability to discover educational abilities for mathematical learning in children's activities' (Thiel 2012, p. 1253).

Therefore, in a common model of preschool teacher's competencies, Fröhlich-Gildhoff et al. (2014) highlight the ability *to analyse and evaluate* situations as one important part of performance competencies. Next to the ability to *analyse and evaluate* concrete situations, the authors emphasise the *role of (self-)reflection* in general for preschool teachers. Through reflection, preschool teachers become aware of their subjectivity, and they learn to take different perspectives and can analyse situations on the basis of theoretical knowledge and knowledge which is based on experiences (Fröhlich-Gildhoff et al. 2011, 2014).

Therefore, in many pre-service and in-service education programmes, such as the Victorian Early Years Learning and Development Framework (Kennedy and Stonehouse 2012), reflection is seen as an important or even the most important component.

Reflection can be distinguished as individual and collective reflection (Berkemeyer et al. 2011, p. 228), which can be divided further in content, object focused or focused on oneself. Research of professional competencies of teachers confirms positive effects of collective reflection on competencies of professionals (Beck et al. 2002; Schuster 2008) and on the development of teaching (West and Staub 2003) and especially on the teaching of preschool teachers (Bleach 2014). Marcos and Tillema (2006) provide a critical overview of empirical research results about reflection and professional development.

The long-term, in-service project 'Children and Adults Explore Mathematics together'—which is linked to the innovative structures in pre-service education within the Karlsruhe BA course 'Childhood Pedagogy'—highlights the need for reflection in professional development. The implementation of reflection is presented in the description of the pre-service course and in-service project in the next paragraph.

Design of the Innovative Pre-service Education Bachelor Course and In-Service Project

Designing Components regarding Different Professional Competencies for BA Course

Often there is a separation of modules focused on theory at university and modules focused on settings outside university. The students' options to gain practical experiences are bound to institutional conditions, e.g. time schedule, etc., which make it difficult to acquire action competencies in early mathematics education. Therefore, the construction of the BA course implements options for acquiring action competencies outside of the regulation of educational institutions, which can also serve as a basis for reflection.

Videotaped (Inter)Actions in a 'Sheltered' Room of Action

Apart from avoiding limitations by educational institutions, the establishment of good conditions for reflection was another module design principle. Pre-service education students are sometimes overwhelmed with the complexity of the possible actions in early mathematics education which inhibits their abilities to reflect on situations of action if theoretical aspects must be considered (Stokking et al. 2003). Therefore, 'sheltered' rooms of action set up particularly for observation and reflection are created in the BA course. This innovative element constitutes a setting where action competencies are not acquired outside university. Rather the practical field 'comes' to the university.

At the university, the sheltered room of action is called 'MachmitWerkstatt'—literally 'join-in-studio'. At this join-in-studio, preschool teachers can play together with their children in prepared playing and exploring environments and explore mathematical aspects together. As in section "Theoretical and Empirical Background" mentioned earlier, it is very challenging for preschool teachers to recognise teachable moments in children's play. So one important aspect of the join-in-studio is to 'provide' teachable moments through a prepared environment.

Another important element of the join-in-studio setting is the video recording of each visit to the studio. As a result of the video recording, the students can observe both children's activities in the playing environments and their own interactions with the children after the visit. Thereby they can develop and improve their diagnostic and reflection competencies. These competencies become the foundation to initiate and support children's mathematical discovery processes. Empirical studies have shown positive effects of the integration of video recording for the analysis of interaction between children and adults (Pianta et al. 2008; Downer et al. 2009). Empirical studies with professionals teaching in school proved also positive effects for their professional development (Nührenbörger 2009; Scherer and Steinbring 2006).

In contrast to the positive aspects mentioned, the prepared learning environment as an artificial situation has some limitations especially for children's learning, which will be discussed in section "Closing Remarks".

Implementation of Videotaped Practical Situations in the BA Course and In-Service Project

The implementation of video-recorded (inter)actions in a sheltered room of actions can be realised by a close connection and interplay with the in-service project 'Children and Adults Explore Mathematics' (Fig. 1).

The goal of the in-service project lies in the evaluation of the in-service education in early mathematics education. The project consists of three phases: (1)

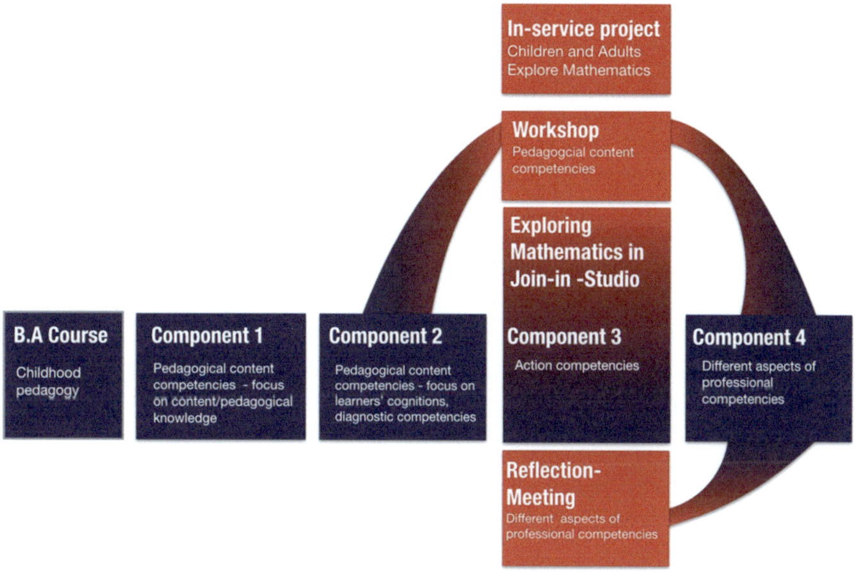

Fig. 1 Connection between in-service project and BA course

workshop for professionals in preschool, (2) children and adults playing and exploring mathematics together in the join-in-studio and (3) reflection meeting on the basis of video documents.

The three phases are conducted within half of a year (time period of a university term). After that period, different mathematical content becomes the focus, and the three phases start again.

Therefore, in each term in the workshop, a different mathematical content is focused on (e.g. counting and seeing, exploring patterns, comparing and measuring). In addition to exposing pedagogical content knowledge, playing environments, in which children can acquire different mathematical abilities, are analysed and created together with the preschool teachers in the workshop. The results of the workshops are reported in a handout for the preschool teachers, so they can implement these tried-and-tested environments in their daily life in kindergarten.

After the workshop, each preschool teacher has the possibility to visit the join-in-studio with the group of kindergarten children they daily work with throughout term time (see Fig. 1). The main focus lies in the possibility to explore mathematics together with their children. Only preschool teachers who attended the workshop are allowed to come to the studio.

As previously mentioned, the join-in-studio also serves as a sheltered room for student teachers to acquire action competencies. Acting in the join-in-studio constitutes the main connection between the in-service project and pre-service course. So not only the preschool teachers are interacting with the children, the student teachers also are interacting with the children in the join-in-studio. Thus, both preschool teachers and student teachers have the possibilities to practise action competencies. As already mentioned, each visit at the studio is video recorded. The student teachers are responsible for the recording. The video-recorded actions and interactions serve later as a basis for the reflection.

The components of reflection constitute a further connection between the in-service course and pre-service project. There are different meetings of reflection for student teachers and for the preschool teachers. In preparation for their reflection meeting, the student teachers are asked to analyse the actions and interactions that occurred during the visits to the join-in-studio. After the analysis, the student teachers choose meaningful video clips for the reflection meetings. Likewise, the video clips are used in reflection meetings with the preschool teachers of the in-service project and for further workshops also. The reflection meeting with the preschool teachers is audio recorded only.

Thus, the three phases of the in-service project interweave with the components of the BA course in different ways.

Circulated Connection of Different Competencies in One Mathematical Content

Apart from the linking of the in-service project and pre-service course and the video-recorded interaction in the join-in-studio, the innovation further emphasises the element of reflection. The reflection component enables a circular model of connections of different phases, which focus on different competencies within one mathematical content. The different competencies can be acquired exemplarily on the basis on one specific mathematical content (Fig. 2).

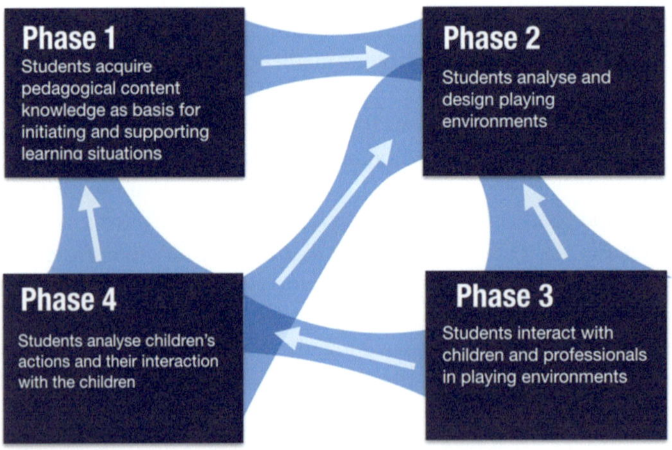

Fig. 2 Connection of competencies within the BA course

In Phase 1, student teachers acquire foundations regarding content knowledge, pedagogical content knowledge, knowledge in developmental psychology and pedagogy with regard to special conditions of preschool settings.

On this basis in Phase 2, the student teachers analyse materials and games and develop playing environments. In doing so, they have to take into consideration a need to balance free and instructed playing. The playing environment shall provide possibilities to accompany child-initiated interactions and prerequisites for sustained shared thinking. By designing playing environments, student teachers have to consider the interplay between instruction and construction.

The implementation into practice takes place in the interactions with alternating groups of children in the join-in-studio at the university. The student teachers are challenged to assess the mathematical potential of playing situations and to create challenging social situations in order to motivate children to play. Due to the alternating groups of children, the student teachers have the chance to intensively and exemplarily focus on one mathematical content both on theoretical and practical perspectives.

During the reflection phases, they can gain a deeper understanding of the mathematical pedagogical content. On the basis of this new insight and extended knowledge, they design and implement playing environments and learning possibilities again. From analysing their own actions, they connect theoretical and practical aspects. Theories can be compared with individual experiences of practical sessions. Mismatches between theories and observations can be used in a constructive way for further design and implementation.

From analysing the videos, a critical distance to their own interactions can be established. In collaborative phases of reflection, student teachers are able to analyse patterns of communication and solution processes of children on the basis of theoretical aspects. The student teachers can also think about alternative (re)actions. New insights can eventually lead to adaptations being made to the playing environments, which provide a connection to Phase 2. Also new theoretical insights can be obtained which connects to Phase 1. The video clips selected by the student teachers can be used to illustrate theoretical aspects in Phase 1 as well.

In the description of the innovative structure, it is postulated that different competencies of preschool teachers can be achieved through the focus on reflection. In the following section, a glimpse is provided of the ongoing analysis of the audio-recorded reflection meetings from the in-service project. It is analysed to determine the aspects of preschool teachers' competencies which are addressed in the reflection meetings.

A Glimpse into the Analysis of Reflection Meetings

The project lasted 4 years and included seven different reflection meetings. The number of preschool teachers at the reflection meetings varied between 10 and 25. Not every preschool teacher who attended the workshop and the join-in-studio took part in the reflection meetings.

Three selected extracts from the recordings are described in detail in order to show how the model of deductive category application of qualitative content analysis was used (Mayring 2007).

Reflective Statements Addressing Pedagogical Content Knowledge

Before reflecting on the video clips, the preschool teachers were asked to report on their impressions in an open discussion. One preschool teacher reported an interesting situation in which the children played buying and selling eggs. In doing so, the children used a rack, which contained 30 eggs with the structure of six rows and five eggs in a row (5×6). Fortunately this situation was recorded clearly on a video recording, and the student teachers made a clip for the reflection meeting with this situation. The preschool teacher commented on the video clip (Fig. 3):

> We played with the eggs, the egg cartons and egg racks. Ina put 30 eggs on the rack. And she started to count them all by 2s, 2, 4, 6, 8, 10, 12 and so on till 30. Then during playing—I don't know why—the children discussed how many eggs half of this egg rack may contain. She did a thing which I never would have realised if I had not been here. They still debated how many eggs will be in half of this egg carton. Then Ina draws an imaginary line in the carton so that there were 3 rows, each with 5 eggs. She laid her hand on 6 six eggs and said, '6', really 6, then on other 6 eggs and said '12' and then on the last 3 eggs and said '15'. Because I didn't understand what was going on, I asked her to explain me again what she did, and then we considered that her explanation will be on the video.
>
> [Wir haben mit den Eiern, den Eierschachteln und -platten gespielt. Ina hat eine Eierpalette gefüllt, auf die 30 Eier passen. Dann beginnt sie zu zählen und dann zählt wirklich in Zweierschritten 2,4,6,8,10,12,14, und so weiter bis 30. Dann während dem Spielen—ich weiß nicht warum—die Kinder haben überlegt, wie viel Eier wohl auf die Hälfte passen. (…). Sie hat erst eine gedachte Linie auf dem Karton gezeichnet, so dass es 3 Reihen mit 5 Eiern waren. Beim Bestimmen der Hälfte der Eier auf der Eierpalette, legt sie ihre Hände zuerst auf 6 Eier, Und sagt dann ohne zu zählen erst mal 6, echt einfach 6 und dann auf andere 6 Eier und sagt 12, und dann auf die letzten 3 Eier und sagt 15. Weil ich nicht gleich verstand, was sie meinte, fragte ich sie, ob sie mir das nochmal erklären kann. Und dann haben wir aufgepasst, dass ihre zweite Erklärung auf Video aufgenommen wird.]

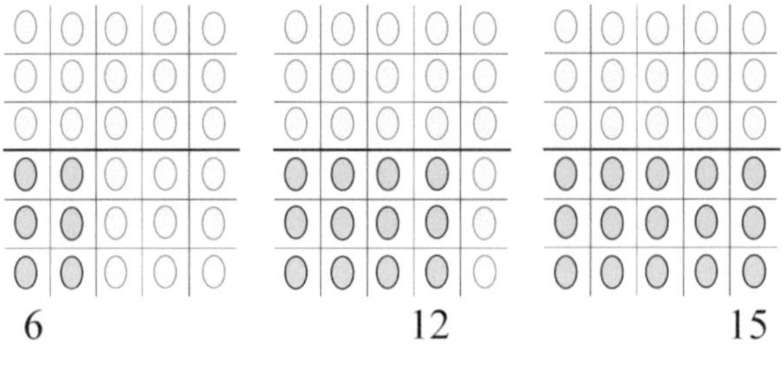

Fig. 3 Ina's process to determine the numbers of eggs

Different perceptions made by the children in regard to the quantity of numbers were reported by the preschool teacher, especially about decomposing a quantity into different parts and using different structures. First, the preschool teacher reports that Ina decomposed the quantity of all eggs into parts of two eggs. Later on, the preschool teacher recognises that Ina decomposed all eggs into two halves. Then she observed that Ina decomposed the quantity of half of the eggs (15) in a structure with 2×6 eggs and three eggs. Next to these different possibilities of decomposition, the preschool teacher also mentioned different processes for judging the quantity like counting in steps by twos and solving a very challenging calculation task like $6 + 6 + 3$.

After this video clip had been shown, the preschool teachers discussed that seeing structures is very individual and that the structures are not just there; they emerge when people look at them. Thus, it can be seen that the reflection meeting pedagogical content knowledge was addressed.

Reflective Statements Referring to Action Competencies

A precondition for action competencies or at least one major part of action competencies is identifying learning possibilities for children. Therefore, identifying learning possibilities or teachable moments in children's actions had a dominant role in the reflection meetings. This can be seen in expressions like 'I noticed especially in how many areas the children find mathematics and how eager they are with it' or 'I never saw so many mathematical situations'.

Acting in specific situations is also a part of action competencies.

> Preschool teacher:
> But I have learned so much from the student teachers; to say: 'Hold back, give the children more possibilities to try themselves'. Not only 'right' and 'wrong' to communicate. And focus on the procedure: 'How do you know this? How could you explain or say this more precisely?' To pick up such questions, that was quite interesting for me, to think in such a way.
>
> [Aber ich hab von den Studentinnen so viel gelernt, zu sagen ,nimm dich zurück, gib den Kindern mehr Möglichkeiten sich auszuprobieren. 'Ja nicht das Richtig und Falsch, was wir vermitteln. So von der Vorgehensweise,woher weißt du das jetzt? Wie kannst du das noch genauer sagen?' Solche Fragestellungen aufzugreifen, das war für mich ganz interessant, so zu denken.]

This preschool teacher focused on the interaction between the children and the adults and identifies possibilities to initiate sustained shared thinking. Many preschool teachers also addressed the interplay between instruction and construction in the interaction with children.

> Preschool teacher:
> So, I then realised, I provide far too much. Not until here [*in the in-service project*] I realised that I should withdraw much more and that I can and should create much more free space, which is very difficult if one is in this job for such a long time. And I just became aware that we should rather help the children, but let them do it themselves—also in regard to mathematical content.

> [Also ich hab dann gemerkt, ich gebe viel zu viel vor. Und hier [*in der Fortbildung*] habe ich dann erst gemerkt, dass ich mich viel mehr zurück nehmen muss, viel mehr Freiräume schaffen kann und sollte, was unheimlich schwierig ist, wenn man schon so lange drin ist. Und es ist mir dann aber immer wieder bewusst geworden, dass man den Kindern eher Hilfe geben soll, aber sie selber machen lassen soll—auch bei mathematischen Sachen.]

One preschool teacher told in the reflection meeting about a situation where she discussed different arrangements of five eggs in an egg carton, which can contain ten eggs in the structure of a ten frame. There were egg cartons with five eggs in a row, but Peter put two eggs in the upper row and three eggs in the lower row. The following discussion can be seen in the videotape:

Preschool teacher	Why can you see easily and quickly that Peter has 5?
Child 1	Because 4 plus 1 is 5.
Preschool teacher	Hmm, can you see how Peter put the eggs in the carton. Is that 4 and 1?
Child 1	Yeees?
Preschool teacher	Hmmmmm, yes, you could see that, too, right. But I've seen something else. Has anyone an idea?

(No child had an idea)

Preschool teacher	How many are in a row?
Child 2	2 and 3
Preschool teacher	Yes, and 2 and 3 is 5, too. Ok.
[Erwachsener	Warum kann man beim Peter gut erkennen, dass es 5 sind?
Kind 1	Weil 4 plus 1 ist 5.
Erwachsener	Hmm, Siehst du wie es Peter gelegt hat? Ist das 4 und 1, was da der Peter da gelegt hat?
Kind 1	Ja?
Erwachsener	Hmmmmm, ja, das kann man auch erkennen. Stimmt. Ich hab was anderes erkannt, wer kann sich vorstellen, was ich erkannt hab. Hat jemand eine Idee?

(Kein Kind hat eine Idee)

Erwachsener	Wie viele sind in einer Reihe?
Kind 2	2 und 3
Erwachsener	Ja, und 2 und 3 ist auch 5. Ok.]

The preschool teacher commented that only after the situation happened, she 'understood how the child could see four and one. Yes, it was the pattern of the dice of four and one. But in this situation I just couldn't see it'. This situation led to a broad discussion with the focus on understanding children's thinking and comprehension as well as about the problem that seeing structures is a very individual act.

Posing questions was another component of the reflection discussions. The preschool teachers jointly discussed appropriate and alternative questions and (re) actions in this situation.

Reflective Statements concerning Attitudes, Beliefs and Motivational and Volitional Tendencies

Preschool teacher:
I was extremely motivated through this (vocational) training to implement mathematics in kindergarten, but at the same time I was positively surprised how much material we were already using, where I was not aware about the mathematical learning opportunities.

[Ich wurde durch die Fortbildung extrem motiviert, Mathematik im Kindergarten umzusetzen, war aber gleichzeitig auch positiv überrascht, wie viel Material wir schon im Einsatz hatten, bei dem mir die mathematischen Lernchancen nicht bewusst waren.]

In this statement, motivational and volitional aspects are addressed by the preschool teacher. Many preschool teachers refer to their own view of mathematics and their experiences in their own school career.

Preschool teacher:
I am also very thankful because it changed something in me. I didn't really like maths in school and I thought children will start early enough with doing maths, so we don't have to bring it as such a concept into the kindergarten, I thought, because the 1×1 and $3+7$, they really learn that early enough in school. For myself, it really changed a lot, because I think that I got out of this that we do mathematics in the daily life and we only have to change our thinking a bit and I really liked it so that I really want to thank you heartily.

[Also ich bin auch sehr dankbar, weil bei mir hat es selber einen Hebel umgelegt. Mathematik war mir ziemlich verhasst in der Schule und ich dachte, die Kinder machen noch früh genug Mathematik, das müssen wir jetzt auch nicht noch unbedingt so als Begriff in den Kindergarten bringen, fand ich, denn das 1×1 und $3+7$, das lernen sie noch wirklich früh genug in der Schule. Für mich selber hat es jetzt sehr viel umgelegt, weil ich denke, ich hier mitgekriegt habe, dass wir eigentlich Mathematik machen im Alltag und wir das einfach nur ein bisschen umdenken müssen und ich fand das so toll, dass ich mich echt herzlich dafür bedanke.]

On the basis of some transcripts, the categories are presented. In summary, the preschool teachers were able to use the shared space of the reflection meetings to discuss *all* the different aspects of professional competencies. Therefore, the postulation that different competencies of professionals can be addressed by reflection can be confirmed.

Closing Remarks

The important role of reflection was analysed on a theoretical and empirical basis for professional development especially in early childhood mathematics education. By a circular connection of different phases concerning different professional competencies and by other innovative aspects like the sheltered room for action and the use of video-recorded interactions, the role of reflections in teachers' learning of pedagogical content knowledge was considered. Still, different aspects of professional development have to be analysed in detail.

Concerning early mathematics education, it must be stated that an artificial sheltered room like the join-in-studio is not an optimal situation for children to learn mathematics. The children's ability to influence their activities can be quite limited if student teachers focus mainly on mathematical aspects. Children should have the chance to acquire mathematical competencies in their play, in their preschool setting or at home in natural learning situations (Gasteiger 2014; van Oers 2014). Nevertheless, many preschool teachers reported that the artificial and sheltered environment helped both the children and the preschool teachers to focus on the activities the children had chosen. This was because there were not so many other activities and possibilities available in the join-in-studio as in their normal preschool institution and so children focused longer on their own play and activities:

> Preschool teacher:
> We realised that children in a sheltered room are sometimes more concentrated on their play or activities, when they are not so distracted by other things which are going on around them. Some children don't get into it so much in their free play time, some need this protection. Here, they are completely concentrated, whereas in the institution, during times of 'free-play', there are rarely occurring such great scenes…
>
> [Wir haben festgestellt, dass die Kinder in einem geschützten Raum manchmal konzentrierten sind beim Spielen und Tun, wenn sie nicht so abgelenkt sind von anderen Sachen um sie herum. Manche Kinder kommen im Freispiel da nicht so richtig rein, manche brauchen das geschützte. Also wenn sie hier sind, dann sind sie voll konzentriert, während im Freispiel in der Einrichtung, kommen selten solche tollen Szenen einfach auf …]

The artificially created situation seemed to help the preschool teachers to acquire wide-ranging knowledge and competencies so that they are now able to identify children's mathematical competencies and teachable moments in their daily routine in the preschool setting. As one preschool teacher said: 'This, I had never perceived in everyday situations, if I would not have been here'. Another preschool teacher stated: 'My view for mathematical situations is now wider. I see more possibilities of exploring mathematics, of supporting children to find solutions'. This last statement illustrates how the different aspects of professional competencies are interweaved. Next to attitudes, beliefs and motivational and volitional tendencies—as seen in statements such as 'my view for mathematical situations is now wider'—growth in pedagogical knowledge was also identified: 'I see more possibilities of exploring mathematics, of supporting children to find solutions'. Seeing 'more possibilities of supporting children to find solutions' is also one part of action competencies. In order to support children constructing new mathematical knowledge, professionals need to recognise the mathematics in children's constructions or, more precisely, in children's statements and actions. Professionals also need to know how to interact with the children in these situations and expand their repertoire of action competencies. So in this statement, different aspects of professional competencies are connected. Furthermore, the complimentary connection of instruction and construction become apparent as well as knowledge of both aspects.

Acknowledgement The in-service project is funded by the *Tschira Stiftung* [Foundation of Tschira], and the implementation of the join-in-studio in the BA course is supported by the *Baden-Württemberg Stiftung* [Foundation of the Land Baden-Württemberg]. The author takes the responsibilities for the content of this publication.

References

Barrett, H. (2005). White paper: Researching electronic portfolios and learner engagement. www.taskstream.com/reflect/whitepaper.pdf. Accessed 31 May 2014.

Baumert, J., & Kunter, M. (2006). Stichwort: Professionelle Kompetenz von Lehrkräften [Keyword: Professional competence of teachers]. *Zeitschrift für Erziehungswissenschaft, 9*(4), 469–520.

Baumert, J., & Kunter, M. (2011). Das Kompetenzmodell von COACTIV [The competence model of COACTIV]. In M. Kunter, J. Baumert, W. Blum, U. Klusmann, S. Krauss, & M. Neubrand (Eds.), *Professionelle Kompetenz von Lehrkräften: Ergebnisse des Forschungsprogramms COACTIV* (pp. 29–53). Münster: Waxmann.

Beck, R., King, A., & Marshall, S. (2002). Effect of videocase construction on preservice teachers' observations of teaching. *The Journal of Experimental Education, 70*(4), 345–361.

Berkemeyer, N., Järvinen, H., Otto, J., & Bos, W. (2011). Kooperation und Reflexion als Strategien der Professionalisierung in schulischen Netzwerken [Cooperation and reflection as strategies of professionalization in school networks]. *Zeitschrift für Pädagogik, 57*(Beiheft 1), 225–247.

Bleach, J. (2014). Developing professionalism through reflective practice and ongoing professional development. *European Early Childhood Education Research Journal, 22*(2), 185–197. doi:10.1080/1350293X.2014.883719.

Chen, J. Q., & McCray, J. (2014). Intentional teaching: Integrating the processes of instruction and construction to promote quality early mathematics education. In U. Kortenkamp, B. Brandt, C. Benz, G. Krummheuer, S. Ladel, & R. Vogel (Eds.), *Early mathematics learning. Selected papers of the POEM 2012 Conference* (pp. 257–274). New York: Springer.

Dewey, J. (1910). *How we think*. Boston: Heath.

Downer, J. T., Kraft-Sayre, M., & Pianta, R. C. (2009). On-going, web-mediated professional development focused on teacher-child interactions: Feasibility of use with early childhood educators. *Early Education & Development, 20*(2), 321–345.

Fröhlich-Gildhoff, K., Nentwig-Gesemann, I., & Pietsch, S. (2011). *Kompetenzorientierung in der Qualifizierung frühpädagogischer Fachkräfte [Competence-orientation in the qualification of professionals in early childhood education]*. Expertise. München: DJI.

Fröhlich-Gildhoff, K., Weltzien, D., Kirstein, N., Pietsch, S., & Rauh, K. (2014). *Expertise—Kompetenzen früh-/kindheitspädagogischer Fachkräfte im Spannungsfeld von normativen Vorgaben und Praxis [Expertise—Competences of professionals in early childhood education in the tension between normative descriptions and practical field]*. Freiburg: Zentrum für Kinder- und Jugendforschung. http://www.bmfsfj.de/RedaktionBMFSFJ/Abteilung5/Pdf-Anlagen/14-expertise-kindheitspaedagogische-fachkraefte,property=pdf,bereich=bmfsfj,sprache=de,rwb=true.pdf. Accessed 31 May 2014.

Gasteiger, H. (2014). Professionalization of early childhood educators with a focus on natural learning situations and individual development of mathematical competencies: Results from an evaluation study. In U. Kortenkamp, B. Brandt, C. Benz, G. Krummheuer, S. Ladel, & R. Vogel (Eds.), *Early mathematics learning. Selected papers of the POEM 2012 conference* (pp. 275–290). New York: Springer.

Ginsburg, H. P., Lee, J. S., & Boyd, J. S. (2008). Mathematics education for young children: What it is and how to promote it. *Social Policy Report, 22*(1), 3–22.

Kasüschke, D., & Fröhlich-Gildhoff, K. (2008). *Frühpädagogik heute. Herausforderung an Disziplin und Profession [Early childhood education today. Challenge for discipline and profession]*. Köln, Kronach: Carl Link.
Kennedy, A., & Stonehouse, A. (2012). *Victorian early years learning and development framework practice principle guide, reflective practice*. Melbourne: Department of Education and Early Childhood Development. www.eduweb.vic.gov.au/edulibrary/public/earlylearning/pracreflective.pdf. Accessed 31 May 2014.
Kolb, D. A. (1984). *Experiential learning: Experience as the source of learning and development* (Vol. 1). Englewood Cliffs, NJ: Prentice-Hall.
Marcos, M. J. J., & Tillema, H. (2006). Studying studies on teacher reflection and action: An appraisal of research contributions. *Educational Research Review, 1*, 112–132.
Mayring, P. (2007). *Qualitative Inhaltsanalyse. Grundlagen und Techniken [Qualitative content analysis. Basics and techniques]* (9th ed., 1st ed., 1983). Weinheim: Deutscher Studien.
Nentwig-Gesemann, I. (2007). Das Konzept des forschenden Lernens im Rahmen der hochschulischen Ausbildung von FrühpädagogInnen [The concept of exploratory learning in the context of university education of teachers in early childhood]. In K. Fröhlich-Gildhoff, I. Nentwig-Gesemann, & P. Schnadt (Eds.), *Neue Wege gehen—Entwicklungsfelder der Frühpädagogik* (pp. 92–101). München: Reinhardt.
Nührenbörger, M. (2009). Lehrer-Schüler-Diskurse im Mathematikunterricht als Gegenstand kollegialer Reflexion—Fallkonstruktionen mathematischer Unterrichtsdiskurse [Discourse of students and teachers in mathematics education as object of reflection of teachers' colleagues—Case studies of mathematical educational discourses]. In M. Neubrand (Ed.), *Beiträge zum Mathematikunterricht* (pp. 131–134). Münster: WTM.
Pianta, R. C., Mashburn, A. J., Downer, J. T., Hamre, B. K., & Justice, L. M. (2008). Effects of web-mediated professional development resources on teacher-child interactions in prekindergarten classrooms. *Early Childhood Research Quarterly, 23*, 431–451.
Scherer, P., & Steinbring, H. (2006). Noticing children's learning processes—Teachers jointly reflect their own classroom interaction for improving mathematics teaching. *Journal for Mathematics Teacher Education, 9*(2), 157–185.
Schön, D. (1983). *The reflective practitioner, how professionals think in action*. New York: Basic Books.
Schuster, A. (2008). *Ich schreibe. Also lerne ich [I'm writing, therefore I'm learning]*. Dissertation. Regensburg: Roderer.
Siraj-Blatchford, I. (2007). Effektive Bildungsprozesse: Lehren in der frühen Kindheit [Effective educational processes: Learning in early childhood]. In F. Becker-Stoll & M. R. Textor (Eds.), *Die Erzieherin-Kind-Beziehung* (pp. 97–114). Berlin: Cornelsen Scriptor.
Siraj-Blatchford, I., & Sylva, K. (2004). Researching pedagogy in English preschools. *British Educational Research Journal, 30*(5), 713–730.
Stipek, D. J., Givvin, K. B., Salmon, J. M., & Mac Gyvers, V. L. (2001). Teachers' beliefs and practices related to mathematics instruction. *Teaching and Teacher Education, 17*, 213–226.
Stokking, K. M., Leenders, F. J., Stavenga-de Jong, J. A., & van Tartwijk, J. (2003). From student to teacher: Reducing practice shock and early drop-out in the teaching profession. *European Journal of Teacher Education, 26*(3), 329–350.
Strehmel, P. (2008). Wovon hängt "gute Bildung" tatsächlich ab? [On what does good education really depend?]. *kindergarten heute,* (1), 8–13.
Sylva, K., Melhuish, E., Sammons, P., Siraj-Blatchford, I., & Taggart, B. (2004). *The Effective Provision of Preschool Education (EPPE) Project: Final report. A longitudinal study funded by the DfES 1997-2004*. London: DfES and Institute of Education, University of London.
Thiel, O. (2012). Investigating the structure, level and development of professional skills of preschool teachers in mathematics. In *ICME-12 pre-proceedings, Seoul* (pp. 1251–1259).

van Oers, B. (2014). The roots of mathematising in young children's play. In U. Kortenkamp, B. Brandt, C. Benz, G. Krummheuer, S. Ladel, & R. Vogel (Eds.), *Early mathematics learning. Selected papers of the POEM 2012 Conference* (pp. 111–123). New York: Springer.

Weinert, F. E. (2001). Concept of competence: A conceptual clarification. In D. S. Rychen & L. H. Salganik (Eds.), *Defining and selecting key competencies* (pp. 45–65). Seattle, WA: Hogrefe & Huber.

West, L., & Staub, F. C. (2003). *Content-focused coaching*. Portsmouth, NH: Heinemann.

Zull, J. E. (2002). *The art of changing the brain*. Sterling, VA: Stylus.

Zull, J. E. (2004). The art of changing the brain. *Educational Leadership, 62*(1), 68–72.

Index

T. Meaney et al. (eds.), *Mathematics Education in the Early Years*,
DOI 10.1007/978-3-319-23935-4

MIX
Papier aus verantwortungsvollen Quellen
Paper from responsible sources
FSC® C105338

If you have any concerns about our products,
you can contact us on
ProductSafety@springernature.com

In case Publisher is established outside the EU,
the EU authorized representative is:
Springer Nature Customer Service Center GmbH
Europaplatz 3, 69115 Heidelberg, Germany

Printed by Libri Plureos GmbH
in Hamburg, Germany